MATERIALS RESEARCH SOCIETY SYMPOSIA PROCEEDINGS

ISSN 0272-9172

Materials for Infrared
Detectors and Sources

MATERIALS RESEARCH SOCIETY SYMPOSIA PROCEEDINGS

ISSN 0272 - 9172

MATERIALS RESEARCH SOCIETY SYMPOSIA PROCEEDINGS

MATERIALS RESEARCH SOCIETY SYMPOSIA PROCEEDINGS

MATERIALS RESEARCH SOCIETY SYMPOSIA PROCEEDINGS

Materials for Infrared Detectors and Sources

Symposium held December 1-5, 1986, Boston, Massachusetts, U.S

EDITORS:

R. F. C. Farrow

IBM/Almaden Research Center, San Jose, California, U.S.A.

J. F. Schetzina

North Carolina State University, Raleigh, North Carolina, U.S.A.

J. T. Cheung

Rockwell International, Thousand Oaks, California, U.S.A.

MRS MATERIALS RESEARCH SOCIETY
Pittsburgh, Pennsylvania

CAMBRIDGE UNIVERSITY PRESS
Cambridge, New York, Melbourne, Madrid, Cape Town,
Singapore, São Paulo, Delhi, Mexico City

Cambridge University Press
32 Avenue of the Americas, New York NY 10013-2473, USA

Published in the United States of America by Cambridge University Press, New York

www.cambridge.org
Information on this title: www.cambridge.org/9781107405639

Materials Research Society
506 Keystone Drive, Warrendale, PA 15086
http://www.mrs.org

First published 1987
First paperback edition 2012

Single article reprints from this publication are available through
University Microfilms Inc., 300 North Zeeb Road, Ann Arbor, MI 48106

CODEN: MRSPDH

ISBN 978-0-931-83755-5 Hardback
ISBN 978-1-107-40563-9 Paperback

Contents

*Invited Paper

*Invited Paper

*Invited Paper

*Invited Paper

Preface

This book contains most of the papers presented at the symposium: "Materials for Infrared Detectors and Sources" held in Boston, Massachusetts, December 1-5, 1986.

This symposium brought together the leading groups from the USA, Europe and Japan working on the preparation and characterization of materials for infrared detectors and sources. Much of the activity in this field is driven by the need for focal-plane array imagers operating in the medium (3-5 micron) and long (8-14 micron) wavelength atmospheric transmission windows. In addition there is now a growing interest in the preparation and exploration of detector and source materials for fiber-optic communications at wavelengths beyond 1.5 microns. The objectives of the symposium were threefold: firstly, to review progress in the key areas of bulk and epitaxial growth technologies for preparation of infrared materials; secondly, to review techniques for characterization of infrared materials; and thirdly, to evaluate the potential of novel materials and structures for infrared detectors as well as new epitaxial technologies for preparation of detector structures.

A one-day joint session with the symposium on "Diluted Magnetic Semiconductors" was held to deal with common approaches in the preparation and characterization of materials. The symposium was well attended and achieved its objectives. In particular it provided a much-needed and timely international forum for wide-ranging, in-depth discussions on all the important infrared materials currently under development.

Symposium co-chairmen:

R.F.C. Farrow   J.F. Schetzina
J.T. Cheung

Acknowledgments

It is our pleasure to acknowledge with gratitude the financial support for this symposium provided by IBM.

In addition, we are very pleased that the symposium was also supported in part by the Office of Naval Research, the Army Research Office, the Air Force Office of Scientific Research, and the Defense Advanced Research Projects Agency.

PART I

Infrared Materials Based on Group III-V Compounds

Epitaxial Semiconductor Structures for the Infrared: Where Are We Now?

Morton B. Panish

AT&T Bell Laboratories
Murray Hill, New Jersey 07974

EXTENDED ABSTRACT

The use of epitaxial structures for the generation, detection, and modulation of light has had as its strongest driving force the development of fiber optic communications systems. Partly for that reason, the semiconductor materials systems that can be most readily optimized for wavelengths (800 - 1600nm) transmittable through state-of-the-art optical fibers, have been the most well developed. It also appears that the communications community has been most fortunate in that the semiconductor systems required, although far more difficult to deal with than Si, are relatively benign compared to those that operate at significantly shorter or longer wavelengths. Two developments that are now more than 15 years old initiated the extensive studies of compound semiconductor epitaxial structures that are now so important. These were the introduction of the heterostructure concept in the study of injection lasers and light emitting diodes (1), and the realization that lattice matched isoelectronic epitaxial layers, consisting in part of solid solutions between the binary compounds, would permit such structures to be grown.

The development of epitaxial technologies for the growth of heterostructures started with the growth of GaAs/Al_xGa_{1-x}As light emitters at about 900 nm. With this semiconductor system it is possible to take advantage of the lattice match that exists between GaAs and Al_xGa_{1-x}As for all values of x. The initially used liquid phase epitaxy method for this system is being largely replaced by molecular beam epitaxy (MBE) (2) and Metal Organic Chemical Vapor Deposition (MOCVD) (3). These epitaxy methods have permitted the growth of an enormous variety of heterostructures with dimensional precision such that both layer thickness and interface abruptness can be controlled to a monolayer.

The advent of optical fibers with low dispersion at 1300nm and low loss at 1550nm spurred the development of epitaxial methods for other III-V systems, most particularly $Ga_xIn_{1-x}As_yP_{1-y}$/InP and $Al_xIn_{1-x}As/Ga_xIn_{1-x}As$/InP, with direct bandgaps that included those wavelengths. Epitaxy in both systems may be achieved with monolayer precision, the former by several varieties of Gas Source MBE (GSMBE) (4,5,6) and the latter by conventional MBE (7), in spite of the need precisely to control the composition to maintain the lattice match with the substrate. These epitaxial methods have permitted a bounteous crop of lasers, light emitting diodes, optical detectors and transistors.

This proliferation of III-V structures and devices has been largely restricted to wavelengths shorter than about 1700nm, most likely because, for many of the systems, the formation of extensive solid solutions between the smaller bandgap III-V compounds is inhibited by miscibility gaps (8) at compositions where there is a lattice match to a convenient binary III-V compound. This unfortunate situation has so far precluded the development of a corresponding range of useful III-V devices that operate at longer wavelengths that are of interest for several major applications. These include, in particular, the potential need for sources and detectors in the 4-5μm range for use with potential ultra-low loss fibers, and for detectors in the 10μm and longer wavelength regions for commercial thermal imaging and for military applications.

It is in these wavelength ranges that II-VI compounds containing Hg become important, particularly Cd_xHg_{1-x}Te/CdTe and more recently superlattice structures of HgTe/CdTe. The Cd-Hg-Te system has the advantage of a close lattice match between CdTe and HgTe, a complete series of solid solutions, and bandgaps that range from essentially zero energy to 1.5eV. All of this has been obvious for years, but the Hg containing II-VI's are difficult to deal with in comparison to the shorter wavelength III-V's. Very little work has been

Mat. Res. Soc. Symp. Proc. Vol. 90. ⁕1987 Materials Research Society done to develop these materials for wavelengths shorter than 10μm or so, although longer wavelength CdHgTe detectors using essentially bulk LPE grown material are well established. There has been a resurgence of interest in these materials in the past few years that dates to Faurie's demonstration (9) that superlattice structures of HgTe-CdTe can be grown by MBE and there is recent work showing that CVD methods can also be applied (10). It is possible that these developments will lead to development of new detectors in the 4–5μm range and at longer wavelengths, for both military and for nonmilitary thermal imaging applications.

While there is potential for detector structures of II-VI compounds grown by MBE and MOCVD to move into the wavelength region at less than about 5μm that had previously been thought of as the province of III-V semiconductors, it should be noted that there is now motion in the other direction at wavelengths of about 10μm and greater. New detector structures, the first of which have been reported by Levine et.al. (11), utilize intersubband transitions in quantum wells and sequential resonant tunneling between quantum wells in GaAs/AlGaAs superlattice heterostructures. These detectors are inherently very fast. The very first ones studied operate at about 10μm and are about ten times worse than good HgCdTe detectors from the point of view of both dark current and efficiency. However, significant improvement should be readily achievable by modification of the superlattice geometry, the device geometry, and by the use of superlattices of other III-V compounds.

REFERENCES

1. I. Hayashi and M. B. Panish, J. Appl. Phys. 41, 150 (1970).

2. "Molecular Beam Epitaxy and Heterostructures," L. L. Chang and K. Ploog, Editors, NATO ASI Series, Series E: Applied Science - No. 87, Martinus Nijhoff Publishers, Boston 1985.

3. R. D. Dupuis, Science 226, 623 (1984).

4. M. B. Panish, Prog. Crystal Growth and Charact. 12, 1 (1986).

5. M. B. Panish, J. Cryst. Growth. In press.

6. W. T. Tsang, J. Electron. Mat. 15, 235 (1986).

7. K. Mohammed, F. Capasso, J. Allam, A. Y. Cho, and A. L. Hutchinson, Appl. Phys. Lett. 47, 597 (1985).

8. Kentaro Onabe, Japn. J. Appl. Phys. 21, L323 (1982).

9. J. P. Faurie and A. Million, J. Cryst. Growth 54, 582 (1981).

10. P. -Y. Lu, C. -H. Wang, L. M. Williams, S. N. G. Chu, and C. M. Stiles, Appl. Phys. Lett. 49, 1372 (1986).

11. B. F. Levine, K. K. Choi, C. G. Bethia, J. Walker and R. Malik Appl. Phys. Lett. In press.

DEFECTS IN Mg-DOPED InP AND GaInAs GROWN BY OMVPE

FRED R. BACHER, H. CHOLAN, AND WALLACE B. LEIGH
Oregon Graduate Center, Department of Applied Physics and Electrical
Engineering, 19600 N.W. Von Neumann Drive, Beaverton, Oregon 97006-1999, USA

ABSTRACT

We report on the defects present in doped InP and GaInAs grown by
organometallic vapor phase epitaxy (OMVPE). The material was grown in an
atmospheric pressure system using group III trimethyl sources, arsine and
phosphine. Bis(cyclopentadienyl) magnesium (Cp_2Mg) was present as a p-type
source of magnesium. Defects in as-grown material were characterized using
photoluminescence (PL), Hall-effect, and deep level transient spectroscopy
(DLTS). Various levels of Mg doping were investigated, ranging from
5×10^{15} to 1×10^{19} cm^{-3}. Radiative defects were observed at 77 K
corresponding to PL emission from conduction band/shallow donor to acceptor
levels including emission at 1.37 eV identified as the shallow hydrogenic
acceptor, and emission lines at 1.3 eV and 1.0 eV in heavily doped material.
Corresponding hole traps in InP:Mg were observed by DLTS having thermal
activation energies of 0.20 and 0.40 eV, the 0.40 eV trap being the dominant
defect in p-type InP. In GaInAs grown near lattice-matched to InP,
radiative emission is also observed from deep centers 100 meV from band edge
emission. This emission is observed to be related to lattice-mismatch of
the ternary with the InP, and is found to be accentuated and broadened in
GaInAs doped with Mg.

INTRODUCTION

A great majority of the present and proposed near-infrared optical and
optoelectronic devices require p-type epitaxial layers of InP, GaInAs,
and/or GaInAsP. Emission devices such as light emitting diodes or solid
state lasers require heavily doped p+ material to insure recombination is
maintained in the active region. Heavily doped p-type material is also
desirable in such applications as photocathodes. A final p+ layer on a
multi-layer epitaxial device insures good ohmic contact as well as lower
series resistance. P-type doping of III-V compounds can often be difficult:
doping during growth from the melt is complicated by unfavorable dopant
distribution coefficients and high vapor pressures. Doping using diffusion
sources requires precise control of diffusion times and temperatures to
insure the proper junction depth. Cadmium and zinc impurities have a
tendency to re-diffuse during any further high temperature processing.
P-type doping during growth via organometallic vapor phase epitaxy (OMVPE)
has the capability to alleviate some of these effects.

At the same time that heavy p doping is needed, it is important that
the p-type layer be relatively free of defects [1]. Frequently, the p+
layer is used as a window for optical transmission, and any stray absorption
by defects is undesirable. Also, deep levels tend to limit diffusion
lengths of carriers, and promote non-radiative recombination. In this
study, defects in Mg-doped InP and GaInAs grown by OMVPE using Trimethyl
group III sources and bis(cyclopentadienyl magnesium), (CP_2Mg) as a
magnesium source have been investigated. Magnesium was chosen as the p-type
dopant because it has a high solubility in III-V compounds while
maintaining a low diffusion coefficient. CP_2Mg was chosen as its source
because it has a favorable vapor pressure for doping in an OMVPE system.
Using CP_2Mg, InP:Mg has been prepared for hole concentrations up to p = 10^{19}
cm^{-3}.

Defects in InP:Mg and GaInAs:Mg were investigated using photoluminescent (PL) and Hall effect, at room temperature and 77 K for both techniques, and by deep level transient spectroscopy (DLTS) performed on Au-semiconductor junctions. Using these techniques, shallow and deep radiative and nonradiative defects related to the Mg impurity were observed. The deep defects are believed to be associated with Mg-native defect or Mg-impurity complexes.

EXPERIMENTAL

Epitaxial films were prepared in a horizontal-tube, RF heated OMVPE grower specifically designed for the preparation of GaInAs and GaInAsP. This growth system to present day has been used to prepare device-quality GaAs and GaInAs on GaAs substrates, and InP and GaInAs lattice-matched on InP substrates. The growth tube is of rectangular cross section having 2.4×4 cm^2. Total gas flow of Pd diffused hydrogen was set a 4 l/min for all growth runs. Sources used are trimethyl indium (TMIn) trimethyl gallium (TMGa) arsine and phosphine. Both of the hydride sources are diluted to 10% in hydrogen. Inlet lines of the system are heated to 60°C to prevent recondensation of the solid Mg and In sources. Typical growth conditions for preparation of III-V materials are given in Table I.

TABLE I. Conditions used for the growth of III-V compounds by OMVPE

	GaAs	InP	GaInAs
Gas Phase Concentration, in Mole Fraction:			
TMIn	--	9.2×10^{-5}	8.2×10^{-5}
TMGa	1.5×10^{-4}	--	$4-5 \times 10^{-5}$
PH3	--	2.2×10^{-3}	--
AsH3	3×10^{-3}	--	8×10^{-3}
Temperature	640°C	600°C	650°C
Growth Rate (microns/min)	0.10	0.05	0.08

Indium phosphide substrates used were Sumitomo sulphur-doped n+. Substrates were etched in 5:1:1 H$_2$SO$_4$:H$_2$O$_2$:H$_2$O solutions at 40°C for 2 min. When the growth conditions of Table I are applied, the result is material with excellent morphologies and layer uniformity.

Photoluminescence spectra were obtained using an Ar$^+$ ion laser focused to a power density of 20 W/cm^2. No surface preparation was done to the samples before PL investigation. Filtered luminescence from the samples was passed through a 1/2 meter monochrometer and was detected using a lead-sulphide detector. Peak heights were corrected for detector response. While no mapping was attempted, the surface was scanned to demonstrate qualitative uniformity.

Ohmic contacts to p-type InP were formed by evaporation of a 90/10 Au/Zn alloy through masks, then followed by a 1 minute anneal in argon at 450°C. For CV and DLTS analysis, an Au dot of 1 mm diameter was evaporated next to the ohmic contact. Deep level transient spectroscopy was performed with a modified Boonton 72 BD capacitance bridge, using the standard 1 MHz test frequency. The output signal of this bridge was subsequently digitized using a storage scope (Tektronix 7D20) with averaging capabilities. The averaged digital waveform is downloaded to a desktop computer (Tektronix 4052) for evaluation of the capacitance transient. Rate windows for this system were set using the conventional two-channel boxcar method as originally demonstrated [2].

RESULTS AND DISCUSSION

Since the main aim of our OMVPE system is the preparation of materials which at various times may include Ga, In, As, or P, GaAs was first prepared on GaAs substrates for characterization, which included a study of equipment performance and the investigation of source purity without the presence of the TmIn source. Table II shows 300 K and 77 K Hall mobiltiy measurements for several GaAs growths. GaAs surfaces were all observed to be specular in nature, with no apparent surface defects. Growth rates for GaAs are typically 0.1 to 0.20 microns/min, and measured background concentrations for samples in Table II are close to 1×10^{14} cm^{-3}. Photoluminescent linewidths for samples in Table I are typically less than 12 meV, and no deep emission is observed. Using separate process systems, which included separate inlet lines, growth tubes and susceptors, we observe the ability to change from GaAs to InP growth with no problematic cross contamination. The results of the GaAs investigation demonstrated the integrity of our OMVPE system and technique.

Growth of Mg-doped InP

Photoluminescent spectra of undoped InP taken at sample temperatures of 300 K and 77 K are shown in Figure 1. PL line widths at 77 K in undoped InP typically were 13 to 15 meV, corresponding to a background concentration ranging from the mid 10^{15} cm^{-3} to the 10^{16} cm^{-3} range. Corresponding 77 K Hall mobilities for similar material grown on Fe-doped substrates were in the 10^4 cm^2/Vs range. Higher residual impurity concentrations in InP as compared to GaAs were attributed to impurities in the starting sources. In the undoped InP, band-to-band transitions at 1.41 eV were the only transitions observed.

Figure 2 represents a series of PL spectra taken from samples grown with various molar flows of CP$_2$Mg present. Molar ratios of (Cp$_2$Mg)/(TmIn) varied from 2.9×10^{-4} to 5.5×10^{-2} for samples of spectra labeled (b) through (e) in Figure 2. These samples were determined as p-type from the sign change of thermoelectric power measurements. Spectra (a) is of undoped InP, for comparison. The band-to-band transition was observed in all but the most highly doped samples, becoming less significant as the Mg concentration increased. A peak at 1.37 eV was observed in all Mg-doped InP, and corresponds to the expected dominant shallow hydrogenic acceptor impurity [3]. After preparing material corresponding to Figures 2(b) and 2(c), subsequent InP growths without Cp$_2$Mg show little sign of Mg-related emission, indicating almost no memory effect of the Cp$_2$Mg source.

There was a shift of photoluminescent emission to lower wavelengths as acceptor concentration was increased, as has also been observed for Zn doping in InP [4]. For group II/group III metal-organic source ratios in the vapor greater than 10^{-3}, a peak at 1.30 eV was observed, as found also in MBE InP:Mg [3]. Moreover, a broad additional PL band appeared centered on 1.0 eV, becoming the dominant defect for the highest doped samples. This broad emission corresponds to a deep Mg-related deep defect in the bandgap, most likely an impurity-impurity or impurity-native defect complex.

In Figure 3, log of the net acceptor concentration achieved in these samples is shown as a function of the Cp$_2$Mg input flux. The solid data

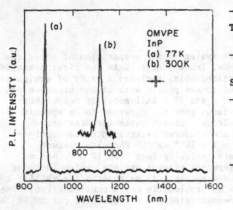

Fig. 1. Photoluminescence spectra
of undoped InP grown by OMVPE.

TABLE II. Hall mobility results of
OMVPE GaAs

Sample No.	μ (cm²/Vs)	
	300 K	77 K
53	6480	38,000
55	5880	86,000
61	5300	88,000
62	5690	88,000

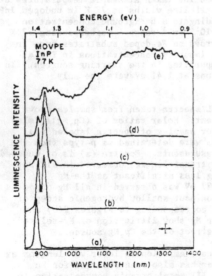

Fig. 2. Photoluminescence spectra
of Mg-doped InP grown at a) 0,
b) 3, c) 5, d) 30, e) 70 × 10⁻⁴
mole fraction Mg.

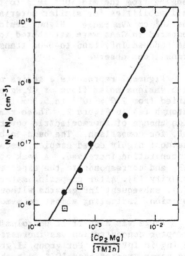

Fig. 3. Net acceptor concentration
versus mole fraction ratio
(CP₂Mg)/(TMIn).

points represent concentrations derived from the reported correlation with PL peak width [4,5]. The open data are measured from CV analysis of evaporated Au-Schottky diodes, and agree within experimental error. The highest achieved doping is slightly less than 10^{19} cm^{-3}, and represents some of the highest doped VPE InP to date. The solid line in Figure 3 is a least- squares-fit to the peak-width data. The slope of the line indicates a quadratic dependence on Mg overpressure, similar to that observed by Nelson and Westbrook [6] in InP prepared by adduct sources.

The data in Figure 2 is representative of material which varies from lightly semiconducting to almost metallic in nature. The lighter doped material was prepared for subsequent CV and DLTS analysis. Gold Schottky barriers were prepared on the semiconductor surface by evaporation through a mask in high vacuum. This formed metal-semiconductor rectification barriers of 1.0 ± 0.2 eV in height, as measured from subsequent CV analysis. Figure 4 shows results of the DLTS investigation of Mg-doped InP. Since the minority carrier injection for Schottky barriers on InP is expected to be negligible, only hole traps are expected to be observed in our p-type samples. These are indicated by negative peaks in Figure 4. Clearly seen are two levels, labeled H(0.20) and H(0.40). The peak labeled H(0.40) is observed in all Mg-doped InP tested so far, and is the dominant level in concentration as observed in these samples. Concentration of H(0.40) and H(0.20) were measured at $2-6 \times 10^{14}$ cm^{-3} and $1-5 \times 10^{13}$, as measured from DLTS peak heights, and corrected for depletion edge effects.

An array of spectra similar to Figure 4 were recorded for various rate windows, and fitted with the corresponding temperatures for peak maxima to the equation for hole emission rate:

$$e_p/T^2 = A\sigma_p \exp E_a/kT \qquad (1)$$

where A is calculated from the equation $A = (v_t N_v/T^2)$. N_v is the effective density of states for the valence band, and v_t the Boltzmann-averaged thermal hole velocity. E_a is the thermal activation energy and σ_p the temperature-independent capture pre-factor. The Arrhenius data for the two traps in Figure 4 are shown in Figure 5. Fitting the data in Figure 5 to Eq.(1), we obtain thermal activation energies, 0.40 ± 0.02 eV and 0.20 ± 0.03 eV, with corresponding capture cross sections of 6 and 4×10^{-16} cm^2, respectively, for the two observed hole traps. The defect labeled H(0.20) is a defect which has been observed by DLTS in both LPE and VPE-grown InP:Zn [7,8]. The dominant peak at 0.40 eV corresponds well to the deep-emission observed in the photoluminescence spectra of Figure 2(e). This emission is tentatively identified as conduction band to H(0.40) emission and the level at H(0.40) ascribed to a Mg-related complex.

Mg-doped GaInAs

GaInAs epitaxial layers were prepared on InP using the growth conditions indicated in Table I. Prior to growth of the ternary, an InP buffer layer of 0.5-0.7 microns was first prepared on the InP substrate. Composition of the ternary layers was determined from PL band edge peak position, but most usually by optical absorption. Optical absorption of the ternary was measured at 10 K for samples mounted face-side down on a glass slide, with the substrate removed via an etching process. Removal of the substrate was necessary as it was found that the substrate absorbs heavily in the optical wavelengths of interest, even though these wavelengths are beyond the InP absorption edge. Etching of the substrate was done in a dilute solution of HCl, which specifically etches all InP, leaving the ternary intact.

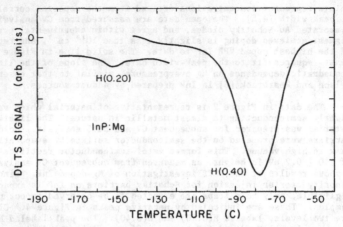

Fig. 4. DLTS spectra of InP:Mg (sample (b) in Fig. 2). The rate window was sec^{-1}.

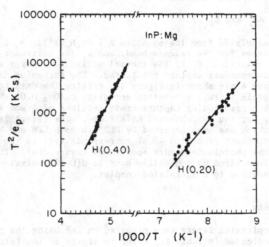

Fig. 5. Arrhenius plot of trap levels H(0.10) and H(0.40).

As radiative centers were observed far from the band-edge emission, and determined as related to impurity-defect complexes for the InP material in Figure 2, similar transitions were observed in GaInAs. In particular, a large radiative center was observed in undoped material, and related transitions in Mg-doped GaInAs. This center was most apparent in material which was not lattice-matched but slightly Ga-rich as indicated from absorption measurements. Figure 6 is a PL spectra of an undoped sample of $Ga_x In_{1-x} As$ for x = 0.50. The composition was confirmed from the measured absorption edge. A large radiative center was observed at ~ 100 meV from the band edge emission. This radiative transition was observed to be more predominant with increased mismatch, becoming the dominant transition for most Ga-rich ternary material.

A radiative transition similar to the type shown in Figure 6 was observed by Yagi et al. [9] in GaInAs grown by liquid phase epitaxy. The luminescent intensity of a broad emission centered around 0.69 eV in that study was observed to be minimized at the ternary composition where the ternary and the InP was lattice-matched at the growth temperature. Thus this defect which was also ~ 100 meV from the band edge emission, was related to defects associated with lattice mismatch at the growth temperature. Similar, although less-intense emission was observed by Goetz et al. [10] and attributed to Fe contamination from the use of Fe doped InP as substrates. In our own material, we believe a radiative level exists at ~ 100 meV which is related to mismatch, and which can be accentuated by the presence of a metal impurity. Figure 7 is a photoluminescence spectra of lattice-mismatched GaInAs doped with Mg. The defect related to lattice-mismatch is observed to be accentuated by the presence of the Mg impurity, and an even broader and deeper emission is observed. This indicates that, similar to the Mg-doped InP containing luminescent activators deep in the bandgap related to defect complexes, in InGaAs a similar defect exists related to lattice-mismatch. This type of luminescent center could have detrimental effects in any radiative devices.

The existence of the defect in Figures 6 and 7 on sulfur-doped material rules out the possibility of contamination by Fe from the substrate as observed by Goetz. However, this does not rule out the possibility of Fe contamination from the Cp_2Mg source via Cp_2Fe impurities. However, this is also unlikely, as Fe was not observed in the DLTS investigation of InP:Mg. More likely, the center is related to strain-related defects, such as may be generated and/or gettered by misfit dislocations.

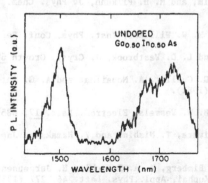

Fig. 6. Photoluminescence spectra of undoped $Ga_x In_{1-x} As$ for x = 0.50.

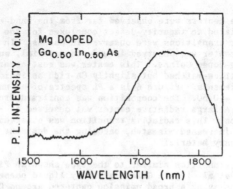

Fig. 7. Photoluminescence spectra of
$Ga_xIn_{1-x}As$:Mg for $x = 0.50$.

Acknowledgements

We wish to acknowledge George Howard for preparation of the OMVPE
material. We also thank R. Kremer and J. S. Blakemore for useful
discussions.

REFERENCES

[1] G. H. Olsen, J. Cryst. Growth **31**, 233 (1975).

[2] D. V. Lang, J. Appl. Phys. **45**, 3014 (1974).

[3] K. J. Bachman, E. Buehler, B. I. Miller, J. H. McFee and F. A. Thiel,
J. Cryst. Growth **39**, 137 (1977).

[4] O. Roder, U. Heim, and M. H. Pilkuhn, J. Phys. Chem. Solids **31**, 2625
(1970).

[5] B. D. Joyce and E. W. Williams, Inst. Phys. Conf. Ser. **9**, 57 (1971).

[6] A. W. Nelson, and L. D. Westbrook, J. Cryst. Growth **68**, 102 (1984).

[7] J. L. Pelloie, G. Guillot, A. Nouailhat and A. G. Antolini, J. Appl.
Phys. **59**, 1536 (1981).

[8] M. Inuishi and B. W. Wessels, Electron. Lett. **17**, 685 (1981).

[9] T. Yagi, Y. Fujiwara, T. Nishino and Y. Hamakawa, Jap. J. Appl. Phys.
22, L467 (1983).

[10] K. H. Goetz, D. Bimberg, K. A. Brauchle, H. Jurgensen, and J. Selders,
M. Razeghi, E. Kuphal, Appl. Phys. Lett. **46**, 277 (1985).

PART II

Materials Requirements for Infrared Detectors, Imagers and Sources

Materials Requirements for IR Detectors and Imagers

M. A. Kinch
Texas Instruments Incorporated
Dallas, Texas 75265

ABSTRACT

The requirements of infrared systems have increased significantly over the years, from the simple linear, low resolution, photoconductive array to the present-day, large area, high-density, photodiode (MIS and metalurgical) arrays, with on-focal-plane signal processing of considerable complexity. The success that has been achieved in meeting the performance goals appropriate for these systems has been due, to a large degree, to significant advances in the relevant materials technologies. The technologies of importance over the last twenty years are briefly reviewed, and the current state of the art, with its dominance by intrinsic alloy materials, is addressed in detail. The limitations of current bulk and epitaxial intrinsic materials technologies are considerable, both from a performance and a producibility point of view, when compared to the quality and quantity of material required by future infrared systems. These limitations are considered together with possible ways to overcome them.

I. INTRODUCTION

Passive thermal imaging systems operating in the earth's atmosphere are designed to sense and display differences in emitted thermal radiation of a scene associated with variations in temperature and emissivity. The spectral density of the photon flux levels emitted by black bodies of various temperatures is shown in Figure 1(a). It is apparent that room temperature emission peaks at a wavelength ~ 12 μm at a flux density level of ~2x10^{17} photons/cm^2-s-μm. The infrared scene consists of minute changes of emitted flux superimposed on this large background flux level and thus the imager is designed to be sensitive to this scene contrast. The variation of scene contrast across the 300°K blackbody spectrum is shown in Figure 1(b), and is seen to increase with decreasing wavelength. The wavelength of operation for terrestrial infrared systems is determined by the spectral output shown in Figure 1(a) and the transmission of the earth's atmosphere, and is essentially limited to two spectral windows located in the 3 to 5 μm, and the 8 to 14 μm spectral ranges. For a temperature difference of 0.1°K Figure 1 indicates that the contrast of a 3 to 5 μm scene is $\Delta\phi/\phi$ = 1/300 superimposed upon a total emitted flux density $\phi_{2\pi}$ = 1.3x10^{16} photons/cm^2-s, whereas in the 8 to 12 μm window $\Delta\phi/\phi$ = 1/600 with a total flux density $\phi_{2\pi}$ = 8x10^{17} photons/cm^2-s.

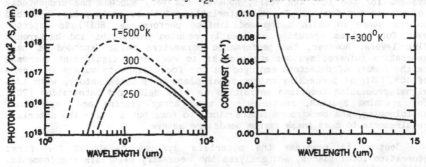

Figure 1 (a) Spectral Radiant Photon Density vs Wavelength for Various T, and (b) Scene Contrast vs Wavelength at T = 300°K

The heart of the infrared imaging system is the detector focal plane, and these detectors fall into two broad categories, namely thermal and photon. Both have been utilized for infrared imagery, but the high sensitivity, high resolution requirement of current and future systems have thus far only been satisfied by materials utilizing photon detection.

Reasonable quantum efficiencies for photon absorption can be achieved in semiconductors either by utilizing 1) transitions between appropriate impurity levels in the bandgap and an adjacent band (extrinsic detection), or 2) transitions between the valence and conduction bands (intrinsic detection). Extrinsic detectors are majority carrier devices whereas intrinsic detectors are essentially dominated by minority carriers. Extrinsic detectors suffer from two severe disadvantages relative to intrinsic devices. Firstly, they are relatively noisy devices from a statistical point of view[1] and hence require a greater degree of cooling to achieve the same sensitivity, and secondly the absorption coefficient although not small still requires detector thicknesses $\sim 10^{-1}$ cm, at meaningful impurity concentrations, to achieve significant absorption. This is to be contrasted with the large absorption coefficients ($\alpha \sim 10^3$ to 10^4 cm^{-1}) available with intrinsic device material. Intrinsic devices thus lend themselves more readily to production-line processing utilizing standard photolithographic techniques.

Early infrared systems employed both extrinsic and intrinsic materials. Hg-doped germanium was developed for operation in the 8-14 μm spectral window, and a variety of intrinsic materials such as InSb, InAs and Pb salts for the 3 to 5 μm window. However the severe cooling requirement ($\sim 28°$K) associated with Ge:Hg for the vitally important long wavelength window led to the development of a variety of narrow bandgap tunable alloy systems [2,3] in the early 1960's, capable of operating at temperatures in the liquid nitrogen (77°K) range. The development of these narrow bandgap alloys set the stage for an explosion in the pervasiveness of infrared imaging in military and industrial systems, resulting in the so-called first generation infrared focal plane, or common module infrared system, in the early 1970's.

This focal plane consists of a linear array of discrete detectors (up to 200 in number) operating at $\sim 77°$K and coupled to the outside world by individual bias and preamplifier leads. The infrared scene is scanned across the array and parallel signals fed out in analog fashion through appropriate electronics to generate a TV compatible real-time visible image. The common module has been the major building block of infrared systems for the last ten years, however its very success has undoubtedly led to its own demise. The characteristics of these devices have advanced to the stage in which background limited performance (BLIP) is expected even for systems operating at greatly reduced bandwidths and background flux levels. However, the performance parameters being demanded of next generation infrared systems are such as to require a significant increase in the number of detectors employed in the focal plane to values in excess of 10^4. Signal processing on the focal plane will be mandatory, with such signal processing functions envisioned as time delay and integration (TDI) for scanning systems, multiplexing, area array staring mode operation, antiblooming, and background subtraction, to name but a few. The materials requirements to meet these system needs are severe.

Section II examines the materials systems developed for first generation focal planes, and analyses the necessary materials requirements. Section III is devoted to a discussion of next generation focal plane materials requirements with particular emphasis being placed on tactical system operation in the 8-14 μm spectral window. Possible options for alleviating the severity of the materials problems will be discussed.

II. FIRST GENERATION FOCAL PLANES

Intrinsic photon detection can be implemented in either of two device forms, namely photoconductive or photovoltaic. Initially the prime candidate materials were HgCdTe and PbSnTe, and both photoconductive[4,5] and photovoltaic[6,7] devices have been fabricated from these materials, with essentially BLIP performance demonstrated for all of them. However the carrier concentrations typical of PbSnTe were not readily conducive to BLIP photoconductor performance and the main thrust in this material concentrated on a photodiode technology, with the development of various diffusion and Schottky barrier techniques. HgCdTe, on the other hand, was readily prepared with n-type doping concentrations $< 10^{15}$ cm^{-3} which was ideal for photoconductors, but the available crystal quality and surface passivation technology, amongst other things, did not lend itself readily at the time to a viable photodiode process. Thus, the prime candidates for first generation scanning infrared systems were HgCdTe photoconductors and PbSnTe photodiodes.

The photovoltaic detector when operating under background limited conditions, offers a theoretical noise advantage[8] over the photoconductor with no supposed penalty of operating temperature. However, the PbSnTe photodiode device structure did not lend itself so well to the producibility aspects required for large scale detector focal plane production. The Pb salts are extremely soft, and readily damaged, and possess enormous expansion coefficients in the 77 to 300°K temperature range. The dielectric constants are extremely high ($\sim$ 500 to 2000 depending on composition and applied electric field) which coupled with the relatively large doping concentrations results in large device capacitances, and hence potential bandwidth limitation problems for scanning focal planes with required bandwidths in excess of 40 kHz. Surface passivation is also not a well understood technology in PbSnTe, and presents a potential device degradation problem for photodiodes on this material (and for photodiodes on any other material that are not fabricated with guard ring structures).

HgCdTe photoconductors, on the other hand, had an adequate materials technology in hand, with the necessary demonstrated electrical, mechanical, and processing characteristics required for large volume focal plane production. Thus despite the inherent noise advantage of the photodiode structure, and the undoubtedly superior crystalline quality of PbSnTe, the decision was made to go with HgCdTe photoconductors. History has proved in fact that this decision was a good one.

A variety of parameters, such as specific detectivity (D*) and noise equivalent power (NEP), have been employed to characterize infrared performance, but most of these concepts are somewhat artificial, and relatively meaningless to anyone other than systems engineers. For infrared detectors and focal planes, particularly when operating in a tactical, high background flux environment, the materials requirements for optimum performance, namely BLIP, are best and most simply understood in terms of detector noise requirements. The sensitivity of the detector and its associated circuitry must be such that detector noise dominates the noise of subsequent electronics, such as preamplifiers. The detector in turn must be operated such that its noise is dominated by background generated minority carriers, i.e., at a temperature where the thermally generated density of minority carriers is negligible.   The final requirement is that the detector be sufficiently fast to accommodate the necessary system bandwidth.

For the n-HgCdTe photoconductor the noise voltage[8] of the detecting element is given by

$$V_N = \frac{2V}{n}\left(\frac{p\tau\Delta f}{v(1+\omega^2\tau^2)}\right)^{1/2},$$ (1)

where V is the applied bias voltage, n the effective electron concentration, p the minority carrier hole concentration, v the detector volume, τ the minority carrier lifetime, Δf the measurement bandwidth, and $\omega = 2\pi f$ the measurement frequency. BLIP operation requires that $p = (\eta\phi\tau/t) > n_i^2/n$, where η is the detector quantum efficiency, ϕ the incident background flux, t the detector thickness, and n_i the intrinsic carrier concentration; it has been assumed that the optically active volume equals the detector volume. Equation (1) then becomes, at low frequencies,

$$V_N = \frac{2V\tau}{nt}\left(\frac{\eta\phi\,\Delta f}{A}\right)^{1/2},$$ (2)

where A is the detector area. There are two limitations on bias voltage[8], namely 1) minority carrier sweepout, and 2) Joule heating in the detector element. The more stringent requirement from a materials point of view is associated with the bias power, limiting values of V typically to ≤ 0.1 V for first generation system geometries. Typical system noise voltages are 2×10^{-9}V/Hz$^{1/2}$, hence, for parameter values $t = 10^{-3}$ cm, $A^{1/2} = 5\times10^{-3}$ cm, $\eta = 0.8$, $\phi = 10^{17}$/cm^2-s, Eq(2) indicates that $(\tau/n) >> 2\times10^{-22}$ s-cm^3 for background photon generated noise to be the dominant noise source. The upper limit on minority carrier lifetime in n-HgCdTe, determined by Auger band-to-band recombination[9], is given by $\tau_A = 2\tau_{Ai}(n_i/n)^2$, and is shown in Figure 2 for x = 0.205 HgCdTe ($\lambda_c = 12$ μm at 77°K) as a function of temperature for various doping concentrations. Thus $(\tau/n) = 2\tau_{Ai}\,n_i^2/n^3 >> 2\times10^{-22}$; hence, at 77°K, where $\tau_{Ai} = 10^{-3}$s, and $n_i = 3\times10^{13}$ cm^{-3}, we have $n << 2\times10^{15}$ cm^{-3}. Any degradation of lifetime by Shockley-Read recombination mechanisms serves only to further limit the required carrier concentration to even lower values. It is interesting to note that the BLIP condition $p = (\eta\phi\tau/t) > n_i^2/n$ for the Auger-limited case translates into $\tau_{Ai} > nt/2\eta\phi = 10^{-5}$s, for $n = 2\times10^{15}$ cm^{-3}. Figure 2 indicates that this condition is satisfied for all temperatures below 110°K.

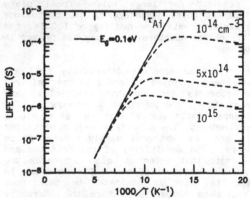

Figure 2   Auger-Limited Lifetime vs 10^3/T for Various Doping Concentrations for 0.1 eV HgCdTe

The above noise analysis has been somewhat oversimplified in that the impact of certain operational conditions has been neglected. These are the effects of 1) ohmic contacts[10] on effective lifetime for detector geometries that are small relative to a minority carrier diffusion length (L_n = 36 μm for τ = 4×10^{-6} s), 2) surface passivation techniques used to minimize both detector 1/f noise and surface recombination velocity effects, and 3) high background fluxes which generate majority carrier densities that are large compared to impurity related electron concentrations, and hence, in turn, limit lifetime values by the Auger-limited mechanism. A consideration of these phenomena has a two fold effect on materials requirements vis-a-vis detector performance. Firstly, the requirements for BLIP operation become somewhat more stringent that predicted by the simple model. A meaningful set of parameters is n < 10^{15} cm-3, and τ > 10^{-6}s, for 12 μm n-HgCdTe photoconductor operation at 77°K, under tactical background operating conditions. Secondly, material uniformity requirements are considerably relaxed. The surface passivation employed is an oxidation process which results in a fixed positive oxide charge at the HgCdTe surface, giving a strongly accumulated surface with typical equivalent values for nt = 1 to 2×10^{12} cm-2. Thus for bulk concentration < 10^{15} cm-3 the detector impedance is dominated by surface shunts which are extremely uniform. The effect of contacts and background flux also tends to dominate minority carrier lifetime values, and effectively damps out any non-uniformities of bulk lifetime for values of $\tau >> 10^{-6}$s. The overall result is one of homogenization for both array impedance and effective lifetime, and hence sensitivity, which obvious benefits with regard to focal plane uniformity.

The materials requirements of n < 10^{15} cm-3, and τ > 10^{-6}s, for first generation HgCdTe photoconductors have been met by material grown by solid state recrystallization (SSR), liquid phase epitaxy (LPE), and metal-organic chemical vapor deposition (MOCVD). The success of SSR material in satisfying these requirements is somewhat surprising in view of its microstructural quality (dislocation densities ~ 10^7 cm-2, and sub-grain boundaries) but is indicative of the forgiving nature of the bulk electrical and photoconductive processes in n-type 12 μm HgCdTe, and their relative immunity to microstructure. This of course was one of the main reasons for the original selection of the HgCdTe photoconductor as the primary first generation focal plane.

III. SECOND GENERATION FOCAL PLANES
     The sensitivity and resolution requirements of next generation infrared systems mandate the use of 10^3 to 10^6 detector elements in the focal plane together with a significant amount of associated signal processing, and rules out any possibility of utilizing an extension of first generation technology. Two significant technologies have emerged to address the overall second generation problem, and both involve the utilization of HgCdTe. The first technology employs metallurgical diodes hybridized to a Si signal processing chip, and the second employs MIS devices, which can be either hybridized to silicon, or fabricated in a monolithic form, e.g., charge coupled devices (CCD), charge injection devices (CID), charge imaging matrix (CIM). Regardless of the technology involved the next generation focal plane concept involves the use of a unit cell that integrates charge (or minority carriers) for a finite period of time prior to subsequent signal processing. It thus becomes meaningful to talk in terms of unit cell (or pixel) well capacity, whose magnitude is determined by required system performance and background flux levels involved. This well capacity can be proportioned between the infrared material and the Si processor, with the infrared materials requirements depending upon the amount of integration carried out there.

HgCdTe is again the infrared material of choice, and the materials requirements are most easily understood in terms of unit cell noise considerations, as was the case for first generation systems. Optimum focal plane performance is achieved when each unit cell is BLIP, and noise considerations for the integrating unit cell are expressed as variances on the number of carriers in the well, which is determined by the length of the integration period and the various current contributions. Minimal infrared materials requirements occur when no integration is attempted in the HgCdTe, as the infrared device is then operated essentially under equilibrium conditions. However in this case BLIP operation depends upon minimizing the noise associated with the Si processor whilst staying within the constraints imposed by focal plane geometry and power budget considerations. We will consider the HgCdTe materials requirements for second generation focal planes utilizing the integrating MIS detector. Requirements for metallurgical photodiodes will manifest themselves as one extreme of this concept.

The integrating MIS detector consists in its simplest form of a metal gate separated from the HgCdTe surface by an insulator. The gate is pulsed to an appropriate voltage V relative to the semiconductor, forming a potential well at the surface of the semiconductor for the collection of minority carrier charge. The device can be represented as an insulator capacitance C_{ox} in series with a depletion layer capacitance C_d, and τ_{int} minority carriers are integrated at the intermediate node, where J is the minority carrier current given by $J = J_d + J_\phi$, J_d is the device dark current and J_ϕ the current due to incident photon flux, τ_{int} the integration time. The maximum charge that can be stored on the node is $Q_{fw} = C_{ox} (V - V_{th})$, where V_{th} represents a threshold voltage. BLIP operation is obtained when $J_\phi > J_d$, and Q_{fw} is sufficiently large that the variance on this node dominates system noise. HgCdTe materials requirements for second generation focal planes are thus driven by considerations of Q_{fw} and J_d.

Well capacity is limited by the maximum voltage V that can be applied to the gate. The surface potential ϕ_s associated with this voltage is limited by tunneling via bandgap states, and the tunnel current is given by[11]

$$(3)$$

$$J_t = 10^{-13} \left[\frac{N_r \phi_s}{E_g} \right] exp \left[- \frac{5.3 \times 10^6 E_g^2}{E} \right] A/cm^2,$$

where N_r represents the density of midgap states (cm-3), E_g is the bandgap (eV) and E the electric field (V/cm). The variation of J_t with E predicted by Eq(3) is shown in Figure 3(a), for the two important atmospheric cutoff wavelengths, namely 5 μm (E_g = 0.25 eV) and 11.5 μm (E_g = 0.108 eV), assuming a value for N_r = 10[13] cm-3, which is characteristic of the best p-HgCdTe material. The right-hand axis of Figure 3(a) indicates the equivalent photon flux generated current assuming η = 0.6. BLIP performance requires that $J_t < J_\phi$, and this criterion is used to define an upper limit on applied electric field, the breakdown field E_{bd}, for the material in question. Tactical background flux levels are shown in Figure 3(a) for a typical f/2 system, and values for E_{db} are 3x10[4] V/cm for a 5 μm cutoff wavelength, and 8x10[3] V/cm for 11.5 μm.

This condition on breakdown electric field is readily translated into one of material quality, by an application of Gauss' law at the semiconductor-insulator interface of the MIS device. Thus for the empty well condition

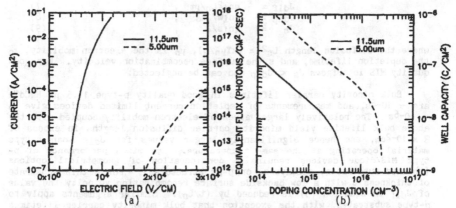

Figure 3 (a) Tunnel Current vs Electric Field and (b) Available Well Capacity vs Doping concentration, for 5 μm and 11.5 μm HgCdTe with $N_r = 10^{13} cm^{-3}$ midgap states, and $C_{ox} = 4 \times 10^{-8}$ F/cm^2

$$C_{ox} (V-\phi) = \varepsilon \varepsilon_0 E_{bd} = qpW \quad , \quad \text{(4)}$$

or,

$$\begin{aligned} C_{ox} V \quad &= \quad \varepsilon \varepsilon_0 E_{bd} + C_{ox}\phi_s \\ &= \quad \varepsilon \varepsilon_0 E_{bd} + C_{ox}\varepsilon \varepsilon_0 E_{bd}^2/2qp \\ &= \quad Q_{fw} + C_{ox}V_{th} \quad , \quad \text{(5)} \end{aligned}$$

where $V_{th} = 2\phi_f + (4\varepsilon \varepsilon_0 qp\phi_f/C_{ox}^2)^{1/2}$, W is the depletion region width, and ϕ_f is the bulk Fermi potential. In a well designed MIS device Q_{fw} is dominated by the E_{bd}^2 term. The dependence of available well capacity on doping concentration given by Equation (5) is shown in Figure 3(b) for the 5 μm and 11.5 μm breakdown field values deduced from Figure 3(a), assuming a realistic value for $C_{ox} = 4 \times 10^{-8}$ F/cm^2. Integrating MIS focal planes require well capacities in the 10^{-8} to 10^{-7} C/cm^2 range (equivalent to 4×10^5 to 4×10^6 electrons/mil^2) for pixel noise to dominate. It is apparent that such well capacities are readily available for good quality 5 μm HgCdTe at realistic doping concentrations in the > 10^{15}cm^{-3} range. However operation at 11.5 μm is a different proposition, due to the narrow bandgap involved, and the relatively low value of $E_{bd} = 8 \times 10^3$ V/cm. Pixel noise and dynamic range considerations indicate a value for Q_{fw} of 2 to 7×10^{-8} C/cm^2 is required at this cutoff wavelength, depending on cell geometry, system requirements, and available Si readout architecture. Figure 3(b) indicates that even for the best available material quality doping concentrations of p = 2.5 to 9×10^{14} cm^{-3} are required. The requirements for n-HgCdTe are even more severe, as the best values for E_{bd} in this material have been found to be ~ 70% that for p-type. Q_{fw} varies as E_{bd}^2, hence the required values can only be achieved for n = 1.25 to 4.5×10^{14} cm^{-3} at 11.5 μm.

Thermally generated dark currents in p-HgCdTe MIS/diode devices are given by

$$J_d = J_{dif} + J_{dep} + J_s \quad , \quad \text{(6)}$$

where the respective components refer to minority carrier diffusion from the bulk, generation from Shockley-Read centers in the depletion region, and generation from fast surface states. These components are given by,

$$J_{dif} = qn_i^2 L_n/p\tau \quad , \tag{7}$$
$$J_{dep} = qn_i W/2\tau_0 \quad , \tag{8}$$
$$J_S = qn_i s/2 \quad , \tag{9}$$

where the diffusion length $L_n^2 = (kT\mu_n/q)$, μ_n is the electron mobility, τ_0 the depletion lifetime, and s the surface recombination velocity. For good quality MIS interfaces $J_S < J_{dep}$ and can be neglected.

Bulk minority carrier lifetimes in good quality p-type 11.5 μm HgCdTe are ~ 10^{-6}s, and measurements of depletion current limited devices give τ_0 ~ 10^{-6}s. The relatively large values of electron mobility coupled with the above bulk lifetime yield minority carrier diffusion lengths in excess of 3×10^{-2}cm, and hence significantly larger values for J_{dif} than n-type material operating at the same temperature. Optimized performance of p-type MIS/diode devices require a consideration of geometrical options (thinned substrates, epilayers, or p-p+ structures). For a thin substrate of thickness t with a low backside surface recombination velocity the value of J_{dif} in Equation (5) is reduced by (t/L_n). Similar arguments apply to n-type substrates, with the exception that bulk minority carrier lifetimes tend to be even higher, in the range 2 to 10 μs.

A calculation of $(J_{dif} + J_{dep1})$ as a function of temperature utilizing Equations (7) and (8) modified for a substrate thickness of 10 μm is shown in Figure 4 for n and p-HgCdTe with an assumed cutoff wavelength of 11.5 μm. A doping concentration of 5×10^{14} cm-3 is assumed in each case, with $\tau_p = 2 \times 10^{-6}$s, $\tau_n = 10^{-6}$s, $\tau_0 = 10^{-6}$s, and an average surface potential of 0.3V. It is apparent that the thermally generated dark current at 77°K is an order of magnitude below that due to background photon flux, and presents no problem at this temperature of operation.

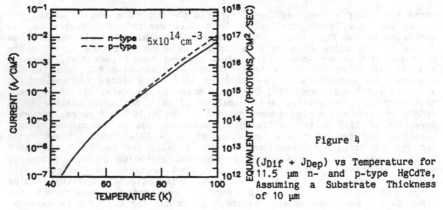

Figure 4

$(J_{dif} + J_{Dep})$ vs Temperature for 11.5 μm n- and p-type HgCdTe, Assuming a Substrate Thickness of 10 μm

Summarizing the MIS materials requirements for the homogeneous alloy for the most stringent case, namely an 11.5 μm cutoff, we can say that BLIP operation of the HgCdTe-Si unit cell can be achieved at 77°K for material with the following properties:

- p-type   2.5 to 9 x 10^{14}cm-3
- n-type   1.25 to 4.5 x 10^{14}cm-3
- Substructure free, low dislocation density ($< 10^5$cm-2), $N_r \leq 10^{13}$cm-3
- bulk minority carrier lifetime > 10^{-6}s
- depletion lifetime > 10^{-6}s

These requirements have been satisfied by a limited quantity of bulk material grown by both SSR and the travelling heater method (THM), but only on a very limited basis. We are not aware of an epitaxial process that can yet achieve these rather stringent goals, although a modified LPE process being developed at Texas Instruments shows promise, and has produced isolated layers that approach requirements.

Similar arguments can be made with regard to metallurgical diode operation in that the MIS device exactly simulates an open circuit one-sided abrupt diode. The thermally generated dark current issues are identical for similar substrates; however tunneling currents are theoretically less for hybrid diode operation between 0 and 0.1 V reverse bias relative to the values of effective reverse bias of 0.5 to 1.0V for pulsed MIS devices. Near-equilibrium diode operation is simulated by near threshold MIS operation, which is indicated in Figure 3(b) by the $Q_{pw} \approx 0$ condition. Hence for the quality of p-type substrate ($E_{bd} = 8 \times 10^3$ V/cm) shown the diode materials requirement on doping concentration is 2 to 4×10^{15}; for n-type it will be lower by a factor of two. However this is only true if the diode formation process results in an abrupt n+-p junction with the p-region of the quality indicated, namely $E_{bd} = 8 \times 10^3$ V/cm (dislocation free, $N_r \leq 10^{13}$ cm-3), otherwise the doping requirement limitations due to tunneling will become more stringent as E_{bd}^2.

The above materials requirements for abrupt junctions in the homogeneous alloy are stringent, requiring approximately state-of-the-art material. This has resulted in the development of heterostructural concepts to alleviate these requirements, namely heterojunctions and superlattices.

Heterojunctions consist of layers of materials with different bandgaps that are typically thicker than the Matthew's criterion[12], and as such any lattice mismatch shows up in the form of misfit dislocations, which can be very detrimental to device performance. The most well-known of these concepts is the heterojunction diode shown in Figure 5 in both its accepted forms, n-p+, and p-n+, for different compositions of HgCdTe. A band diagram for a corresponding homojunction is included for comparison. Apart from the obvious difference of a thermal dark current reduction from one side of the junction, some reduction in tunnel current is also achieved. A consideration of the geometry of tunnel transitions from midgap states to the various bands, in equilibrium at zero bias, indicates that the n+-p structure is the more optimum for inhibiting tunnel current, even though compositional grading is mandatory to eliminate the barrier to minority carrier flow in the conduction band. Extreme care must be exercised in matching the compositional grading and impurity distribution profiles in these junctions such that barriers do not appear in the p (narrow bandgap) regions of the device[13]. The one major drawback of the heterojunction diode concept is that the maximum electric field occurs at the p-n junction interface which is adjacent to the narrow bandgap material (within a fraction of a depletion region), thus limiting the degree of tunnel current reduction that can be achieved, particularly under reverse bias. This problem can be overcome by a judicious choice of compositional grading in the neighborhood of the p-n junction, although this will in turn mandate stringent planar uniformity requirements on doping concentration and compositional profiles. MIS heterojunction concepts are evolving which will involve photon detection in a low electric field narrow bandgap region, and charge integration in a high electric field wide bandgap region

24

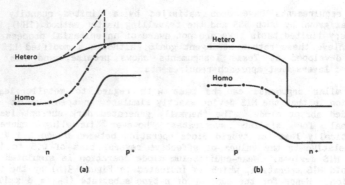

Figure 5 Heterojunction Diode a) n-p$^+$, and b) p-n$^+$ Geometries

The critical issues with regard to heterojunction devices are 1) the role played by misfit dislocations, 2) doping profile control, 3) compositional grading control if required, and 4) interface planarity. The misfit dislocation issue may well necessitate the use of quaternary alloys to eliminate the lattice mismatch problems.

Infrared superlattices on the other hand consist of quantum wells that are thin compared to the Matthew's criterion, and hence no misfit dislocations occur within the active volume of the structure, although individual layers will be in a state of elastic strain determined by the degree of lattice misfit. Dimensional quantization provides energy levels within the quantum well determined by the well thickness. Alternating barrier layers provide coupling between individual quantum wells, and the energy levels are broadened into bands. It becomes meaningful then to talk in terms of an effective mass in the superlattice direction, and this effective mass (inversely proportional to the bandwidth) is determined directly by the width of the barrier layers. Thus to first order the bandgap and effective mass of the structure are independent variables, which is very attractive from a tunnel current point of view. Initially it was thought that infinitely large masses, and hence very small tunnel currents could be employed for narrow bandgap structures, but a consideration of device electric fields, both in the superlattice direction and in the plane of the layers[14], indicates that one is limited to effective mass values not greatly in excess of the quantum well material band-edge mass. These considerations indicate that the gray tin bandstructure offers the greatest promise for implementation as an infrared detector, and investigations are underway at various laboratories on the HgTe/CdTe system. The tunnel current of Equation (3) contains an implied $m^{*1/2}$ term in the numerator of the exponent[11] and as the doping concentration requirement for tunnel breakdown varies as E_{bd}^2 it will be relaxed directly as m^*. Thus operation at 11.5 μm with a HgTe/CdTe superlattice (in which $m^* \sim 0.04\ m_0$) will theoretically require doping concentrations that are larger by a factor of 5 relative to the homogeneous HgCdTe alloy case.

The critical issues for the infrared superlattice are 1) layer interdiffusion, which will completely negate the effective mass argument above, 2) band structure, which is not yet completely understood, 3) doping concentrations which as yet are too high, typically $\sim 10^{16}$ cm^{-3}, and 4) defect concentrations due to low temperature non-equilibrium growth, which cannot be eliminated by post-growth annealing due to the layer interdiffusion problem.

IV. SUMMARY

Intrinsic detector materials, and in particular the alloy HgCdTe, have achieved a position of dominance in the infrared detector and imaging fields. The needs of current generation focal planes utilizing HgCdTe photoconductors have been met by a variety of growth techniques, including SSR bulk, and LPE and MOCVD epitaxial processes. Next generation focal plane materials requirements depend strongly on the operational wavelength range. Meaningful system operation at 3 to 5 µm is readily achievable with today's bulk or LPE material, however the 8 to 12 µm spectral range poses significant materials problems. Bulk growth by THM and SSR has yielded limited quantities of suitable material but it is not yet clear whether these techniques can meet the predicted large volume production requirements. No homogeneous alloy epitaxial process has yet demonstrated adequate performance, even for the somewhat relaxed requirements of metalurgical diode operation, however, modified LPE shows promise. Heterostructure concepts offer the potential for considerably relaxing the basic materials requirements for next generation infrared detectors, although at the expense of structural complexity, and indeed some grown heterojunction diode arrays have already exhibited excellent performance.

V. REFERENCES

1. D. Long, Infrared Phys. 7, 169 (1967).
2. W. D. Lawson, S. Nielson, E. H. Putley, and A.S. Young, J. Phys. Chem. Solids 9, 325 (1959).
3. J. O. Dimmock, I. Melngailis, and A. J. Strauss, Phys. Rev. Letters 16, 1193 (1966).
4. M. A. Kinch, S. R. Borrello and A. Simmons, Infrared Phys. 17, 1327 (1977).
5. M. B. Reine, A. K. Sood and T. J. Tredwell, "Semiconductors and Semimetals," Vol. 18, edited by R. K. Willardson and A.C. Beer (Academic Press, New York, 1981) p. 201.
6. I. Melngailis and T. C. Harman, Appl. Phys. Letters 13, 180 (1968).
7. M. R. Johnson, R. A. Chapman and J. S. Wrobel, Infrared Phys. 15, 317 (1975).
8. M. A. Kinch and S. R. Borrello, Infrared Phys. 15, 111 (1975).
9. M. A. Kinch, M. J. Brau and A. Simmons, J. Appl Phys. 44, 1649 (1973).
10. M. A. Kinch, S. R. Borrello, B. H. Breazeale and A. Simmons, Infrared Phys. 17, 137 (1977).
11. M. A. Kinch, J. Vac. Sci. Technology 21, 215 (1982).
12. J. W. Matthews and A. E. Blakeslee, J. Cryst. Growth 27, 118 (1974).
13. P. Migliorato and A. M. White, Solid-St. Electron. 26, 65 (1983).
14. M. A. Kinch and M. W. Goodwin, J. Appl. Phys. 58, 4455 (1985).

DEVELOPMENT OF HgCdTe FOR LWIR IMAGERS

JOSEPH L. SCHMIT
Honeywell Physical Sciences Center, 10701 Lyndale Ave. So.
Bloomington, MN 55420

ABSTRACT

This paper provides a historical perspective on the emergency of HgCdTe as the material of choice for long wavelength infrared (LWIR) imagers. The need for devices which see room temperature objects through the atmospheric window actually drove the development of this material. The lack of elemental or compound semiconductors having the desired wavelength response forced the choice of the alloy semiconductor, HgCdTe. The development of this material in several countries and companies beginning in the late 1950's is traced. The crystal growth methods used to grow HgCdTe have included melt growth techniques such as Bridgman, zone-melting, quench-anneal and slush-growth. The solution growth techniques include growth from HgTe-rich, Te-rich and Hg-rich solutions. Vapor phase growth has included evaporation, sputtering, molecular beam epitaxy (MBE) and metal-organic chemical vapor deposition (MOCVD). No perfect method has yet been developed, but several have provided material for the large area arrays needed for modern imagers.

NEED FOR 0.1 eV BAND GAP

Paul Kruse has given an excellent historical overview for the development of HgCdTe [1]. The military need for a thermal imaging capability at longer wavelengths drove the development of HgCdTe. In the late 1950's infrared detectors were either lead salts or InSb responding to 3-5um radiation or doped Ge which requires cooling to 30K. Figure 1 indicates one reason for interest in the longer wavelengths. It is a plot of radiant emission from a black body at three temperatures as a function of wavelength. Note that a 3000K black body (such as the filament of a light bulb) emits radiation in the visible range. Since our eye can detect radiation in the visible, we can see a hot filament and also things reflecting its radiation. However, the output from a 300K black body is 10,000 times lower and its peak intensity occurs at 10um, where our eye cannot see. Clearly, we need detectors operating near 10um if we expect to "see" room temperature objects such as people, trees and trucks without the aid of reflected light.

Usually the desire to see in the dark also includes the desire to see at long distances through air, therefore an additional constraint is added by the water and CO_2 absorption present in air. Figure 2 is a plot of the transmission through one mile of air as a function of wavelength [2]. Specific absorption bands of water and carbon dioxide are indicated. For longer distances, the transmission outside of the 3 to 5 and the 8 to 14 um "windows" drops nearly to zero. More detailed plots are available but this is sufficient to indicate one reason why people want detector response to be at specific wavelengths. The desired detector should thus have a band gap near 0.1eV and operate at 77K or above.

Elemental and compound semiconductors do not exist with the right band gaps to operate in the desired ranges. Figure 3 is a plot of cut-

28

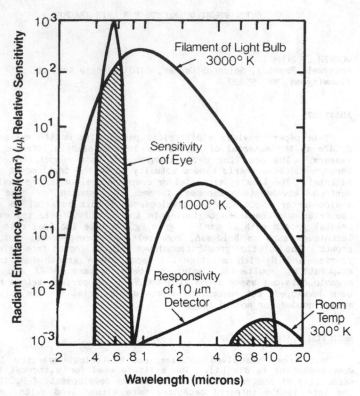

Fig. 1. Energy emitted by a black body and spectral sensitivity of two detectors versus wavelength.

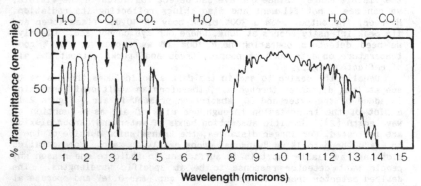

Fig. 2. Transmission versus wavelength for 1 mile of air. Water and carbon dioxide absorption bands are indicated.

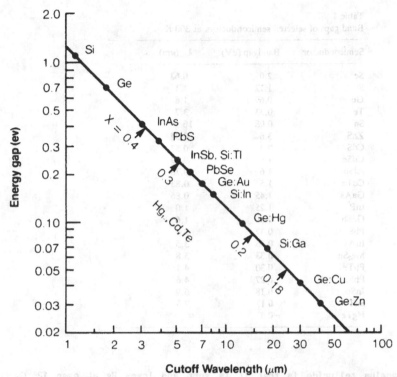

Fig. 3. Energy gap versus cutoff wavelength for elemental, compound, alloy and extrinsic detectors.

off wavelength as a function of energy for several semiconductors. The straight line is the relationship: $E = hc/\lambda = 1.24/\lambda$ where E is in eV and λ is in um. This plot includes elemental semiconductors such as Si and compound semiconductors such as PbSe, as well as extrinsic and alloy semiconductors. The data for the elemental, compound and alloy materials is for 77K while the data for the extrinsic materials are for lower temperatures. This plot does not include all possible materials because the plot becomes too crowded below 5 um. Table 1 is a more complete list of elemental and compound semiconductors [3]. Two facts stand out: (1) Each semiconductor has a specific energy gap which defines the wavelength it detects best. (2) None of the materials listed has a gap in the important 8-14 um window.

In 1959 a group at RSRE in England published a paper indicating that the alloy system $Hg_{1-x}Cd_xTe$ was a semiconductor over a wide composition range [4]. $Hg_{1-x}Cd_xTe$ is an alloy of HgTe and CdTe and has band gaps between those of the end compounds. Since HgTe is a semimetal with a negative band gap, the alloy covers the energy range of zero to 1.6 eV, or the spectral range 0.8 um to ∞. Figure 4 is a plot of the band gap versus composition at 80K. The lattice constants of HgTe and CdTe differ by only 0.3%. A composition of 20% CdTe is desired to give 10 um detectors, and figure 3 indicates the convenient range of band gaps available. The biggest disadvantage of mercury

Table 1
Band gaps of selected semiconductors at 300 K

Semiconductor	Bandgap (eV)	λ_{co} (μm)
Se	2.0	0.62
Si	1.12	1.1
Ge	0.67	1.8
Te	0.33	3.7
Sn	0.08	16
ZnS	3.6	0.34
CdS	2.4	0.52
CdSe	1.8	0.69
AlSb	1.6	0.78
CdTe	1.5	0.83
GaAs	1.45	0.86
InP	1.25	1.0
GaSb	0.80	1.6
PbS	0.37	3.4
InAs	0.35	3.5
Mg_2Sn	0.33	3.8
PbTe	0.30	4.1
PbSe	0.27	4.6
InSb	0.18	6.9
SnTe	0.17	7.3
HgTe	−0.3	∞

cadmium telluride is that it is soft and loses Hg at over 125°C.
However, its combination of a wide choice of band gaps, direct band
gap for intrinsic photodetection and a technology developed to provide
donor and acceptor concentrations below 10^{15}cm^{-3} have led to its wide
use in IR sensor systems world-wide. One of its advantages, wide
choice of band gap energy, has also led to a problem. It is necessary
to control composition closely to control the cut-off wavelength,
especially for long wavelength devices. The lack of knowledge of the
band gap was a formidable problem in the early 1960's. Studies were
begun in at least the U.S. [5,6], England [7], France [8], Poland [9],
and the USSR [10].

GROWTH OF HgCdTe

The high vapor pressure over HgCdTe presented a problem generally
not fully acknowledged in published reports: the quartz ampules used
to contain the materials often exploded.

Melt growth techniques

Bridgman growth was one of the first methods attempted with
HgCdTe. It consists of holding a melt above its melting temperature
and then lowering it slowly through a temperature gradient. It is
ideally suited to congruent melting materials in which the vapor
pressures of the constituents are negligible. However, neither of
these conditions are met with HgCdTe. I discuss details of the P-T
phase diagram later. Harman at MIT Lincoln Labs [11] solved the

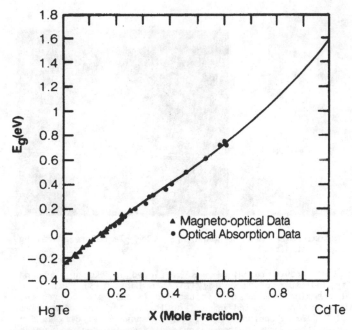

Fig. 4. Energy gap of $Hg_{1-x}Cd_xTe$ at 80K versus composition. The gap is negative and therefore effectively zero for IR detectors below 0.15 mole fraction CdTe.

explosion problem by using a lower temperature side arm. Parker and Kraus at Texas Instruments [12] used high pressure stainless steel containers to prevent the ampules from exploding while being quenched. Boules grown by the Bridgman technique often contain large radial and axial compositional gradients, due to segregation across a liquid-plus-solid phase [13]. The modification devised by Kruse [14] has come to be known by various names, including the quench-anneal (QA) process. In the QA process, a quartz capsule of liquid HgCdTe is quenched across the liquid-plus-solid zone, forming a dendritic mass having good long range homogeneity. This is then annealed below the solidus temperature and converted to a homogeneous crystal by solid state recrystallization. Figure 5 shows two low power microphotographs of a slice of $Hg_{1-x}Cd_xTe$ after quenching, but before annealing. The dendrites have been stained by a 30 s etch in $1:1:1=HCl:HNO_3:H_2O$ which blackens higher x material. Note that the dendrite spacing is on the scale of a few tenths of a millimeter. These photos are from 1965, since more recent ingots are always annealed before removal from the growth capsules. Figure 6 is a plot of the composition of one of these samples measured using an electron beam microprobe. The measurements were taken about every millimeter and thus missed many of the high x dendrites; however, it is clear that the sample is very inhomogeneous over a short range. After annealing for 2 weeks at ~650°C, the samples are homogeneous over several inches to better than the accuracy of the microprobe which is ±0.01 mole fraction CdTe. The repeatibility of the composition from run to run is <0.005 mole fraction CdTe for the QA process.

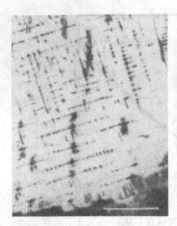

Fig. 5. Photomicrographs of QA grown $Hg_{0.8}Cd_{0.2}Te$ before annealing. Markers represent 1 mm.

Another melt growth technique was developed by Harman, of the Massachusetts Institute of Technology Lincoln Labs. [15]. It is a variation of freezing from a large volume and is called slush growth. This technique has been scaled up and optimized for the commercial production of HgCdTe by Cominco, Inc. Figure 7 shows a T-x diagram for HgCdTe showing the lens shaped liquid-plus-solid region, which prevents congruent solidification of solid having the same composition

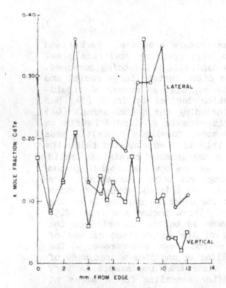
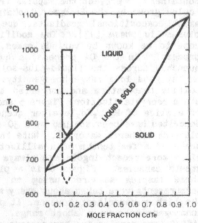

Fig. 6. Composition versus distance across a QA grown $Hg_{1-x}Cd_xTe$ sample before annealing.

Fig. 7. T-x phase diagram of $Hg_{1-x}Cd_xTe$ including a sketch indicating the slush-growth method.

as the liquid. In the slush growth technique, a capsule of HgCdTe of lower x than the desired composition is first held at position 1 to homogenize the melt. It is then quickly moved to position 2 where the tip is quenched. The capsule is kept in position 2 for several weeks, while solute (CdTe) diffuses to the solid-liquid interface and precipitates as HgCdTe. The very slow growth rate yields crystals which are very uniform across the face. However, the composition varies along the length of a crystal, unless CdTe is replenished in the melt.

Some work was done at Honeywell in the late 1960's to develop zone melting to grow HgCdTe [16]. By keeping the molten zone short and constant with time, boules were grown with regions of constant x extending over a large fraction of the boule length. However, these crystals always contained radial x-gradients of the order of 1 mole %. Figure 8 is a plot of the composition of a typical zone melted ingot versus length.

A more recent melt growth technique [17] has provided material of excellent compositional uniformity. Dozens of samples have been grown with carrier concentrations below $5 \times 10^{13} cm^{-3}$ and mobilities above $2 \times 10^5 cm^2$/Vs indicating lack of compensation.

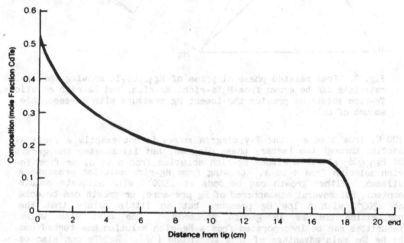

Fig. 8. Plot of composition vs length of a zone melted ingot.

Solution growth techniques

Liquid phase epitaxy (LPE) can be done from any solution supersaturated with HgCdTe. Three choices for the solution have been used. Figure 9 is a plot of four interrelated phase diagrams for $(Hg_{1-x}Cd_x)_{1-y}Te_y$: the ternary tie-line diagram, the T-x diagram for the psuedobinary, the T-y diagram through Te and Hg and a P_{Hg}-y diagram. The ternary diagram gives no temperature information, but indicates the compositions in equilibrium with $Hg_{0.6}Cd_{0.4}Te$. The alloy can be grown from three different solutions, with the composition and temperature as independent variables. The T-x diagram indicates that $Hg_{0.6}Cd_{0.4}Te$ can be grown from a HgTe-rich solution at about

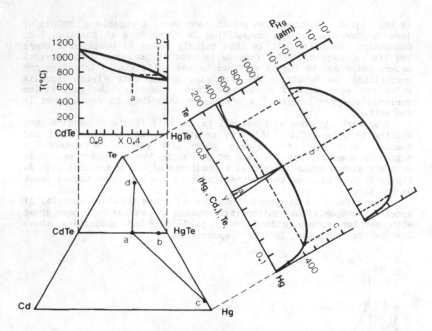

Fig. 9. Four related phase diagrams of $Hg_{1-x}Cd_xTe$ showing that x=0.4 materials can be grown from HgTe-rich, Hg-rich, and Te-rich solutions. Te-rich solutions provide the lowest Hg pressure with a reasonable amount of solute.

$700^{\circ}C$, from b to a. The T-y diagram shown is not exactly a vertical section through the ternary phase diagram, but illustrates the growth of $Hg_{0.6}Cd_{0.4}Te$ from either Hg-rich solution from c to a, or from Te-rich solution from d to a. Growing from Hg-rich solution presents a dilemma. Either growth can be done at $\sim500^{\circ}C$ with adequate solute content, but several atmospheres of Hg pressure; or growth can be done at $\sim300^{\circ}C$ with a low Hg pressure but so little solute that the composition changes as growth proceeds. The ease with which impurities can be incorporated from a Hg-rich solution has turned out to be the main advantage of this technique [18]. HgCdTe can also be grown from a Te-rich solution with adequate solute content and a low Hg pressure. The P-y diagram indicates that at $500^{\circ}C$, the Hg pressure over a Te-rich solution will be ~0.1 atm. This minimal pressure is maintained either by providing a Hg source upstream in the H_2 carrier gas stream or by enclosing the solution under a cover with a source of Hg. The liquidus temperature of the solution can be controlled independently of the x-composition, by controlling the y-composition (excess Te). Slider systems have provided layers uniform in composition to ±0.002 mole fraction across the surface and into the layer [19]. Figure 10 is a profile of such a layer as measured using an electron beam microprobe. The composition is constant with distance from the surface within the 1 mole % precision of the instrument.

Several companies are now making state of the art detectors and arrays, using material grown by LPE. The excellent lateral homogeneity has been the main reason for its success [20]. Electron concentrations have been achieved in the low $10^{14}cm^{-3}$ range. The biggest remaining problem is the terraced surface morphology, which often occurs on LPE films and limits its use for multiple layer structures having different composition, electrical type and doping concentrations.

Vapor phase techniques

Several vapor phase growth techniques have been proposed in order to reach the next level of sophistication. I discuss three techniques here: isothermal closed-spaced epitaxy (ICE), molecular beam epitaxy (MBE), and metal organic chemical vapor deposition (MOCVD).

In the ICE technique, a CdTe substrate is held a few mm above a source of Te in a capsule with a controlled Hg pressure in an isothermal furnace [21]. The driving force for the transport of Te is the chemical potential difference between the source and the substrate. As Te reaches the substrate, it combines with Hg to form HgTe, which stops further deposition until the HgTe interdiffuses with the CdTe from the substrate. Thus the surface composition is independent of growth time, but the thickness increases with growth time and the composition gradient decreases with growth time as shown in figure 11. To get a different composition, one must grow with a different composition source or grow at a different temperature. This technique was developed in the late 1960's but was abandoned for use in photoconductive (PC) detectors due to the x gradient. The ICE technique is again being investigated with the addition of a temperature gradient to try to reduce the compositional gradient.

MBE is a very simple process done with expensive equipment. The name implies that beams of molecules such as CdTe are deposited epitaxially on a substrate. In fact, beams of atoms or molecules of only one element are evaporated onto a substrate in a very high

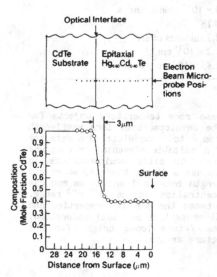

Fig. 10. Compositional profile through an LPE film grown from a Te-rich solution.

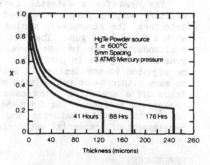

Fig. 11. Compositional profiles through VPE films grown by ICE for different lengths of time.

vacuum. In the case of HgCdTe the vacuum is not as good as for other materials, because a flux of excess Hg atoms must be supplied to prevent dissociation of the grown material. Research at LETI in Grenoble has been successful in producing HgCdTe with high mobility and low carrier concentration by MBE [22]. A novel MBE process has been developed by Rockwell International. The technique, known as LADA (laser assisted deposited), uses a laser pulse to evaporate the source, instead of using thermal Knudsen cells. Results using this technique are similar to those using the conventional MBE approach. Table 2 lists some of the parameters for the best layer reported at LETI.

Another vapor phase technique being pursued is MOCVD. It holds the potential of being a lower cost, higher throughput technique, but it is at a less developed stage. A group at RSRE was first of many to report growth of HgCdTe by MOCVD [23]. They studied the pyrolysis of individual metal organic compounds to determine the optimum compounds. The best material parameters have been reported by the GEC group in England [24]. They report a donor concentration of $2.2 \times 10^{14} cm^{-3}$ at 77K with a mobility of $4.9 \times 10^5 cm^2$/Vs indicating a lack of compensation. Material of this quality should fill the requirements for detector arrays suitable for the IR imagers now being produced. The hope is that it will do so at a cost sufficiently low that IR systems will proliferate into the commercial market.

Some of these growth techniques and the resulting material properties have been explained in more detail previously [25].

Table 2
$Hg_{1-x}Cd_xTe$ by MBE

Parameter	Value
T growth	180°C
Hg flux	5×10^{17} atoms/cm$^2 \cdot$s
x	0.18
Rocking curve	3.7 minutes of arc
n	1.2×10^{15} cm^{-3}
μ	1.85×10^5 cm^2/V$\cdot$s

CONCLUSION

The need for a detector to sense room temperature objects for military mapping purposes spurred the development of long wavelength detectors. The atmospheric window from 8 to 14 um dictated a detector with a 0.1eV band gap. The lack of a suitable elemental or compound semiconductor led to the development of alloy semiconductors in general and HgCdTe in particular. It has a direct band gap which can be adjusted to any desired IR wavelength between 1 and 20 um and can be made either n- or p-type with concentrations in the $10^{14} cm^{-3}$ range. Trade off's are still being made between the desired properties and the growth methods. The fundamental properties are well understood, the technology is maturing and the future looks bright for the development of commercial heterostructure arrays.

REFERENCES

1. P.W. Kruse, Semiconductors and Semimetals, 18, Edited by Willardson and Beer, (Academic Press, New York, 1981) p. 1-5.

2. P.W. Kruse, L.D. McGlauchlin and R.B. McQuistan, Elements of IR Technology (Wiley, London, 1962) p. 164.

3. J.L. Schmit, Crystal Growth of Electronic Materials, edited by E. Kaldis (Elsevier Sci. Publishers, 1985) chapter 20, p. 281.

4. W.D. Lawson, S. Nielsen, E.H. Putley and A.S. Young, J. Phys. Chem. Solids 9, 325 (1959).

5. T.C. Harman, A.J. Strauss, D.H. Dickey, M.S. Dresselhaus, G.B. Wright and J.G. Mavroides, Phys. Rev. Lett. 7, 403 (1961).

6. P.W. Kruse, M.D. Blue, J.H. Garfunkel, and W.D. Saur, Infrared Phys. 2, 53 (1962).

7. J.C. Woolley and B. Ray, J. Phys. Chem Solids 13, 151 (1960).

8. F. Bailly, G. Cohen-Solal and Y. Marfaing, C.R. Acad. Sci. Paris 257, 103 (1963).

9. R.R. Galazka, Acta Phys. Polon. 24, 791 (1963).

10. B.T. Kolomiets and A.A. Mal'kova, Fiz. Tverd. Tele 5, 1219 [English transl.]: Sov. Phys. Solid State 5, 889 (1963).

11. T.C. Harman, Physics and Chemistry of II-VI Compounds, edited by M. Aven and J.S. Prener, (Wiley, New York, 1967) P. 784 .

12. S.G. Parker and K. Kraus, U.S. Patent 3 468 363 (1969).

13. B.E. Bartlett, P. Capper, J.E. Harris and M.J.T. Quelch, J. Crystal Growth 46, 623 (1979).

14. P.W. Kruse and J.L.Schmit, U.S. Patent 3 723 190 (1973). Filed October 9, 1968; held under secrecy order until 1973.

15. T.C. Harman, J. Electron. Mater. 1, 230 (1972).

16. C.J. Speerschneider, unpublished work.

17. D.A. Nelson, W.M. Higgins, R.A. Lancaster, R.P. Murosako and R.G. Roy, Proc. IRIS, Vol. 29, p. 389-398 (1984), unclassified.

18. M.H. Kalisher, J. Crystal Growth 70, 365 (1984).

19. J.L. Schmit, R.J. Hager and R.A. Wood, J. Crystal Growth 56, 485 (1982).

20. S.H. Shin, M. Chu, A.H.B. Vanderwyck, M. Lanir and C.C. Wang, J. Appl. Phys. 51, 3772 (1980).

21. O.N. Tufte and E.L. Stelzer, J. Appl. Phys. 40, 4559 (1969).

22. J.P. Faurie and A. Million, Appl. Phys. Letters 41, 264 (1982).

23. S.J.C. Irvine and J.B. Mullin, J. Crystal Growth 55, 107 (1981).

24. M.J. Hyliands, J. Thompson, M.J. Bevan, K.T. Woodhouse and V. Vincent, J. Vac. Sci. Technol. A(4), 2217 (1986).

25. J.L. Schmit, J. Crystal Growth 65, 249 (1983).

MODELING OF HgCdTe HETEROJUNCTION DEVICES

KEN ZANIO AND KEN HAY
Ford Aerospace & Communications Corporation
Ford Road, Newport Beach, CA 92658-9983

ABSTRACT

A model for generating the composition and doping profiles from growth and diffusion parameters was developed for heterostructure devices. Poisson's equation was applied to these structures to predict barriers in the conduction band to minority carrier flow for long wavelength HgCdTe infrared detectors prepared by LPE techniques. Spectral response and quantum efficiency measurements illustrate the presence of these barriers and support the use of this model in predicting barrier formation.

A. INTRODUCTION

Heterojunction technology has significantly increased the resistance area products (RA) of long wavelength HgCdTe infrared detectors prepared by LPE techniques. Often increases in the RA of these heterojunction devices are associated with decreases in the quantum efficiencies. Electronic band models[1,2] have predicted potential barriers in HgCdTe which block minority carrier flow to the electrical junction and decrease the quantum efficiency. These models use the complementary error function to determine the composition profiles. This approach prevents the calculation of band diagrams for material having more complex composition profiles such as those obtained in the LPE growth of HgCdTe by the tipping technique. Even for the simple case of cap layers with flat composition profiles, previous models arbitrarily place the relative position of the metallurgical and electrical junctions, assume abrupt shapes of the doping profiles, and assume the interdiffusion coefficient in the transition region between the cap and base layers to be independent of composition. Because the resolution of the measurement of the actual composition and doping profiles is limited at best to a few thousand angstroms in LPE HgCdTe, further modeling is necessary to realistically predict the actual shape of the band diagrams, especially at these reduced dimensions. In this paper we incorporate growth and diffusion in an iterative model to predict the shape of the composition and doping profiles. We then use these profiles to calculate the band diagrams and the height of the barriers to minority carrier flow. A range of practical crystal growth conditions for low barrier heights are obtained when transition and depletion widths are comparable and doping levels in the n-type caps are either higher than or comparable to the doping levels in the p-type absorber layers. Our device results support this model.

B. GROWTH AND DIFFUSION MODEL

In the AlGaAs heterostructure system the composition profiles at the interfaces are abrupt because the interdiffusion coefficients are small. However in HgCdTe rapid interdiffusion occurs between Cd and Hg at LPE growth temperatures (375°C - 475°C) and must be considered in any electronic band structure model. The profiles are further complicated by a significant increase in the interdiffusion coefficient with increasing Hg content.[3] Often in previous models the p-n junctions have been arbitrarily located with respect to the composition profiles, and the donor and acceptor concentrations have been assumed to be abrupt. The resulting vague definition of the metallurgical interface and the uncertain position of the p-n junction leads to uncertain shapes of the band diagrams.

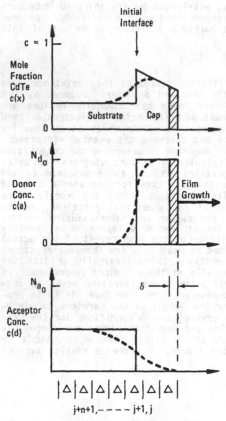

Figure 1. Schematic of the iterative growth and diffusion model with (- - -) and without (——) diffusion.

The general approach (Figure 1) for establishing a composition profile is to (1) grow in a time step, Δt, an incremental film of $Hg_{1-x}Cd_xTe$, δ, on the initial substrate, (2) using the new growth surface as the origin, slice the new substrate into thin slices of Δ, and (3) iterate the composition changes due to diffusion in Δt from the first time period to the next time period. The unknown composition $c(x)_j^{n+1}$ at time $(n+1)$ and at position j is directly determined from $c(x)_{j-1}^n$, $c(x)_j^n$, and $c(x)_{j+1}^n$ at the previous time n. The procedure is repeated until a film of desired thickness is grown. The iteration uses the following finite difference representation for Fick's second law:

$$c_j^{n+1} = r_j^n c_{j-1}^n + \left(1-2r_j^n\right) c_j^n + r_j^n c_{j+1}^n + \frac{\Delta t}{4\Delta^2}\left(c_{j+1}^n - c_{j-1}^n\right)\left(D_{j+1}^n - D_{j-1}^n\right) \quad (1)$$

$$j = 1,2,3,\ldots,n+1,$$

$$n = 0,1,2,3,\ldots,N.$$

Here j represents the jth slab of thickness Δ and N represents the number of iterations to grow a film of thickness $N\Delta$ in time $N\Delta t$. Distance and time are related by $r_j^n = D_j^n \Delta t/\Delta^2$, where D_j^n is the interdiffusion coefficient at position j in period n.

The expression for the interdiffusion coefficient[3] from 400°C to 600°C is:

$$D_{Hg_{1-x}Cd_xTe}(micron^2/sec) = 3.15 \times 10^{10} \times 10^{-3.53x} \times \exp(-2.24 \times 10^4/K) \qquad (2)$$

The diffusion coefficient varies about two orders of magnitude per hundred degrees and is about three and one-half orders of magnitude larger in HgTe than in CdTe.

The approach for establishing the donor and acceptor profiles are similar except the diffusion coefficients for the donors and acceptors are assumed to be independent of x. The donor and acceptor concentrations for the jth slab at time n+1 are respectively:

$$c_j^{n+1}(d) = r_j^n c(d)_{j-1}^n + \left(1 - 2 r_j^n\right) c(d)_j^n + r_j^n c(d)_{j+1}^n \quad \text{and} \qquad (3)$$

$$c_j^{n+1}(a) = r_j^n c(a)_{j-1}^n + \left(1 - 2 r_j^n\right) c(a)_j^n + r_j^n c(a)_{j+1}^n. \qquad (4)$$

The composition and doping profiles were calculated over a wide range of parameters for a wide band gap n-type cap on a narrow band gap p-type absorber. Fig-

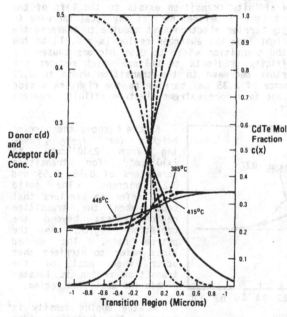

ure 2 shows the composition profiles for n-type caps having an x-value of 0.34 generated for 15 minutes at temperatures of 445°C, 415°C and 385°C at a deposition rate of 8 microns/hr on a p-type absorber layer having an x-value of 0.21. These conditions result in 15/85 composition transition widths, $W_{15/85}$, of 0.98, 0.55 and 0.27 microns. The 15/85 transition width is defined as the distance over which the composition changes from 15% to 85% of the total composition difference. The acceptor and donor concentrations are 5 x $10^{15}cm^{-3}$. The donor diffusion coefficient is for In and was determined from SIMS measurements to be 1 x 10^{-5} micron²/sec at 430°C. The acceptor diffusion coefficient is associated with native defects and is assumed here to be 8 x 10^{-5} microns microns²/sec.

Figure 2. Composition and dopant profiles for an n-type cap (right) deposited on a p-type base (left) at three different temperatures.

C. ENERGY BAND MODEL

A program was developed to calculate the energy bands as a function of depth from given doping and composition profiles in HgCdTe heterojunctions. The technique implemented[4] solves Poisson's equation utilizing Fermi-Dirac statistics which are necessary to correctly model systems with degenerate regions, such as n-type HgCdTe due to the extremely small electron effective mass. Only the the equilibrium condition is considered. The common anion rule is assumed causing the total band gap discontinuity to fall across the conduction band. The cap layers were assumed thick enough to achieve charge neutrality at the two extremes. A Newton-Raphson iteration technique was used to solve for the electrostatic potential, which satisfies Poisson's equation, to an accuracy of 10^{-4} kT.

The band structures were calculated over a wide range of parameters for a wide band gap n-type cap on a narrow band gap p-type absorber. The 77°K band structure for each of these cases in Figure 2 is plotted in Figure 3. The position of equivalent free electron and hole concentration, p/n, is located about 0.2 microns to the left of the Matano interface. This offset is primarily due to the higher density of states in the valence band than in the conduction band. For a transition width of 0.98 microns a barrier of 26 meV is located to the left of the p-n junction. This barrier is present because a major portion of the electron affinity transition exists to the left of the depletion region. The barrier height is defined here as any local increase in potential energy that a minority carrier electron must acquire to traverse the conduction band from left to right from the absorber region (x = 0.21) to the cap (x = 0.34). Decreasing the transition width to 0.55 microns causes the electrostatic and electron affinity gradients to balance which reduces the barrier to within kT. A further decrease in the transition width to 0.27 microns results in the appearance of a 35 meV barrier on the right hand side of the p-n junction. This is due to a more extreme electron affinity gradient between the cap and base layers.

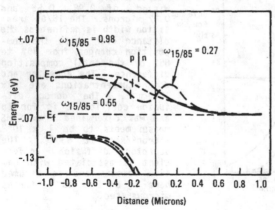

Figure 3. Energy band diagrams for $N_a = N_d = 5 \times 10^{15} \mathrm{cm}^{-3}$ as the transition width is changed.

Figure 4 shows the barrier height for $N_A = N_D$ ranging from $2 \times 10^{15} \mathrm{cm}^{-3}$ to $2 \times 10^{16} \mathrm{cm}^{-3}$ for transition parameters of 0.98, 0.55 and 0.27 microns. The solid lines refer to barriers that occur when the transition region exists beyond the depletion regions in the p-type base. The dashed lines refer to barriers that occur when most of the transition region is located within the depletion region.

As the doping density is reduced, the depletion width increases allowing structures with wider transition regions to become more acceptable. For example, the barrier height

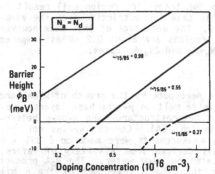

Figure 4. Barrier heights for equivalent donor and acceptor concentrations.

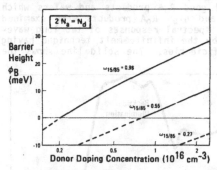

Figure 5. Barrier heights for donor concentrations twice the acceptor concentrations.

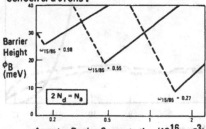

Figure 6. Barrier heights for acceptor concentrations twice the donor concentration.

for a 0.98 micron transition decreases from 45 meV to 10 meV as the doping density decreases from 2×10^{16} cm^{-3} to 1.5×10^{15}cm^{-3}. In contrast material with a 0.27 micron transition region becomes unacceptable when the doping level is reduced below about 7×10^{15}cm^{-3}. The band gap gradient becomes much larger than the electrostatic gradient at the metallurgical junction resulting in large barriers and the onset of the Anderson discontinuity. Material having the intermediate transition width of 0.55 microns shows the widest range of acceptable doping levels with a small barrier. From about 2×10^{15}cm^{-3} to 1×10^{16}cm^{-3} the barrier is below 2 kT. Figure 5 shows the barrier heights for the same transition regions and doping range but for $2N_a=N_d$. In general, the curves shift downward making the wider transition regions more acceptable and the narrow transition regions less acceptable. However, the average band gap in the depletion region is narrower than for $N_a=N_d$, which enhances recombination and tunnelling currents. For the case where the acceptor doping is twice the donor doping (i.e., $2N_d=N_a$) the barrier heights again decrease as the doping decreases. However, with further decreases in the doping concentration the barriers cannot be reduced to an acceptable value. This phenomenon is a consequence of the extension of the depletion region further into the more lightly doped cap layer that separates the electrostatic and electron affinity gradients and results in higher barriers. Figure 6 shows that decreases in the transition region will result in the reduction of the barrier height. This case is restrictive from the viewpoint of controlling materials parameters. The advantage of this case however is that the band gaps in the depletion regions are about 0.2 eV as compared to the 0.11 eV to 0.15 eV range for the $N_a=N_d$ and $2N_a=N_d$ cases.

D. EXPERIMENTAL RESULTS

Heterostructures were prepared at Ford Aerospace by the growth of wide band gap n-type cap layers from a Hg-rich infinite melt on p-type base layers. Devices were prepared by mesa etching our heterostructures and n-on-p heterostructures provided by Fermionics (Chatsworth, CA). Current versus voltage, spectral response and quantum efficiency measurements were made on 6 mil x 6 mil, 8 mil x 8 mil, and 15 mil x 15 mil devices from a variety of wafers. Approximately fifty devices were fabricated on each wafer. The R_oA product of 8 mil devices were generally larger than the R_oA product of the 6 mil devices. However, since in general no correlation between R_oA and size was found between the 8 mil and 15 mil devices, their R_oA products were characteristic of the bulk and not the surfaces. Wafers that had devices with good quantum efficiencies (40-60%) and good R_oA products and wafers which had poor quantum efficiencies (<10%) and high R_oA products were examined in further detail. Figure 7 shows 77°K spectral responses of two long wavelength heterojunction devices prepared by the infinite melt technique having both good and poor external quantum efficiencies. The solid line shows the

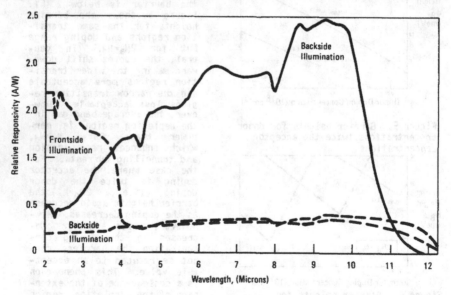

Figure 7. Spectral responses at 77°K for "barrier" (- - -) and "normal" (_____) HgCdTe heterojunction devices.

typical response of devices having efficiencies of about 40% and R_0A products of about 40 Ω-cm^2. The dashed lines show front and backside illuminated responses typical of devices from another wafer. These devices had poor (2-8%) quantum efficiencies at long wavelengths but high R_0A (105-165Ω-cm^2). For illumination through the cap there is also a short wavelength component of the response whose cutoff is about 3.8 microns and is characteristic of the composition of the n-type cap. Minority carrier electrons generated in the p-type base region by long wavelength photons are blocked from the junction by the barrier. However, minority carrier holes generated in the n-type cap from shorter wavelength photons are not blocked by a barrier in the valence band. Sputter XPS profiles by Rocky Mountain Analytical (Golden, CO) of this barrier device in Figure 8 shows the transition width to be about 1.4 microns. Devices fabricated from Fermionics' wafers consistently had both good quantum efficiencies and R_0A. Figure 8 also shows a Sputter XPS profile from a Fermionics heterostructure wafer. The transition width here is about 0.4 microns. The solid lines in Figure 8 are a fit of the model to sputter XPS data. The dashed lines refer to the model but for no interdiffusion.

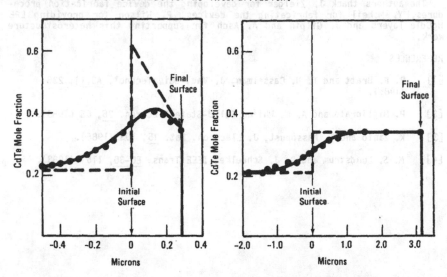

Figure 8. Fit of model (_____) to sputter XPS data (• • • •) for "barrier" (right) and "normal" (left) HgCdTe heterojunction devices.

E. DISCUSSION

Reference to model plots as in Figure 4, Figure 5 and Figure 6 can guide the preparation of heterostructure devices with reduced barrier heights. These figures apply to growth methods where the cap composition is constant but could be used as guide lines for cases of gradual changes in composition. Given that p-type substrates with an acceptor concentration of 5×10^{15}cm^{-3}

and an x-value of 0.21 were available and n-type caps with an x-value of 0.34 and a donor concentration of $10^{16} cm^{-3}$ could be obtained, appropriate temperatures and times must still be determined to obtain heterojunctions with acceptable barriers. Figure 5 shows that for this case a transition width of about 0.5 microns would be required. The transition region can be approximated by $2\sqrt{D_{HgCdTe} \cdot t}$ where the interdiffusion coefficient refers to the base region and t is the time of growth. This approximation also assumes that the growth rate is high enough for the cap thickness to be at least several times the transition width. Assuming no subsequent annealing, a growth time of 5 minutes would require a diffusion coefficient of 2.1×10^{-4} microns^2/ sec. From Equation 2 this corresponds to a growth temperature of about 450°C. Longer growth times would require lower growth temperatures. The growth time is one of the most uncertain parameters because the nucleation and growth of the layer at a precise time is difficult to control. Better control of such critical parameters will further improve the quality of HgCdTe long wavelength HgCdTe heterojunctions prepared by LPE.

The authors thank J. Zielger for developing the device fabrication procedures, Y. McNeil for fabricating the devices, P. Wilson for providing LPE HgCdTe layers and J. Gilpin and A. Asch for supporting this heterostructure work.

REFERENCES

[1]   P. R. Bratt and T. N. Casselman, J. Vac. Sci. Technol. A3(1), 238 (1985).

[2]   P. Migliorato and A. M. White, Solid-State Electron. 26, 65 (1983).

[3]   K. Zanio and T. Massopust, J. Electron. Mat. 15, 103 (1986).

[4]   M. S. Lundstrum and R. J. Schuelke, IEEE Trans. ED-30, 1151 (1983).

PREPARATION AND APPLICATIONS OF LEAD CHALCOGENIDE DIODE LASERS

DALE L. PARTIN
Physics Department, General Motors Research Laboratories, Warren, MI
48090-9055

ABSTRACT

Lead chalcogenide diode lasers are useful for spectroscopic and fiber optics applications in the mid-infrared (2.5-30 μm) wavelength range. These devices have previously required cryogenic cooling (<100 K) for CW operation. This limitation has been overcome through the use of a new, lattice-matched alloy system, $Pb_{1-x}Eu_xSe_yTe_{1-y}$, as well as the introduction of advanced, quantum well active region device structures grown by molecular beam epitaxy (MBE). Operating temperatures have been increased to 175 K CW (at 4.4 μm) and to 270 K pulsed (at 3.9 μm). Thermal leakage currents out of the device active region appear to be limiting device performance. This has led to the study of band offsets in PbEuSeTe/PbTe heterojunctions as well as to exploration of alternative high energy band gap alloys of PbTe with Ge, Yb, Ca, Sr, and Ba. The status of this work and examples of ultra-high resolution studies done with these tunable laser sources will be included.

INTRODUCTION

Lead-salt diodes provide tunable laser sources in the 2.5 to 30 μm wavelength range [1-5]. This wavelength range can be covered with either the PbCdS, PbSSe, and PbSnSe material systems or with PbGeTe and PbSnTe. The devices often have multimode emission and limited CW operating temperatures (<80 K), although CW operation up to 120 to 130 K has been reported [1,4,6,7]. These devices were often fabricated by diffusion into bulk crystals, although in some cases double heterojunction lasers were grown by liquid phase epitaxy, hot wall epitaxy or molecular beam epitaxy [1,4].

Three years ago, we reported on the growth of a new lead-salt semiconductor, lead-europium-selenide-telluride ($Pb_{1-x}Eu_xSe_yTe_{1-y}$), grown by molecular beam epitaxy [8]. The energy band gap is controlled by the europium concentration and lattice-matching to PbTe substrates is accomplished by adjustment of the selenium concentration. We have done preliminary studies of double heterostructure and quantum well active region devices made from this material system [9-15]. This work has stimulated others to explore the MBE growth of $Pb_{1-x}Eu_xSe/PbSe$ lattice-mismatched

Mat. Res. Soc. Symp. Proc. Vol. 90. ‹ 1987 Materials Research Society single quantum well diode lasers, and these devices have operated up to 163 K CW (235 K pulsed) [16]. We have also recently shown that PbSnTe active region devices can be used with PbSnYbTe confinement layers for longer wavelength operation [11]. The wavelengths that can be covered with diode lasers made from these materials systems are shown in Fig. 1, along with the wavelengths at which some gaseous molecules can be studied spectroscopically. Many of the materials systems shown may give high device operating temperatures when grown under optimum conditions by advanced epitaxy techniques. However, despite their short history, PbEuSeTe/PbTe devices have so far given the highest operating temperatures. The next section of this paper concerns the experimental techniques used to fabricate them. This is followed by sections on PbEuSeTe device properties, new materials, and tunable diode laser applications.

EXPERIMENTAL TECHNIQUES

The MBE growth system and procedures used in this study were similar to those previously reported for double heterostructure PbEuSeTe diode lasers [10]. The dopant and composition (x) profiles of $Pb_{1-x}Eu_xSe_yTe_{1-y}$ diode lasers are shown in Figs. 2a and 2b, respectively. The selenium concentration was adjusted to obtain lattice-matching between the PbEuSeTe layers and the PbTe substrate [8]. This laser structure has a PbTe single quantum well active region of thickness $L_z = 300$ Å. The $Pb_{1-x}Eu_xSe_yTe_{1-y}$ confinement layers have x = 0.018 near the active region, yielding an

TUNABLE SEMICONDUCTOR LASERS

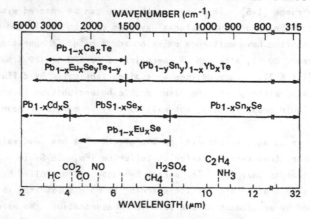

Fig. 1   Wavelength coverage of lead salt materials systems and molecular absorption lines. All possible systems are not shown for clarity.

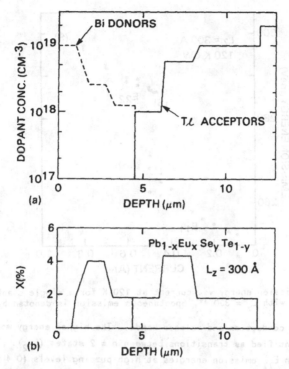

(a)

(b)

Fig. 2 (a) Dopant profile and (b) europium concentration vs. depth for a $Pb_{1-x}Eu_xSe_yTe_{1-y}$ large optical cavity single quantum well (LOCSQW) diode laser.

increase in energy band gap of 103 meV at 80 K [9]. The europium concentration was increased farther from the active region to form a separate optical cavity structure, since the index of refraction of PbEuSeTe decreases with increasing europium concentration [8]. Mesa stripe geometry diode lasers were fabricated as previously reported using an anodic oxide for electrical insulation [15]. The stripe widths for these lasers were 16 to 22 μm, and the cleaved cavity lengths were 325 to 450 μm long.

DIODE LASER PROPERTIES

The CW emission energy of a laser with a 300 Å wide active region is shown in Fig. 3 as a function of current at 120 K. The laser is generally single mode from laser threshold at 0.085 A up to about 0.35 A. The laser mode structure then breaks down, and at 0.47 A a new mode with much higher emission energy is observed. The lower emission energy modes are identified by "E_{11}", i.e., they are caused by transitions between the n = 1 quantum

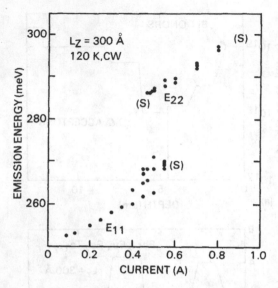

Fig. 3 CW emission energy vs. current at 120 K for a single quantum well
laser with L_z = 300 Å. Spontaneous emission is denoted by (S).

states of the conduction and valence bands. The higher energy modes are
similarly identified as transitions between n = 2 states (E_{22}). The rather
wide spread in E_{11} emission energies at high pumping levels (0.4 to 0.5 A)
is not well understood. Weakly allowed E_{12} transitions are apparently not
involved [17]. However, for the most part, the increase of E_{11} or E_{22}
emission energies with diode current is caused by joule heating of the laser
active region and the consequent increase in energy band gap normally
observed in lead-chalcogenide semiconductors. At lower heat sink
temperatures, the E_{11} transition persists up to higher currents. At higher
heat sink temperatures, the E_{11} threshold current increases and the current
at which the E_{22} mode appears decreases until the laser switches to the E_{22}
mode at threshold.

An important aspect of understanding the operation of these single
quantum well PbEuSeTe lasers is the energy band offset at a PbTe/PbEuSeTe
heterojunction. Initially, attempts were made to determine these offsets by
careful modeling of diode laser emission energies [14]. However, the near
symmetry of the conduction and valence bands at the L-point band extrema in
PbTe and the fact that the heavy hole band is non-degenerate with the light
hole band make the calculated laser emission energies rather insensitive to
the band offsets. The only clear result is that neither band offset can be
less than about 0.1 ΔEg, where ΔEg is the increase in energy band gap at the
heterojunction. In more recent studies of 70 Å wide PbTe single quantum

wells with $Pb_{0.9}Eu_{0.1}Se_{0.096}Te_{0.904}$ boundary layers, magnetoresistance and Hall measurements indicate that the valence band offset is approximately 0.9 ΔEg [18]. It should be noted that the energy gap of $Pb_{1-x}Eu_xSe_yTe_{1-y}$ increases rapidly and linearly with x up to x ~ 0.05, beyond which it increases relatively slowly. This effect is especially prominent at low temperatures [19]. Thus, the band offsets for PbEuSeTe heterojunctions with x ≲ 0.05, which are appropriate for quantum well diode lasers, might not directly follow from band offset determinations at higher europium concentrations. However, the present data indicate that the conduction band offset at PbTe/PbEuSeTe heterojunctions is relatively small, suggesting that electron leakage current out of a PbTe quantum well active region may be dominating the threshold current at higher operating temperatures. At 270 K, this would indicate that the barrier for electron leakage is ~KT [14].

One approach to reducing device threshold currents is therefore to design structures which minimize electron leakage currents out of a PbTe quantum well active region. As a preliminary effort in this direction, a side optical cavity single quantum well (SOCSQW) structure was grown. As shown in Fig. 4, this is an asymmetrical structure in which the quantum well

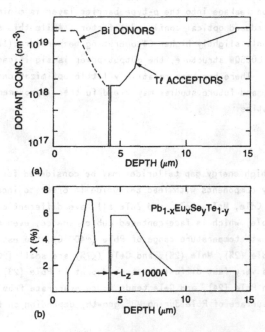

Fig. 4 (a) Dopant profile and (b) europium concentration vs. depth for a side optical cavity single quantum well (SOCSQW) diode laser.

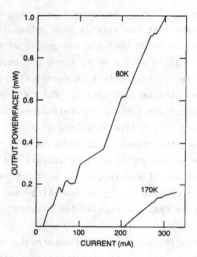

Fig. 5 Single-ended CW output power vs. diode current at 80 K and 170 K
heat sink temperatures.

active region is displaced to one side of an otherwise conventional large
optical cavity single quantum well (LOCSQW) structure [15]. Thus, the
barrier to electron leakage into the p-type barrier layer is minimized at
the expense of a reduced optical confinement factor. While this structure
has so far given only slightly higher CW operating temperature (175 K) than
the corresponding LOCSQW structure, the output power is significantly
improved (Fig. 5). There has been relatively little optimization of this
structure to date, and future studies may yield further improvements in
device characteristics.

NEW MATERIALS

A number of high energy gap tellurides may be considered for
pseudobinary alloy components with lead telluride in order to increase its
energy band gap. CdTe, MnTe, GeTe, and ZnTe all have different crystal
structures than PbTe, which is face-centered cubic. Hence, even at the high
end of the MBE growth temperature range of PbTe (~400°C), the estimated bulk
solubilities of CdTe (2%), MnTe (2%), and GeTe (≲7%) are small [20-22].
Also, cadmium is a very fast diffusing n-type dopant in PbTe [23], Mn atoms
tend to cluster in PbTe [24], and GeTe tends to re-evaporate from or surface
segregate on the surface of PbTe during MBE growth, depending on the growth
temperature [25].

These low solubilities have led us to consider alloys formed between
PbTe and high energy gap tellurides which have the fcc crystal structure.

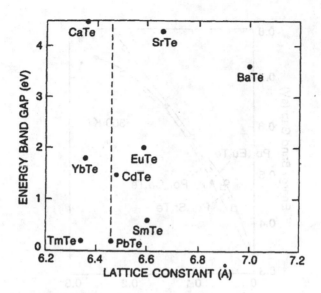

Fig. 6   Energy band gap vs. lattice constant for several face-centered-cubic
telluride compounds and for CdTe.

Some possible binaries are shown in Fig. 6. Of the four semiconducting rare
earth-monotellurides, TmTe is ruled out by its relative small energy gap.
SmTe has a marginally larger energy gap than PbTe, but is also known to be a
donor in PbTe [26]. EuTe and YbTe appear to form a continuous series of
solid solutions with PbTe [27-29] However, our studies of MBE grown PbYbTe
indicated that it cannot be doped heavily p-type, apparently because of
instability in the Yb valence state: $Yb^{+2} + Yb^{+3} + e$ [30,31]. This same
instability may explain the donor action of Sm in PbTe. While YbTe can be
used in combination with PbSnTe to make very long wavelength diode lasers,
it is basically only EuTe-PbTe alloys which have proved useful for making
heterojunction diode lasers involving rare earth elements. This is caused
by the very high stability of Eu in the +2 valence state in these
materials [33]. The properties of PbEuSeTe lattice-matched heterojunction
diode lasers have been described in the preceding section.

     The alkaline earth tellurides are an alternative to the rare earth
tellurides for high energy gap alloys with PbTe (Fig. 6). There was a brief
comment in some early work that CaTe and BaTe are essentially insoluble in
PbTe [34]. This conclusion was apparently based on attempts to grow bulk
crystals of the alloys at relatively high temperature. More recently, PbCaS
and PbSrS films were successfully grown by MBE [35].

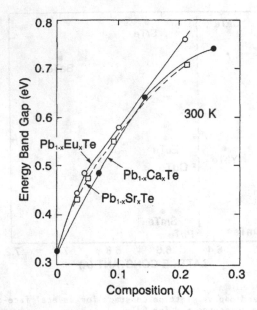

Fig. 7 Energy band gap at 300 K vs. composition for PbTe-based ternary
compounds.

We have recently studied the properties of MBE grown PbCaTe, PbSrTe,
and PbBaTe [36,37]. Several materials properties were explored, including
the dependence of energy band gap on alloy composition (see Fig. 7). The
curve for PbBaTe is not included since BaTe appears to have a small
solubility in PbTe. From the figure, it is seen that increases in energy
band gap are roughly comparable for PbEuTe, PbCaTe, and PbSrTe. Thus, their
relative merits for diode lasers may be determined by factors such as energy
band offsets, which have not yet been determined for PbCaTe/PbTe and
PbSrTe/PbTe heterojunctions. In other recent work, PbCdSSe/PbS lattice-
matched double heterojunction diode lasers have been grown by MBE with
pulsed operation up to 200 K at an emission wavelength of 3.27 μm [38].
Despite the problem with Cd diffusion into PbTe mentioned earlier, there may
be a possibility of using CdTe/PbTe heterojunctions for diode laser
applications. As shown in Fig. 6, such heterojunctions would be nearly
lattice matched at room temperature (Δa/a = 0.3%), and give a large change
in energy band gap. Such heterojunctions have been grown by hot wall
epitaxy [39,40], ionized cluster beam epitaxy [41,42], and molecular beam
epitaxy [43].

TUNABLE DIODE LASER APPLICATIONS

The fundamental characteristics of lead salt diode lasers which make them useful for various applications are listed below.

- Coverage of the wavelength range 2.5-30 μm.
- Ease of tunability by changing the heat sink temperature.
- Capability of rapid, repetitive wavelength modulation by modulating the diode current.
- Tuning range of one diode over 700 cm^{-1} CW.
- Single mode tuning >5 cm^{-1}.
- Line width $\lesssim 10^{-4} cm^{-1}$ ($<10^{-6} cm^{-1}$ has been observed) [44].
- Single mode power of ~1 mW.
- CW operation up to 175 K, pulsed to 270 K.
- High device reliability.

While not all of these features are attainable in the entire wavelength range that can be covered, the trend in the last few years has been for rapid advances in device performance. This has mainly resulted from the impact of advanced epitaxy techniques, such as MBE, to grow lattice-matched heterojunctions, and from the use of new materials systems, such as PbEuSeTe.

The main areas of application are:

- Ultra-high resolution molecular spectroscopy.
- *Trace gas detection and pollution monitoring.*
- Heterodyne detection [45].
- Isotope selective gas detection [46].
- Surface gas adsorbates.
- Fiber optic communications.

Of these areas, perhaps the most potentially exciting are the last three. Isotope selective gas detection is currently of interest for elements which do not have radioisotopes with adequately long half-lives (e.g., O, N) or where the use of available radioisotopes is precluded by biomedical applications (e.g., C). In these circumstances, tunable diode laser spectroscopy has much higher sensitivity and selectivity than alternative techniques such as mass spectroscopy, especially in cases where different molecules have the same mass (e.g., $^{14}N_2$ and $^{12}C^{16}O$, or $^{12}C^{17}O$ and $^{13}C^{16}O$). A recent innovation in this field is the use of a dual path measurement cell in which one path length is adjustable (Fig. 8) [47]. This is extremely useful for determining with precision the concentration of a very small concentration of one isotope in the presence of a very large concentration of another isotope. The narrow linewidth and large tuning range of a single laser mode are essential for this application.

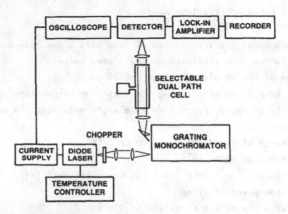

Fig. 8 Schematic diagram of the diode laser isotopic analysis system. The
short path cell was used to measure the more abundant isotopic
species. The long path cell was used to measure the less abundant
isotopic species.

Electroreflectance vibrational spectroscopy (EVS) is a new technique to
study molecules adsorbed on surfaces [48,49]. It is highly sensitive and
selective, and may offer the potential to study surface adsorbates at
atmospheric pressure. The wide tunability (several hundred wavenumbers) of
lead salt diode lasers and single mode, high power emission are essential in
this application.

Finally, optical fibers made from heavy metal fluorides and other
compounds are currently under development which have the potential for
achieving optical losses in the 10^{-2}-10^{-3} dB/km range at $3 \gtrsim \lambda \lesssim 10$ μm
wavelengths if their impurity and defect concentrations can be reduced.
This would allow longer transmission distances to be achieved without diode
laser repeaters than is attainable with silica fibers and diode lasers
operating at short (0.8 - 1.5 μm) wavelengths. The long emission
wavelengths, relatively high output power and high operating temperatures of
lead salt diode lasers are essential in this application.

REFERENCES

1. H. Preier, Appl. Phys. 20, 189 (1979).
2. T. C. Harman, J. Phys. Chem. Solids 32, 363 (1971).
3. A. R. Calawa, J. Luminescence 7, 477 (1973).
4. H. Holloway and J. N. Walpole, Prog. in Cryst. Growth and Charact. 2,
    49 (1979).
5. K. J. Linden, SPIE Conf. on Tunable Diode Laser Development and
    Spectroscopy Applications 438, 2 (1983).
6. H. Preier, M. Bleicher, W. Riedel, and H. Maier, J. Appl. Phys. 47,
    5476 (1976).

7. K. J. Linden, K. W. Nill, and J. F. Butler, IEEE J. Quant. Electron. QE-13, 720 (1977).
8. D. L. Partin, J. Electron. Mater., 13, 493 (1984).
9. D. L. Partin and C. M. Thrush, Appl. Phys. Lett. 45, 193 (1984).
10. D. L. Partin, Appl. Phys. Lett. 43, 996 (1983).
11. D. L. Partin, Optical Engrg. 24, 367 (1985).
12. D. L. Partin, Appl. Phys. Lett. 45, 487 (1984).
13. D. L. Partin, R. F. Majkowski, and D. E. Swets, J. Vac. Sci. Technol. B 3, 576 (1985).
14. D. L. Partin, Superlattices and Microstructures 1, 131 (1985).
15. D. L. Partin, J. Heremans, and C. M. Thrush, Superlattices and Microstructures (to be published).
16. P. Norton, G. Knoll, and K. H. Bachem, J. Vac. Sci. Technol. B 3, 782 (1985).
17. B. Lax (private communication).
18. J. Heremans, D. L. Partin, M. Shayegan and H. D. Drew, Proc. 18th International Conference on the Physics of Semiconductors, O. Engstrom (ed.), World Scientific Publishing Co. PTE, Ltd, Singapore, Aug. 11-15, 1986.
19. W. C. Goltsos, A. V. Nurmikko, and D. L. Partin, Sol. St. Comm. 59, 183 (1986).
20. A. J. Rosenberg, R. Grierson, J. C. Wooley, and P. Nikolic, Trans. Met. Soc. AIME 230, 342 (1964).
21. V. G. Vanyarkho, V. P. Zlomanov, and A. V. Novoselova, Inorg. Mat. 6, 1352 (1970).
22. D. K. Hohnke, H. Holloway, and S. Kaiser, J. Phys. Chem. Solids 33, 2053 (1972).
23. E. Silbeg, Y. Sternburg, and N. Yellin, J. Solid State Chem. 39, 100 (1981).
24. D. G. Andrianov, N. M. Pavlov, A. S. Savelev, V. I. Fistul and G. P. Tsiskarishvili, Soc. Phys. Semi. 14, 711 (1980).
25. D. L. Partin, J. Vac. Sci. Technol. 21, 1 (1982).
26. G. T. Alekseeva, M. N. Vinogradova, K. G. Gartsman, A. Yu. Zyuzin, Kh. R. Mailina, L. V. Prokofeva, and L. S. Stilbans, Sov. Phys. Sol. St. 27, 1953 (1985).
27. R. Suryanarayanan and C. Paparoditis, J. Phys. 29, C4 (1968).
28. C. Paparoditis and R. Suryanarayanan, J. Cryst. Growth 13/14, 389 (1972).
29. R. Suryanarayanan and C. Paparoditis, Colloq. Int. C.N.R.S., No. 180, 149 (1969).
30. D. L. Partin, J. Vac. Sci. Technol. B1, 174 (1983).
31. D. L. Partin, J. Electron. Mater. 12, 917 (1983).
32. D. L. Partin, Optical Engrg. 24, 367 (1985).
33. A. Jayaraman, in: Handbook on the Physics and Chemistry of Rare Earths, ed. K. A. Gschneidner, Jr. and L. Eyring, North Holland Publ., New York, p. 575 ff (1979).
34. L. M. Rogers and A. J. Crocker, J. Phys. D 4, 1016 (1971).
35. H. Holloway and G. Jesion, Phys. Rev. B 26, 5617 (1982).
36. D. L. Partin, B. M. Clemens, D. E. Swets, and C. M. Thrush, J. Vac. Sci. Technol. B 4, 578 (1986).
37. D. L. Partin, C. M. Thrush, B. M. Clemens, (to be published).
38. N. Koguchi, T. Kiyosawa, and S. Takahashi, Fourth Int'l Conf. on Molecular Beam Epitaxy, York, U.K., 9 Sept. 1986.
39. A. Lopez-Otero and W. Huber, Surface Science 86, 167 (1979).
40. J. Humenberger, H. Sitter, W. Huber, N. C. Sharma, and A. Lopez-Otero, Thin Solid Films 90, 101 (1982).
41. T. Takagi, H. Takaoka, and Y. Kuriyama, Thin Solid Films 126, 149 (1985).
42. I. Yamada, H. Takaoka, H. Usui, and T. Takagi, J. Vac. Sci. Technol. A 4, 722 (1986).

43. J. Yoshino, H. Munekata, and L. L. Chang, Seventh U.S. MBE Workshop, Amer. Vac. Soc., Cambridge, MA, 20 Oct. 1986.
44. C. Freed, J. W. Bielinski, and W. Lo, Appl. Phys. Lett. 43, 629 (1983).
45. J. M. Hoell, Jr., C. N. Harward, and W. Lo, Optical Engrg. 21, 320 (1982).
46. D. Labrie and J. Reid, Appl. Phys. 24, 381 (1981).
47. P. S. Lee and R. F. Majkowski, Appl. Phys. Lett. 48, 619 (1986).
48. D. K. Lambert, Proc. Soc. Photo-Optical Inst. Eng. 438, 158 (1983).
49. D. K. Lambert, J. Vac. Sci. Technol. B 3, 1479 (1985).
50. T. Miyashita and T. Manabe, IEEE J. Quant. Elec. QE-18, 1432 (1982).
51. D. C. Tran, G. H. Sigel, Jr., and B. Bendow, J. Lightwave Technol. LT-2, 566 (1984).

EPITAXIAL IV-VI SEMICONDUCTOR FILMS

T. K. CHU† AND A. MARTINEZ†**
† Naval Surface Weapons Center, 10901 New Hampshire Ave., Silver Spring, MD 20903-5000
** Physics Department, The American University, Washington, DC 20016

ABSTRACT

Epitaxial films of IV-VI semiconductors and their alloys form the basis of an infrared detector technology that offers advantages in material stability as well as spectral versatility. These films are prepared by epitaxial hot-wall techniques and their material properties are essentially the same as those of bulk crystals. Because of their stability, multi-layer growths of the materials can be achieved in a straight-forward manner. To date, multi-color detectors and small scale two-color detector arrays have been demonstrated successfully. A brief review of the growth method and the growth characteristics is given. Recent advances in superlattice research, especially those of interest to electro-optical devices, will be discussed. These include persistent photoconductivity and sub-bandgap optical transition.

INTRODUCTION

The IV-VI semiconductors form an important class of infrared materials. Typical applications include laser diodes, detectors and optical filters [1,2]. Their use for infrared detectors dates before the second world war; and photoconductive PbS and PbSe detectors still occupy an important segment of the commercial detector market. This photoconductive detector technology is based on polycrystalline films deposited by wet chemical processes. Studies of epitaxial thin films of the IV-VI compounds started in the early sixties [3]. Their use as a basis for photovoltaic infrared detectors followed in the early seventies. Among the technological semiconductors, the IV-VI compounds are unique in being stable molecularly and having high vapor pressures (or low melting temperatures). This combination makes them suitable for the thermal evaporation techniques. Thus, most of the earlier epitaxial work has been done with the hot wall method or flash evaporation method. However, because of interests in heterostructure diode lasers, quantum wells and superlattice, molecular beam epitaxy method has also been used in recent years.

In the following, a brief discussion of the hot wall thermal evaporation method is given. Results on the growth characteristics of a IV-VI compound and II-IV-VI alloys are presented. Advances in multicolor photovoltaic infrared detectors, for which the IV-VI semiconductors are especially suitable, will also be given. Our recent work on superlattices of the IV-VI compounds will also be described.

HOT WALL EPITAXY (HWE)

Hot wall epitaxy was used as early as 1967 for the growth of IV-VI semiconductors [4], mostly the Pb-salts and their alloys. The essence of the hot wall method is that the source material and the surface of deposition (i.e. that of the substrate) are contained within the same enclosure. The wall of the enclosure is heated usually to the same temperature or to a higher temperature than the source. Film growth presumably takes place near thermodynamic equilibrium condition. As the type and concentration of charge carriers in the IV-VI semiconductors are governed by deviation for perfect stoichiometry, the HWE technique in principle allows a degree of control over these material parameters equal to that of the starting materials. An additional measure of control by means of an auxilliary source, either of the group IV or group VI element, can also be incorporated into the HWE apparatus. Indeed, high quality IV VI semiconductor films have been grown on alkali halide and on group II fluoride substrates. For PbTe, carrier mobilities of $\sim 4 \times 10^2 \ m^2 \ v^{-1}s^{-1}$ (4.2 K) has been reported [5]. We have grown PbTe and In-doped PbSnTe films with carrier concentrations $\sim 5 \times 10^{22} \ m^{-3}$ and $\sim 5 \times 10^{23} \ m^{-3}$; and mobilities at 4.2 K of 70 $m^2v^{-1}s^{-1}$ and 30 $m^2v^{-1}s^{-1}$ respectively. An extensive review of the HWE method has been previously given by Lopez-Otero [5]. Reviews of IV-VI material properties have also been published [6].

Despite its success with IV-VI compounds there remain certain important questions on the nature of film growth under HWE. It has been argued [1] that HWE is basically a vapor transport method, and that film growth takes place under near thermodynamic equilibrium is not clearly established. Evidence for the above argument is also suggested by the growth of certain pseudo-binary alloys: Weiser et al. [7] has reported a significant change in composition during the evaporation process with various PbSnTe compounds. Our growth experiments with PbSe and a PbCdS alloy indicates that near thermodynamic equilibrium condition is not necessary for good film growth, and that kinematic considerations may be more important.

The growth apparatus is in an "open" hot wall configuration, where there is a space of 2-4 cm between the source oven and the substrate, as is reported previously [8]. The preparation of the source material was similar to that reported: elemental materials were vacuum sealed in a quartz ampoule, thermally reacted, melted in an oven, and quenched. Before loading the materials, the inside surface of the ampoule was coated with a thin layer of carbon for the prevention of reaction between the metals with quartz and for the reduction of possible oxides on the surfaces of the elemental materials. Substrate materials are BaF_2, air-cleaved with cleavage plane perpendicular to the <111> direction.

The PbSe films were grown with source temperature from 500 to 550 C and substrate temperature from 300 to 400 C, with the majority of films grown at 350 C. The source material is p-type. The films are also p-type with carrier concentration $\sim 3 \times 10^{17} \ cm^{-3}$. Figure 1 shows the Hall mobility (μ) of the films grown at substrate temperature 350 C. It is seen that the mobility at 77 K falls within a range of 2 - 3 $m^2v^{-1}s^{-1}$. Low mobilities usually can be correlated with poor substrate quality: especially when a large cleavage step runs across the Hall specimen, causing discontinuity in the film. This problem becomes more acute for the thinner samples, the mobilities of which are biased toward lower values. For comparison, the mobility of one specimen ($\mu_{293K} \sim 1 \ m^2v^{-1}s^{-1}$ and $\mu_{77K} \sim 3 \ m^2v^{-1}s^{-1}$) was also measured at 4.2 K and yielded a value of 40 $m^2v^{-1}s^{-1}$. This is among the highest reported for PbSe [5,9]. It should be emphasized that films were grown with the closed HWE in Ref. 5, and with effusion cells in Ref. 9.

With the substrate kept at 350 C the growth characteristics under various combinations of source temperature (T_s) and oven-to-substrate

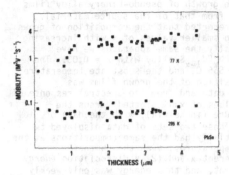

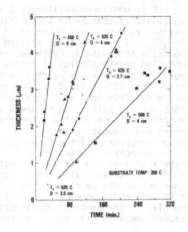

Figure 1: The Hall mobility of PbSe films grown by an "open" hot wall apparatus on BaF$_2$ substrates at 77 K (upper points) and 293 K (lower points) as a function of film thickness. The substrate temperature was 350 C, and source temperatures 500 C, 525 C and 550 C. Films with the highest mobilities were grown with source temperature at 550 C.

Figure 2: Thickness of grown PbSe films versus growth time at several combinations of source temperature (T_s) and oven-to-substrate distance (D).

distance (D) are given in Figure 2. With one exception where D = 3.5 cm film growth rates were steady and increased with increasing T_s. In fact the growth rate is in direct proportion to the vapor pressure of PbSe, indicating the evaporated material comes out of the oven in a material beam. (A rough estimate gives a divergence of ~ 18° for a source oven length of 5 cm and a diameter of 1.9 cm.)

The rate of deposition (R) can be approximated by

$$R < \alpha \cdot v\rho_\gamma \tag{1}$$

where v is the velocity of efflux (from the source) ρ_γ is the ratio of the density of the gas to that of the solid and α is the sticking coefficient of the gaseous material to the solid film. Assuming ideal gas behavior and single molecular gas particles, and using simple kinematic arguments, the rate of growth is calculated to be 1 μm/hr^{-1} for T_s = 525 C and α = 1. Obviously, this calculated rate is an overestimate; however, the actual growth rates are even higher: e.g., with T_s = 525 C and D = 2 cm, R = 3.1 μm/hr^{-1}. Barring gross mistakes in temperature measurements or in vapor pressure data [3], the conclusion can be drawn that the PbSe molecules are clustered: estimate of the cluster size can be made from Eq. (1). Again assuming ideal gas behavior, it is quite straightforward to show that R $\alpha \sqrt{M}$, where M is the mass of the gas particle. As the actual growth rate is ~ 3 times higher than the calculated value, it can be concluded that the cluster is composed of nine or more PbSe molecules.

The fact that the growth rate for D = 3.5 cm was erratic, and that the rate for D = 2.7 cm is lower than that for D = 4 cm indicate that other factors, such as turbulence in the material beam also affect the growth of these films.

It is mentioned earlier that the growth of pseudo-binary alloy films can result in changes in composition from that of the source material. Previously, Jensen and Schoolar [8] reported that the composition of grown $Pb_{1-x}Cd_xS$ (x = 0.06) films shifted to smaller values of x with increasing substrate temperature. With essentially the same apparatus we have conducted growth studies of another $Pb_xCd_{1-x}S$ alloy with x = 0.03. The source temperature was maintained at 565 C, and the substrate temperature varied from 350 to 475 C. The composition of the grown films was determined from optical transmission data and from the spectral response of photo-voltaic cells. In Figure 3 values of x are plotted versus the reciprocal of the substrate temperature (in Kelvin), with the data of Jensen and Schoolar also included. These two sets of data displayed two common features: (1) the films essentially had the same compositions as the source materials at low substrate temperatures, even though these temperatures were different for different x and (2) a dissociation energy could be associated with the Cd content, and this energy was only weakly dependent on the Cd concentration. It should be added that these changes in Cd concentration were not due to changes in the source material. For example, in the present investigation the growths at the lowest substrate temperature were carried out at the end of the growth sequence.

From the above observations one can construct a qualitative picture of the growth of the PbCdS alloys (for Cd content < 6%): they vaporize congruently. The vapor molecules are probably in clusters, similar to PbSe as discussed above, so that the compositions of the source materials are maintained. Differential dissociation occurs during deposition, resulting in the change in composition with temperature. From the slope of the

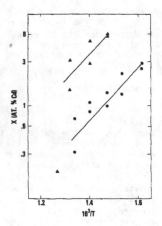

Figure 3: The composition of grown $Pb_{1-x}Cd_xS$ films as function of reciprocal substrate temperature (in Kelvin). x for the source material in this investigation is 3 At % (●), and for Ref. (8) at 6 At % (▲). The points shown indicate the highest and the lowest values for the samples measured. The source temperature for both investigations is 565 C.

straight lines in Figure 3, the dissociation has a characteristic energy of 0.58 ± 0.15 eV. It is interesting to note that this value is the same order of magnitude as the difference in the sublimation energies of PbS and CdS, 0.25 eV.

INFRARED DETECTORS

Photovoltaic detectors based on IV-VI films have been under development since the early 70's. Recent advances have been made with metal-semiconductor (p-type) barrier devices rather than p-n junction devices [1,10]. Detectors with near background limited operation for the 3 to 5 μm wavelength range have been demonstrated [11,12].

One of the advantages of the IV-VI semiconductors as infrared materials is the relatively large selection of alloy systems so that specific spectral responses can be chosen. Because of their material stability and ease of growth, multi-level and contiguous growths of films of various compositions can be achieved in a straightforward manner. Thus, multi-color detectors can be made with a single-chip fabrication process. Figure 4 is a photograph of a four color device in various stages of fabrication. The four colors are, at 77 K: 1-4 μm (PbS), 4-5 μm ($PbS_{0.5}Se_{0.5}$ on PbS), 4-7 μm (PbSe on PbS) and 5-7 μm (PbSe on $PbS_{0.5}Se_{0.5}$ on PbS). The detectivities of these devices are typically 1×10^{11} $cmHz^{1/2}w^{-1}$ at 4 μm, 5×10^{10} $cmHz^{1/2}w^{-1}$ at 5 μm and 3×10^{10} $cmHz^{1/2}w^{-1}$ at 7 μm.

Figure 4: Photograph showing various stages of the fabrication of a four color detector. On the left is the three IV-VI semiconductor layers: one full circle (grown on the substrate) with two semicircles on top grown successively. The semicircles are at right angles to each other. The middle picture shows the Pb junctions (dark dots, diam. 0.06 cm) and BaF_2 insulating layers (gray rectangles). The picture on the right is the final product with Au contacts deposited. Four four-color detectors can be made in a 1 cm x 1 cm substrate. The film structures for the pictures in the middle and on the right consist of full-area films on the substrates instead of circles for the first layer of film.

From an application view point one interesting aspect of these detectors is that they have maximum detectivity at ~ 140 K [1]. This is due to the combination of two factors: the junction resistance being constant for T < 140 K [13] and the decrease in background radiation noise. At 195 K the detectivity of our detector is typically a factor of 2 - 3 lower than the detectivity at 77 K. Thus they should be of practical use in thermoelectrically cooled systems. Figure 5 gives an illustration of the spectral response of three of the four-color detectors at 195 K.

The fabrication of the Pb-metal barrier detector has been benefited greatly by the discovery of Cℓ on the IV-VI semiconductor surfaces, and subsequently chlorination treatment (before Pb deposition) [14]. As a result, the fabrication reliability as well as the detectivity of the detector was improved substantially: variation in diode reverse saturation current was within a factor of 2 or 3 over the area of the substrate (1 cm x 1 cm) and detectivity improved by a factor of 5 (on a best effort basis) compared to an unchlorinated sample. Preliminary Auger and XPS studies

64

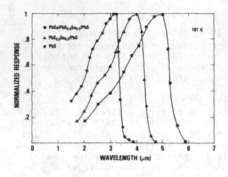

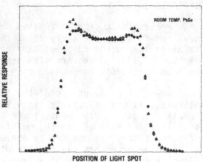

Figure 5: The spectral response of three of the four colors at 197 K. The detectivities are typically 8×10^{10} cmHz$^{1/2}$w^{-1} at 3.2µm (PbS), 4×10^{10} cmHz$^{1/2}$w^{-1} at 4.3µm (PbS$_{0.5}$Se$_{0.5}$) and 1×10^{10} cm$^{1/2}$w^{-1} at 5.3µm (PbSe).

Figure 6: The photoresponse of two PbSe diodes under laser spot scan. The widths of the diodes are ~ 0.01", and the laser spot is ~ 0.0015". One of the diodes (▲) is deposited on a cleavage step. The photoresponse is degraded somewhat in the vicinity of the step, as shown by the irregularity on the right-hand side of the scan.

indicate that the chlorine exists in an ionic state, possibly as an oxychloride within a surface layer thickness of ~ 20 Å. However, the uniformity of such a layer has not yet been determined; thus, the chlorine may have facilitated an MIS structure or it may have reacted preferentially with possible metal inclusions on the film surfaces [15]. In either case improvement in diode characteristics and detector performance is achieved. Benefits of the chlorination process can be illustrated by Figure 6 in which results of laser spot scans on two adjacent diodes are shown. One of these diodes was deposited on top of a large cleavage step of ~ 1µm. It is demonstrated clearly that diode junctions can be formed even on highly defective surfaces.

To-date, in addition to multicolor detectors we have fabricated small scale two color arrays (32 element/color, element size 0.002" x 0.002") with essentially the same technology.

IV-VI SUPERLATTICES

IV-VI superlattice research lags behind that of the III-V and group IV semiconductors. However, the usefulness of the HWE [16] (in addition to MBE [17]) technique for superlattice growth has been demonstrated. We have grown superlattice and multi-quantum well structures with the closed hot wall version and the open hot wall version. So far, we have observed negative magneto-resistance, non-linear field dependence of the Hall resistance, persistent photoconductivity and sub-bandgap transition in structures based on PbS and PbSe. Some of the sample characterization [18] as well as experiment results have been reported previously [19]. We present here our observation of sub-bandgap transition. It was detected with the photovoltaic effect on a p-type PbS/PbSe structure. The structure consisted of 20 periods of alternating PbS and PbSe layers. The thickness of each layer was ~ 500 Å. It was grown on a BaF$_2$ substrate with a buffer layer of PbSe of thickness 0.2 µm. The photovoltaic cell was a Pb metal junction made the same way as the infrared detectors mentioned above.

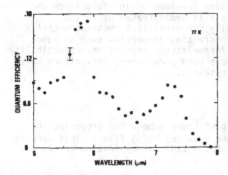

Figure 7: The photoresponse of a photovoltaic cell made on a PbS/PbSe superlattice. The shift of the peak response to a wavelength shorter than the cutoff for PbSe (7µm) is due to the small thickness of the sample.

The sub-bandgap transition is illustrated in Figure 7 where the quantum efficiency of the photovoltaic cell at 77 K is plotted versus wavelength. For the sake of clarity only the photoresponse at longer wavelengths are given. The bandgap energies of PbS and PbSe are 0.31 eV and 0.18 eV respectively at this temperature. The photoresponse observed at wavelengths > 7 µm is therefore considered sub-bandgap. At this point the nature of this transition can not be established clearly. A few possibilities will be discussed below.

The first possible explanation considered was a type I' band alignment, where $\left| \Delta E_c - \Delta E_v \right| \neq \Delta E_g$ and $\left| \Delta E_c \text{ or } \Delta E_v \right| < \Delta E_g$. This is proven incorrect because of the alignment is of type I, determined from photoresponse studies of heterostructures of PbS/PbSe [20]. Effects due to alloying as a result of interdiffusion [21] can be discounted also because of PbS and PbSe form a completely miscible alloy system. Defect states are also unlikely because such long wavelength responses have not been observed in PbSe photovoltaic cells. A Moss-Berstein shift due to transfer of carriers from the PbS to the PbSe layers is possible but would not result in a dip in the photoresponse (at ~ 7 µm) as observed. Other possibilities included band bending within individual layers and band-bending at the depletion layer. Both would reduce the effective bandgap and separate the photocarriers spatially. However, a simple estimate gives values of < 0.1 meV for band bending in the layers, which is smaller than the observed shift of ~ 10 meV. Even though band bending at the depletion layer gives approximately the correct energy shift, it does not provide an explanation for the dip in photoresponse. The last consideration is interface states arising from the matching of energy bands. This is an interesting possibility that merits further experimental and theoretical investigations.

CONCLUSION

In this paper we have emphasized one important aspect of the IV-VI semiconductors: their material stability. The material growth methods, the growth characteristics and the multicolor infrared detectors described above all attest to the fact that high quality device grade semiconductor materials can be prepared with simple equipment requiring low cost investments. Their potential for future development in the science and technology of electro-optical materials is demonstrated by recent activities in superlattices, especially the persistent photoconductivity and sub bandgap transition. Other interesting properties such as band inversion in the tin chalcogenides, ferroelectric transition, compatibility with the group V semimetals, and simple band structures, etc. offers many more interesting aspects for future research.

ACKNOWLEDGEMENT

The authors acknowledge the help of Elmer Gubner for preparing the source materials and the growth of the PbSe and PbCdS films. This work is supported by NSWC Independent Research and Exploratory Development Funds, and by the Office of Naval Research.

REFERENCES

1. H. Holloway and J. N. Walpole, Prog. Crystal Growth Charact. $\underline{2}$ 49 (1979).
2. D. L. Partin, Paper R2.4 Material Research Society Meeting, Dec. 1986.
3. J. N. Zemel, Solid State Surface Science, Vol. 1, pp. 291-403, (M. Green, Editor) Dekker, New York, 1969.
4. P. Hudock, Trans. Metall. Soc. AIME 239 339 (1967).
5. A. Lopez-Otero, Thin Solid Films, 49 3 (1978).
6. See, for example, R. Dalven, Solid State Physics, $\underline{28}$ 179, 1973, and G. Nimtz and B. Schlicht, in Narrow Band Semiconductors Springer Tracts in Modern Physics 98, (G. Höhler, Editor), Springer-Verag, New York 1983.
7. K. Weiser, K. A. Klein, and M. Ainhorn, Appl. Phys. Lett. $\underline{34}$ 607 (1979).
8. J. D. Jensen and R. B. Schoolar, J. Elect. Materials 7 239 (1978).
9. D. K. Hohnke and S. W. Kaiser, J. Appl. Phys. 43 897 (1974).
10. T. K. Chu, A. C. Bouley and G. M. Black, Proc. Intern. Soc. Opt. Eng. 409, 89 (1983).
11. A. C. Bouley, T. K. Chu and G. M. Black, Proc. Intern. Soc. Opt. Eng. 285 26 (1981).
12. A. C. Bouley, J. D. Jensen, G. M. Black and S. Foti, Proc. Intern. Soc. Opt. Eng. $\underline{246}$ 2 (1980).
13. W. Maurer, Infrared Phys. $\underline{23}$ 257 (1983).
14. T. K. Chu, A. C. Bouley and G. M. Black, Proc. Intern. Soc. Opt. Eng. 285 33 (1981).
15. K. Duh and H. Preier, Thin Solid Films $\underline{27}$ 247 (1975).
16. See for example, M. Kinoshita and H. Fujiyasu, J. Appl. Phys. 51 5845 (1980), and H. Clemens, E. J. Fantner and G. Bauer, Rev. Sci. Instrum. 54 685 (1983).
17. D. L. Partin, J. Vac. Sci. Tech. $\underline{21}$ 1 (1982).
18. The Growth and Characterization were presented at the Meeting of The American Physical Society, Las Vegas, 1986. (Bull. Am. Phys. Soc. $\underline{31}$ 522)
19. A. Martinez, T. K. Chu and R. S. Allgaier, Proc. Intern. Conf.: Appl. of High Mag. Field in Semi. Phys. 1986, to be published.
20. T. K. Chu, D. Agassi and A. Martinez, to be published.
21. F. Santiago, A. Martinez, and T. K. Chu, Presented at the 33rd Symposium of the Am. Vac. Soc., Oct. 1983, to be published.

PART III

Issues in Bulk Crystal Growth of Infrared Materials and Structure-Property Relations

ISSUES IN THE GROWTH OF BULK CRYSTALS OF INFRARED MATERIALS

K. J. BACHMANN AND H. GOSLOWSKY
Department of Chemistry, North Carolina State University, Raleigh, North
Carolina 27695-8204.

ABSTRACT

Selected issues in the growth of bulk single crystals for applications
in infrared optoelectronics are reviewed including an overview over materials choices, bulk III-V crystal growth, and the growth of II-VI, IV-VI
and I-III-VI$_2$ compounds and alloys. The most important issues are the
control of purity, perfection, stoichiometry, and uniformity during crystal
growth and the control of the surface properties in wafer fabrication.
Specific examples are given to illustrate problems related to these issues
and to discuss approaches to their solution.

1.   INTRODUCTION

The infrared (IR) wavelength region extends from λ = 780 nm to infinity
and is customarily divided into the near (0.78 < λ < 3 μm), intermediate (3
< λ < 30 μm) and far (λ > 30 μm) regimes. Consequently semiconductors that
are suitable for the fabrication of IR devices have band gaps 0 < E_g < 1.6
eV. Several classes of compounds have members in this energy region, e.g.
the tetrahedrally coordinated group IV semiconductors and their III-V, II-
VI, II-IV-V$_2$ and I-III-VI$_2$ isoelectronic analogs as well as the NaCl structure IV-VI compounds.

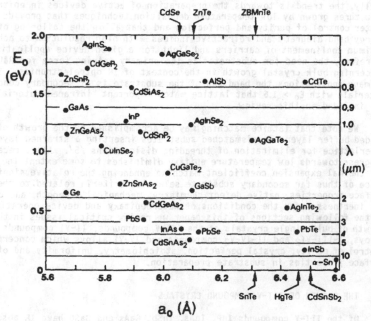

Figure 1 shows a plot E_g vs. a-axis lattice parameter for these compounds. The data refer to the smallest gaps at room temperature between which continuous band gap and lattice constant tailoring is possible within the existence ranges of solid solutions. For applications in the near IR the III-V compounds and alloys are usually preferred because of their direct band gaps and well behaved extrinsic doping properties. The intermediate and far IR are covered by the II-VI and IV-VI alloys, in particular the $Cd_xHg_{1-x}Te$ and $Sn_xPb_{1-x}Te$ systems that exhibit at room temperature zero gap at x = 0.1 and x = 0.59, respectively. Landau level absorption and emission in the range 70 μm < λ < 500 m has been reported for bulk InSb /1/ that, also, is an important detector material for wavelengths in the intermediate IR extending in the form of alloys to the 8-10 μm region /2/.

We note that although solid solubility over the entire range of compositions 0 < x $\leq$ 1 has been assumed for most of the pseudobinary III-V alloys in the early literature, in the past decade, extended regions of immiscibility in solid state have been identified on the basis of classical thermodynamic calculations for many III-V systems /3/. Also, ordering at distinct compositions below a critical temperature has been predicted on the basis of more comprehensive quantum chemical calculations /4/ and there exists mounting experimental evidence that these predictions hold for a number of ternary III-V alloys. Therefore, the validity of band gap tailoring and lattice matching through alloying requires scrutiny establishing the existence of the solid solutions at room temperature. Of course, where solid solutions are preserved at room temperature in metastable quenched-in state, their use for device fabrication, is perfectly legitimate provided that this can be done without compromise in reliability.

Bulk single crystals of the above "infrared materials" are needed for the fabrication of both substrate wafers for epitaxy and bulk effect devices, e.g. photoconductive detectors, electrooptic modulators, etc. Generally, the trend is towards the preparation of active devices in epitaxial structures grown by low temperature deposition techniques that provide for better control of purity and perfection and that allow the tailoring of the energy band diagrams and refractive indices with high resolution to achieve optimum confinement of carriers and light for a given device application. Therefore, the need for substrates is the primary driving force for R&D concerning bulk crystal growth in the context of IR optoelectronics. Since in many applications the band gap of the substrate is of secondary concern materials with E_g > 1.6 that lattice match important "infrared materials" are included in this review.

We note that lattice matching may be accomplished by the growth of graded buffer layers on mismatched substrates inserting a strained layer superlattice for elimination of threading dislocations /5/. Also, recent progress towards low temperature epitaxy diminishes to some extent the need for thermal expansion coefficient matching enhancing the relative significance of thus far secondary substrate selection criteria related to the surface properties, native defect chemistry, mechanical strength, and chemical inertness under the conditions of heteroepitaxy and device operation. In the following sections of this paper we review critical issues in the growth of bulk single crystals of the III-V compounds, II-VI compounds and alloys, I-III-VI$_2$ and II-IV-V$_2$ compounds and IV-VI alloys which concern the control of purity, crystal perfection, stoichiometry, uniformity and of the surface properties in substrate preparation.

2. THE GROWTH OF III-V COMPOUND CRYSTALS

Of the III-V compounds InP, InAs, InSb, GaAs and GaSb have IR absorption edges and are used in the manufacturing of IR devices. An important issue in the fabrication of bulk single crystals of these materials is purity requiring a considerable effort in the preparation of ultrapure starting materials and vigilance in maintaining purity during synthesis, bulk crystal growth and wafer fabrication. Differences in solubility, ion exchange, electrorefinement, distillation and zone refining are the primary principles of materials purification and produce elemental starting materials for compound synthesis of nominally 6N purity, i.e. total impurity contents in the sub ppma range, as discussed in more detail in ref.6. The analytical characterization of the composition and levels of residual impurities in semiconductor grade materials is by itself a critical area of research because it pushes the limits of contemporary analytical chemistry. The methods employed in trace analysis usually rely on electronic transitions of the impurity atoms, mass spectrometric analysis or radioactive isotope decay measurements that achieve sensitivities in the ppba to ppta range. The techniques are complimentary in both sensitivities and interferences requiring generally a combination of several analytical tools to establish a reliable characterization. Of the methods based on electronic transitions atomic absorption spectroscopy (AAS) is most convenient and achieves excellent specificity and sensitivity in single element analysis. For multielement analysis inductively coupled plasma emission spectroscopy (ICPS) is a superior choice. Sensitivities in the ppta range for several transition metal impurities have been achieved recently by ICP using atomic fluorescence spectroscopy as detection method /7/. Plasma sources, also, are utilized in modern mass spectrometric trace analysis /8/ replacing together with laser vaporization techniques /9,10/ the traditional spark source mass spectrometric analysis technique which is still utilized by many suppliers for the characterization of semiconductor grade materials. Although SIMS has been used widely in bulk semiconductor characterization it has been rarely utilized in the analytical evaluations of bulk metals and dielectrics /11/. In our opinion, the method has considerable merit in this application when ultrapure materials are studied where matrix effects are negligible and absolute concentrations can be measured reliably against a set of standards. Table I shows the impurity concentrations in a batch of commercial 6N pure red P determined by neutron activation analysis and, for comparison, the vendor supplied analysis. Seven elements, not listed by the vendor are found at substantial concentration, and some of these impurities, e.g. Hg, certainly are of concern in the context of subsequent III-V synthesis. Although we are painfully aware of the shortcomings of analytical chemistry in semiconductor characterization we believe that it is nevertheless essential for careful synthesis and crystal growth work since it provides valuable clues in the tracking of contamination sources and it contributes to the understanding of the complex interactions between impurities and native defects in compound semiconductors. Of course, the more detailed knowledge in the latter area is obtained by analytical evaluations of the finished crystal or device where higher sensitivity is achieved by advanced techniques of semiconductor characterization /12/.

Of the methods of crystal growth zone melting, the horizontal Bridgman technique, vertical gradient freezing, and liquid encapsulated Czochralski pulling are most frequently utilized in work on III-V compounds. Zone melting has produced InSb crystals of a net carrier concentration of ~10^{13} cm^{-3} /13/, but no progress has been made beyond this limit for the past 25 years. Ultrapure InSb is important for the development of IR lasers and an improvement of the purity to the 10^{12} cm^{-3} range would be required for the realization of advanced InSb cyclotron emission devices challenging the imagination of the crystal growth research community.

The horizontal Bridgman method has been used successfully for many years to produce highly doped GaAs crystals for applications in the manufacturing of optoelectronic devices. Its use is diminishing as compared to LEC pulling because of the increasing interest in nominally undoped semiinsulating

Table I. Neutron activation analysis of the impurity concentration C_N in commercial 6N pure red phosphorus and vendor supplied analytical data C_V.

Impurity	C_N (ppmw)	Relative Error	C_V (ppmw)
Ag	<0.0005	-	<0.1
Al	2272	+5%	NL
As	8.95	+3%	0.3
Ca	<500	-	<1
Cl	45	+8%	NL
Co	0.073	+7%	NL
Cu	<5	-	0.06
Fe	<5	-	<0.1
Hg	2.9	+10%	NL
Na	21.5	+1%	ND
Sb	0.43	+1%	NL
Sc	0.003	+12%	NL
Ti	18	+10%	NL
Zn	<10.5	-	<0.1

NL Not listed by vendor, ND not detected.

GaAs for IC fabrication by ion implantation. The primary difficulty in HB growth of SI GaAs is the reduction of a high silicon doping background which requires the development of alternative containers or, at least, of alternative liner materials.

Vertical gradient freezing (VGF) has been developed with considerable success for the growth of InP and GaAs and is, in our opinion, an excellent choice for the production of larger diameter crystals in the future. Nearly perfectly flat solid/liquid interface shape has been attained by judicious engineering of the heat flow conditions as demonstrated by IR transmission for axially cut InP crystals /14/. The primary remaining flaws are type I striations /15/. In our opinion, it should be possible to suppress these striations by the provision of a vertical magnetic field in the VGF apparatus.

Application of magnetic fields that increase the effective viscosity of the melt /16,17/ has been utilized successfully in suppressing type I striations under the conditions of LEC pulling of Si /18,19/ and GaAs /20,21/. Also, it has been shown to reduce the incorporation of impurities originating at the container walls /22/. A problem that is harder to control under the conditions of LEC pulling as compared to VGF are radial temperature gradients leading to strain that determines the density and distribution of extended defects /23/. Provision of enhanced radial heat input into the B_2O_3 encapsulant layer /24/ and increasing its thickness have been used to grow perfect GaAs crystals /21/. However, although reducing the temperature gradients in III-V crystal growth prevents strain it may complicate the control of twinning /25/ which is one of the least understood, most cumbersome problems of compound semiconductor crystal growth. Therefore, other avenues of controlling extended defect formation have been explored, in particular isoelectronic doping, which has been successfully applied to both InP /26,27/ and GaAs /28/ where careful mapings by IR transmission reveal the built-in strain and show a clear improvement not only in the strain distribution, but also in the concentration and distribution of the EL2 trap that is more uniform for In doped GaAs /29-31/. Unfortunately, as we learn to suppress the formation of dislocations we may encounter more aggravated

conditions regarding the formation of type II striations /32/, i.e. striations that are due to the decoration of risers and/or valleys at solid/-
liquid interfaces exhibiting step bunching phenomena. Figure 2 shows an
example of a valley trace running almost perpendicular to the type I striations in a crystal of InP_yAs_{1-y} grown by zone leveling /33/.

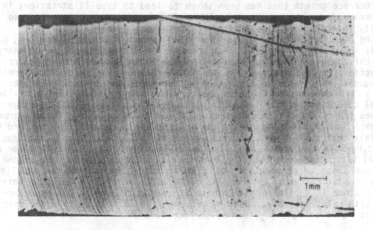

Figure 2. Type I and Type II Striations in InP_yAs_{1-y}.

3.   THE GROWTH OF II-VI COMPOUNDS AND ALLOYS

The growth of the II-VI compounds and alloys differs in several important aspects from the growth of the III-V compounds since they exhibit much
smaller thermal conductivities, exhibit stronger association in liquid
state, vaporize under favorable conditions congruently at relatively high
rate, exhibit solid state phase transformations and tend to self compensate
making extrinsic doping a far more complex endeavor than in the case of the
well behaved III-V compounds. Variants of zone melting and the Bridgman
method employing stoichiometric or off-stoichiometric liquidus compositions, solution growth by solvent evaporation, the Piper-Polich method,
chemical transport and solid state recrystallization have been used for
many years in II-VI crystal growth, but difficulties due to the formation
of low angle boundaries and twinning still exist. ZnTe /34,35/, CdTe
/36,37/, and $Cd_xHg_{1-x}Te$ /38/ have been grown successfully from Te rich
liquidus compositions. In the case of ZnTe the zinc blende to wurzite
structure transformation at 1425°C complicates the growth of large single
crystals and even more aggravating difficulties due to solid state transformations exist in the case of chalcopyrite structure $I-III-VI_2$ compounds that
often undergo more than one solid state transformation during cooling from
the growth temperature to room temperature. For CdTe precipitate formation
due to the strongly retrograde solid solubility range on the Te-rich side of
the binary phase diagram leads to difficulties in the control of composition
since impurities segregated in the precipitates may be released during
subsequent annealing as part of device processing and Hg diffusing to Te
precipitates at heterojunctions forms metallic inclusions that are detrimental to the properties. Growth by solvent evaporation /39/ and the traveling
heater method /40/ from Cd-rich liquidus compositions have been explored to
combat this problem and lower compensation levels have been achieved by the
latter method as compared to crystals grown from Te-rich solutions. A
penalty that one usually has to pay for high purity and perfection is slow

growth rate, of the order of 1 mm/day, which is a severe limitation in a production oriented environment.

Seeded growth and control of the solid/liquid interface shape avoiding concave regions at the container walls are required to improve the yield of crystals of device quality. Seeding is important to grow out twins and to avoid terrace growth that has been shown to lead to type II striations in melt grown semiconductor crystals /15/. The latter are due to preferred impurity incorporation at macrosteps that change in velocity due to bunching phenomena in the course of crystal growth /15,32/. Therefore, even if the type I striations, that are due to modulations in the boundary layer thickness and that run parallel to the solid/liquid interface, are suppressed type II striations may persist destroying the spatial uniformity of composition. Also, the control of the shape of the solid/liquid interface is difficult to implement in practice for materials exhibiting smaller thermal conductivity than the container materials /41/. Adiabatic zones proposed for interface shape control under the conditions of Bridgman growth /42/ have been found to aggravate the distortion of the melting isotherm for materials that show large increases in thermal conductivity upon melting, e.g. $Cd_xHg_{1-x}Te$ that has K_L/K_S values ranging from ~2 for x = 1 to 7 for x = 0.05 /41,43/. To reduce limiting container effects on the control of the interface shape we propose the use of porous container structures and liners. Access to a microgravity environment could be exceedingly helpful in this context since it should permit the use of coarse SiO_2 or BN basket structures with relatively wide openings between ribbons that confine and shape the melt which is held together by surface tension.

In view of the difficulties in the growth of CdTe crystals two alternative approaches to the provision of substrates for $Cd_xHg_{1-x}Te$ epitaxy are presently under evaluation: 1. The utilization of substrates outside the II-VI family that can be grown reliably with large diameter and high perfection; 2. The growth of exactly lattice matching II-VI alloys with improved structural and/or surface properties. Examples of the first category are the use of GaAs, InSb and sapphire as substrates onto which CdTe buffer layers are grown by low temperature epitaxial techniques, e.g. MBE. Of the III-V substrates InSb is preferred because it closely matches the lattice constant of CdTe, $\Delta a/a = 5 \times 10^{-4}$ at room temperature. Excellent structural integrity of the CdTe buffer layers grown by MBE on InSb substrates has been reported /44/, but difficulties related to the surface cleaning of InSb and the possible incorporation of In into the epilayers require further attention. Sapphire is chemically inert, has excellent mechanical strength and can be grown in the form of large boules from the melt. Device structures with favorable properties have been fabricated /45/. However, the large mismatch in lattice constants is a concern and requires further work optimizing the design of buffer layers and exploring alternative ceramic crystals providing for better matching.

There exist several II-VI alloy systems that cover the range of lattice constants at the CdTe-HgTe pseudobinary with minor substitutions, e.g. $Zn_xCd_{1-x}Te$, $CdSe_yTe_{1-y}$ and $Cd_{1-x}Mn_xTe$. $Zn_xCd_{1-x}Te$ has the highest hardness of these alloys. However, twinning remains to be a problem, and the solidus-liquidus separation on the CdTe-ZnTe pseudobinary is sufficiently wide to require zone leveling for the control of segregation. Within the zinc blende structure range of compositions $CdSe_yTe_{1-y}$ shows less tendency to twinning while $Cd_xMn_{1-x}Te$ has advantageous surface properties and exhibits in the range 0 < x < 0.25 a relatively small solidus liquidus separation /46/ that allows the convenient growth of this material by the Bridgman method. Figure 3 shows the broadening of the photoreflectance (PR) spectra as a function of composition for $Cd_{1-x}Mn_xTe$ in the range x < 0.4. In accord with the phase diagram, for moderate Mn concentrations, the broadening is dominated by alloy scattering, and becomes affected by extended

defects at x ≲ 0.25 /47/. TEM studies show that the microstructure of
the alloy changes in this composition
range exhibiting microtwinning at the
higher concentrations /48/.

Studies of the native oxides,
sputter cleaning and the low temperature annealing kinetics by XPS, PR,
and photoluminescence /49/ show that
for $Zn_xCd_{1-x}Te$ and $CdSe_yTe_{1-y}$ the
native oxide consists primarily of
tellurium oxide while for $Cd_{1-x}Mn_xTe$
a mixed manganese tellurium oxide
exists. These oxides may be removed
by sputter cleaning which is a preferred cleaning technique because of
its compatibility with subsequent
heteroepitaxy by MBE. As shown in
fig. 4 the sputtering damage for
comparable of $Cd_{1-x}Mn_xTe$ is substantially less than of pure CdTe and
a 30 min anneal at 200°C recovers the
PR line shape of $Cd_{1-x}Mn_xTe$ while
irreversible changes persist in CdTe.
Also, the PL spectra of sputtering
damaged $Cd_{1-x}Mn_xTe$ show only a very
weak enhancement of subbandgap luminescence features that are more pronounced for pure CdTe /49/ and that
dominate the PL spectra of $Zn_xCd_{1-x}Te$
and $CdSe_yTe_{1-y}$ /50/. Therefore,
$Cd_{1-x}Mn_xTe$ appears to be a superior
material as compared to $Zn_xCd_{1-x}Te$ in
the context of surface cleaning.
However, while improved $Cd_xHg_{1-x}Te$
epitaxy has been established for
$Zn_xCd_{1-x}Te$ substrates /51/ a similar
study of $Mn_xCd_{1-x}Te$ is still out-
standing and is required for a fair
assessment of the relative utility of
these two materials in substrate applications.

4. OTHER MATERIALS AND CONCLUSIONS

In contrast to the zinc blende
structure III-V and II-VI compounds the
rocksalt structure IV-VI compounds and
alloys contract upon freezing. This
may lead to the formation of voids in
directional solidification and appropriate measures must be taken to avoid
this problem. Czochralski pulling has
been employed successfully in the
growth of $Pb_{1-x}Sn_xTe$ and is probably
the most versatile technique for the
growth of low defect density material
/52/.

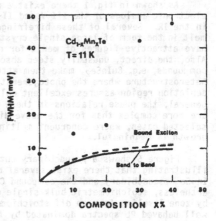

Fig.3 Broadening of the PR spectra of $Cd_{1-x}Mn_xTe$ as a function
of X.

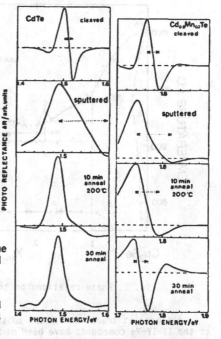

Fig. 4 PR spectra of CdTe and
$Cd_{1-x}Mn_xTe$ prior and after sputtering damage and annealing.

As shown in fig. 1 there exist a number of chalcopyrite structure isoelectronic analogs of the III-V and II-VI compounds with absorption edges
in the IR. Several of these birefringent materials can be grown from the
melt in the form of large single crystals, can be doped p- or n-type and
have attractive figures of merit for non-linear optical applications.
Also, the direct, unusually steep absorption edges of several of these
compounds, e.g. $CuInSe_2$, make them attractive for thin film photovoltaic
heterostructure where the photogeneration of minority carriers within the
depletion region assures excellent collection efficiency. However, in
general, the phase relations in these ternary I,III,VI and II,IV,V systems
are more complex than for the above discussed binary materials, and only in
selected cases, where congruent melting is approached, R&D on bulk crystal
growth is meaningful.

Figure 5 shows a pseudobinary cut $Cu_2Se-CuInSe_2$ of the Cu,In,Se system
illustrating that there exist several maximum melting points at the Cu_2Se-
In_2Se_3 pseudobinary and the compound $CuInSe_2$ does not melt congruently. Nevertheless, stoichiometric bulk single crystals in this compound can be grown
by zone leveling from an off stoichiometric liquidus composition and exhibit
well behaved PL spectra dominated by free exciton emission /53/. Also, there
exist wide homogeneity ranges and several solid state transformations at
elevated temperatures that generally complicate bulk crystal growth that
represents clearly a more difficult task than for the parent II-VI compounds.

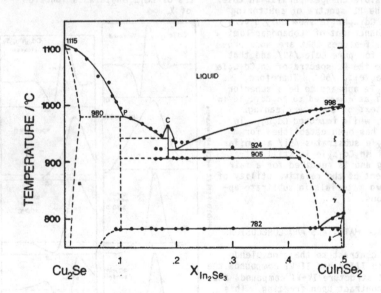

Fig. 5 Phase relations on the $Cu_2Se-In_2Se_3$ pseudobinary.

Attempts at transition metal substitutions on the group II sublattice
of the II-IV-V_2 compounds have been made to explore alternative DMS materials, but have failed thus far to produce evidence for substantial solid
solubility as is expected from the behavior of the parent III-V compounds. In
our opinion more interesting opportunities exist in the exploration of pseudobinary alloys between the I-III-VI_2 compounds and transition metal chalco-

genides. Bulk alloy crystals among the I-III-VI$_2$ compounds themselves already have been grown, in some cases with amazingly sharp luminescence properties for alloy compositions close to the middle of the pseudobinaries /54/. We note that Mn substitution in CdTe results in a significant increase in d$_{14}$ and in the non-linear susceptibility for this material /55/. In view of high figurs of merit for non-linear optics applications of some of the I-III-VI$_2$ compounds this makes the exploration of transition metal substituted I-III-VI$_2$ compounds an interesting topic. Also, we predict that compensating substitution of InSb by Cd+Sn will result in a decrease of its energy gap as demonstrated previously by alloying of InP with its II-IV-V$_2$ isoelectronic analog that shifted the long wavelength cut-off and emission wavelength of CdSnP$_2$/InP detectors and LEDs deeper into the IR /56/. To which extent Cd+Sn substituted InSb alloys exist and the long wavelength limit of the absorption edges attainable with this system are presently unknown, but warrant exploration.

In summary, progress has been made in the control of extended defects in large bulk single crystals of the III-V compounds. However, there exist still problems in the control of purity and of the distribution of shallow dopants and traps that require further attention. This must include research on methods of starting materials purification that is essential not only for progress in bulk IR materials synthesis and crystal growth, but is equally important for the improvement of the purity of sources for MBE, LPE, CVD and OMCVD. Control of stoichiometry and of the thermal conditions that determine the interface shape and the thermal strain are general areas that need continuous work and inventive thinking. They are even more important in the context of chalcogenides, i.e. the II-VI, IV-VI and I-III-VI$_2$ compounds that exhibit low thermal conductivities and, in some cases, relatively large changes in the thermal conductivity upon melting. Although Zn$_x$Cd$_{1-x}$Te has been proven to be an advantageous exactly lattice matching substrate for Cd$_x$Hg$_{1-x}$Te epitaxy and possesses improved structural and mechanical properties, Cd$_{1-x}$Mn$_x$Te exhibits attractive resistance to sputtering damage and fast low temperature annealing kinetics warranting thus further evaluation in the context of Cd$_{1-x}$Hg$_x$Te epitaxy. Also, in our opinion, an enhanced level of exploratory research our novel IR materials and methods of bulk crystal growth would be desirable.

Acknowledgments: The work presented in this paper has been supported, in part, by NSF Grant DMR 8414580, SERI Subcontract XL4-04041-1, and NASA Grant NAG 1-354.

References

1. B. D. McCombe, R. Kaplan, R. J. Wagner, E. Gornik and W. Muller, Phys. Rev. B13, 2536 (1976).

2. P. K. Chiang and S. M. Bedair, Appl. Phys. Lett. 46, 383 (1985).

3. G. B. Stringfellow, J. Crystal Growth, 65, 173 (1983).

4. A. A. Mbaye and A. Zunger, Proc. 7th Int. Conf. on Ternary and Multinary Compounds, Snowmass, CO, 1986, to be published.

5. J. W. Matthews and E. Blakeslee, J. Crystal Growth 27, 118 (1974).

6. K. J. Bachmann and M. S. Su, Proc. 1st Int. Conf. on the Processing of Semiconducters, Santa Barbara, CA, 1986, The Engineering Research Foundation, N.Y., to be published.

7. E.B.M. Jansen and D. R. Deniers, Analyst, 110, 541 (1985).

8. J. W. Coburn and W. W. Harrison, Appl. Spectroscopy Rev., 17, 95 (1981).

9. J.A.J. Jansen and A. W. Witmer, Fresenius Z. Anal. Chem., 309, 262 (1981).

10. D. S. Simons, Springer Ser. Chem. Phys. 36, A. Benninghoven, J. Okano, R. Shimizu and H. W. Werner, eds., Springer Verlag, Berlin, 1984, p. 158.

11. N. H. Turner, B. I. Dunlap and R. Colton, Anal. Chem., 56, 373R (1984).

12. G. Stillman, Proc. 1st Int. Conf. on the Processing of Semiconductors, Santa Barbara, 1986, to be published.

13. K. F. Hulme and J. B. Mullin, J. Electronics and Control 3, 160 (1957).

14. W. A. Gault, E. M. Monberg and J. E. Clemans, J. Crystal Growth, 74, 491 (1986).

15. E. Bauser, Festkorperprobleme, XXIII, P. Grosse, ed., Springer Vielag, Braunschweig, 1983, p. 141.

16. W. B. Thompson, Phil. Mag. Ser. 7, Vol. 42, 1417 (1951).

17. S. Chandrasekhar, Phil. Mag. Ser. 7, Vo. 43, 501 (1952); Vol. 45, 1177 (1954).

18. K. Hoshikawa, Japan J. Appl. Phys. 21, L545 (1982).

19. K. M. Kim, J. Electrochem. Soc. 129, 427 (1982).

20. K. Terashima, T. Katsumata, F. Orito and T. Fukuda, Japan J. Appl. Phys. 23, L302 (1984).

21. H. Kohda, K. Yamada, H. Nakanishi, T. Kobayashi, J. Osaka and K. Hoshikawa, J. Crystal Growth 71, 813 (1985).

22. K. Terashima, F. Orito, T. Katsumata and T. Fukuda, Japan J. Appl. Phys., 23, L485 (1984).

23. A. S. Jordan, R. Caruso and A. R. Von Neida, Bell Syst. Tech. J. 59, 593 (1980).

24. T. Shimada, K. Terashima, H. Nakajima and T. Fukuda, Japan J. Appl. Phys. 23, L23 (1984).

25. S. Tohno and A. Katsui, J. Crystal Growth 74, 362 (1986).

26. G. Jacob, M. Duseaux, J. P. Farges, M.M.B. van den Boom, and P. J. Roksnoer, J. Crystal Growth 62, 417 (1983).

27. A. Katsui, S. Tohno, Y. Homma and T. Tanaka, J. Crystal Growth 74, 221 (1986).

28. G. Jacob, Semi-Insulating III-V Materials, Evian 1982, S. Makram-Ebeid and B. Tuck, eds., Shiva Publ., Nantwich, UK 1982, p. 2.

29. P. Dobrilla, J. S. Blakemore, A. J. McCamant, K. R. Gleason and R. Y. Koyama, Appl. Phys. Lett. 47, 602 (1985).

30. P. Dobrilla and J. S. Blakemore, Appl. Phys. Lett. 48, 1303 (1986).

31. P. Dobrilla and J. S. Blakemore, J. Appl. Phys. 58, 208 (1985).

32. Y. C. Lu and E. Bauser, J. Crystal Growth, 71, 305 (1985).

33. K. J. Bachmann, F. A. Thiel and S. Schreiber, Jr. Progr. Crystal Growth and Characterization 2, 171 (1979).

34. R. Triboulet, Y. Marfaing, A. Cornet and P. Siffert, J. Appl. Phys. 45, 2759 (1974).

35. R. N. Bhargava, J. Crystal Growth, 59, 15 (1982).

36. T. Taguchi, J. Shirafuzi and Y. Inuishi, Japan J. Appl. Phys. 17, 1331 (1978).

37. R. Triboulet and G. Neu, 2nd Internat. Conf. II-VI Compounds, Aussois, France, 1985, Abstr. CW3, p. 73.

38. T. Taguchi, S. Fujita and Y. Inuishi, J. Crystal Growth, 45, 204 (1978).

39. J. B. Mullin, C. A. Jones, B. W. Straughan and A. Royle, J. Crystal Growth, 59, 135 (1982).

40. R. Triboulet, 2nd Int. Conf. on II-VI Compounds, Aussois, France, 1985, Abstr. PL4, p. 123.

41. T. Jasinski and A. F. Witt, J. Crystal Growth, 71, 295 (1985).

42. T. W. Fu and W. R. Wilcox, J. Crystal Growth, 48, 416 (1980).

43. R. J. Naumann and S. Lehoczky, J. Crystal Growth, 61, 707 (1983).

44. S. Wood, J. Greggi, Jr., R.F.C. Farrow, W. J. Takei, F. A. Shirland and A. J. Noreika, Westinghouse R&D Center Technical Document 83-9D2-EPCAD-P1, Pittsburgh, PA, December 21, 1983.

45. E. R. Gertner, W. E. Tennant, J. D. Blackwell and J. P. Rode, 2nd Int. Conf. on II-VI Compounds, Aussois, France, 1985, Abstr. CJ1, p. 87.

46. R. Triboulet and G. Didier, J. Crystal Growth, 52, 614 (1981).

47. K. Y. Lay, H. Neff and K. J. Bachmann, Phys. stat. sol. (a), 92, 567 (1985).

48. J. Narayan, K. Y. Lay and K. J. Bachmann, unpublished results.

49. H. Neff, K. Y. Lay, B. Abid, G. Lucovsky and K. J. Bachmann, J. Appl. Phys. 60, 151 (1986).

50. K. Y. Lay, H. Neff and K. J. Bachmann, unpublished results.

51. T. Maekawa, T. Saito, M. Yoshikawa and H. Takigawa, MRS Symp. Proc. Vol. 56, J. M. Gibson, G. C. Osbourn and R. M. Tromp, eds., Pittsburgh, PA, 1986, p. 109.

52. E. D. Bourret and A. F. Witt, J. Crystal Growth, 63, 413 (1983).

53. P. Lange, H. Neff, M. Fearheiley and K. J. Bachmann, Phys. Rev. B31, 4074 (1985).

54. P. Lange, H. Neff, M. Fearheiley and K. J. Bachmann, J. Electron. Mater. 14, 667 (1985).

55. L. Kowalczyk, 2nd Int. Conf.. on II-VI Compounds, Aussois, France, 1985, Abstr. PW9, p. 233.

56. K. J. Bachmann, E. Buehler, J. L. Shay and J. H. Wernick, US Patent 3,922,533; Nov. 25, 1975.

BULK CRYSTAL GROWTH OF $Hg_{1-x}Cd_xTe$
FOR AVALANCHE PHOTODIODE APPLICATIONS*

T.NGUYEN DUY, A.DURAND, J.L.LYOT
Société Anonyme de Télécommunications, 41 Rue Cantagrel 75631 PARIS CEDEX 13

ABSTRACT

High gap $Hg_{1-x}Cd_xTe$ (MCT) crystal is grown by a solvent method using a travelling heater zone. The use of the solvent zone permits a low temperature and low mercury pressure growth of MCT in a large composition range. In the Cadmium rich alloy range the MCT material exhibits a large spin orbit coupling leading to a resonance at $Eg = \Delta 0$. Due to this particular resonance effect the ionization coefficient of hole is higher than that of the electron, resulting in a low exess noise factor in the avalanche photodiodes whose bandgap energy is close to the spin orbit splitting $\Delta 0$.

INTRODUCTION

The $Hg_{1-x}Cd_xTe$ alloy system is presently the most widely used semiconductor for infrared detection. Due to the complete miscibility between the two binaries CdTe and HgTe and to their small lattice mismatch, less than 0.3 %, the CMT alloy can be synthethized in its whole x range leading therefore to a tailored bandgap material covering a wide range of I.R. detection corresponding to different transmission windows. The 8-12 μm and 3-5 μm atmospheric windows are the most frequently used and the related materials with $x \simeq 0.2$, 0.3 are extensively studied.

But so far as the x compositions higher than 0.5 are concerned the material as well as the resulting components are not yet fully explorated, for example the high x values corresponding to 2-2.5μm, 1.55μm, 1.3μm usable for fiber optics applications. Therefore the purpose of this work is to present some preliminary properties of high gap MCT material and the related photodiodes.

THE GROWTH OF $Hg_{1-x}Cd_xTe$.

Several methods of bulk metallurgy exist for MCT in the low x range : Quench Anneal [1], Czochralski, Slush[2]. All these methods us a stoichiometric melt of CdTe and HgTe whose temperature is given by the CdTe HgTe phase diagram. At high x values the melt temperature is greatly increased rendering these methods unusable. Typically with an x value of 0.7 a temperature up to 1000°C is necessary to melt the alloy, leading to a mercury pressure of 100 atm. To overcome the problem of high temperature and high pressure a new metallurgical process based on the growth from solution is elaborated : the travelling heater method. This method has been sucessfully used for various materials such as Si[3], Si-Ge[4], GaP[5], CdTe[6], HgTe[7] and also CMT alloy[8]. It allows a growth temperature far below the liquidus point : 700°C at 0.5 atm of mercury compared to 1000°C 100 atm.

The growth method is schematically presented in fig.1. It is based on the ternary phase diagram CdTe - HgTe - Te[9] where Te is the solvent, CdTe and HgTe the solutés. Three regions are shown : the source material, the molten zone and the crystallized material $Hg_{1-x}Cd_xTe$. A thermal profile is applied to the ternary zone in order to maintain this zone liquid during all the growth with an appropriate gradient. Due to the lowering movement of the ampule through a stationary heater a temperature difference appears between the upper part and the lower part of the molten zone. This temperature difference ΔT constitutes the driving force of the crystal growth.

---

* Work partly supported by the DAII

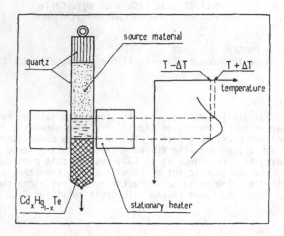

Figure 1 : Schematic diagram of the THM process

ΔT depends on the thermal profile of the heater and on the lowering rate of the ampule.

It can be pointed out that the main characteristic of this method is that the composition of the grown MCT is the same as that of the dissolve source material. Matter transport occurs by diffusion and convection through the solvent zone under the influence of the ΔT driving force. Therefore the solvent dissolves at its upper part a small quantity of the source material, transferring it to the bottom where the lower temperature promotes the crystallization of the grown material. This process of growth is similar to the LPE one, the only difference is that the liquid zone in the THM is continuously supplied by the source material in order to keep its composition constant during the growth.

This growth is governed by the following parameters : the solid composition X_S, the liquid composition X_L, and the temperature of crystallization T related by the Gibbs phase rule $F (X_L, X_S, T) = 0$. It can be concluded that:

- At defined temperature T and constant composition of the liquid the crystallized material has a constant composition.
- At constant composition of the liquid zone the composition of the crystallized material depends on the growth temperature.

The THM presents several specific advantages versus conventional stoichiometric methods.

- The crystal growth occurs at temperatures far below the melting point and consequently the mercury pressure during the growth is greatly reduced. Therefore it is possible to crystallize a wide range of MCT using the same method. Large dimensions ingot is rendered possible up to 40 mm and also 2 inches diameter.

- The growth is governed only by two independent parameters. Typically X_S and T are choosen to grow a desired composition. It is also possible to obtain the desired composition at different temperatures in order to control the stoichiometry of the crystallized material.

- The third advantage is the purification effect related to the use of a solvent zone. It is also possible to dope the material by adjusting the impurity concentration in both the molten zone and the source material according to the segregation coefficient.

- And finally the THM allows the growth of single crystal ingot by using a seed located under the molten zone.
The experimental process can be summerized as followed :
- preparation of the molten zone : an appropriate proportion of CdTe + HgTe + Te, deduced from the ternary diagram, is melt for several hours to assume a good homogeneity and quenched.
- The feeding source material consists of a CdTe + HgTe cylinder the cross sections of which are in the ratio of the desired composition. The cylinder after cleaning and etching is introduced in the quartz ampule over the molten zone.
- The growth occurs in a cylindrical furnace heated by Joule effect at a rate of 2 μm mn^{-1}.

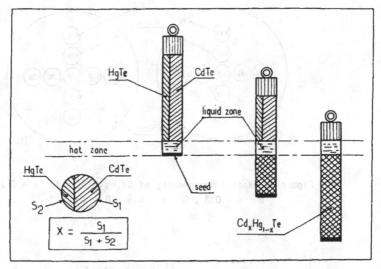

Figure 2 : Crystal growth of Cd$_x$Hg$_{1-x}$Te by THM

. Results and discussion.
All of the composition range can be obtained by changing X$_S$ and T. Typically the following compositions are obtained :

X	0.20	0.3	0.5	0.6	0.7
λ μm	8-12	3-5	2.5	1.6	1.3

Currently the diameter of the ingot ranges from 20 to 40 mm and the length is 60 mm.
The composition homogeneity of the ingot and of the wafer is measured by optical transmission.
The typical longitudinal and radial homogeneity is shown in fig.3,4.
In the range of high x composition the variation of the optical cut-on is less than 0.01 μm at 300°K.
The electrical properties of undoped MCT but presenting p or n type related to stoichiometric defects are presented in the following table with different x values.

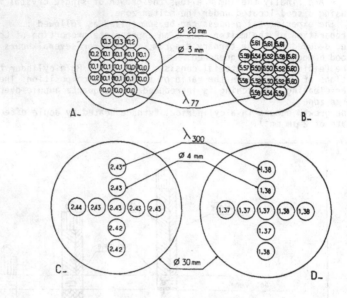

Figure 3 : Radial homogeneity of $Cd_xHg_{1-x}Te$: A : x = 0.2 ;
B : x = 0.3 ; C : x = 0.5 ; D : x = 0.7

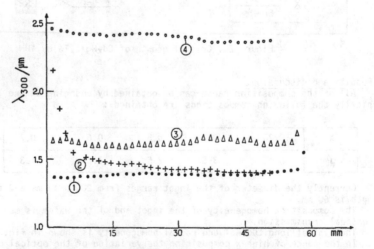

Figure 4 : Composition profiles of $Cd_xHg_{1-x}Te$ ingots :
1   $Cd_{0.7}Hg_{0.3}Te$
2   idem with a non equilibrated molten zone
3   $Cd_{0.6}Hg_{0.4}Te$
4   $Cd_{0.5}Hg_{0.5}Te$

x	T	P-type Concentration(p) cm^{-3}	μ_e $cm^2V^{-1}s^{-1}$	N-type Concentration(n) cm^{-3}	μ_n $cm^2V^{-1}s^{-1}$
0.2	77°K	$3.10^{16} - 4.10^{17}$	500 - 200	2.10^{14}	140 000
0.3	"	$10^{16} - 2.10^{17}$	"	"	60 000
0.5	"	$10^{15} - 2.10^{16}$	100 - 50	3.10^{14}	20 000
0.5	300°K	$2.10^{15} - 9.10^{16}$	400 - 50	6.10^{14}	18 000
0.6	"	$2.10^{15} - 5.10^{16}$	100 - 50	5.10^{14}	2 000
0.7	"	$10^{15} - 3.10^{16}$	"	"	1 000

The P type concentration depends on the growth temperature and/or the post annealing temperature.

The N type concentration can be achieved after the growth at appropriate temperature or after conversion from P $\rightarrow$ N by heat treatment under controlled mercury pressure.

Different carrier concentration can also be obtained by intentionnal P or N impurities doping during the growth.

Concerning the crystalline aspect, usually the ingots resulting from conventionnal growth contain several grains the size of which depends maintly on the composition x : the higher is x , the smaller is the grain size. The ingot diameter seems to have negligible influence on the grain size. Assuming that the growth process is comparable to that of LPE, single orientated CdTe crystal is used as a seed to initiate the crystallization of the ingot. For low x value ($x \leqslant 0.3$) single crystals up to 30 mm in diameter and 60 mm length are currently obtained. For high x value ($x > 0.5$) single crystal is presently limited to 20 mm diameter. The $\langle 100 \rangle$ or $\langle 111 \rangle$ orientation or the nature of the seed CdTe or HgCdTe does not seem to have any significative influence on the crystallization. On some ingots, small twins may be observed at the periphery of the wafer. But their size remains quite small (< 1 mm) and the twins are not continuous throught the ingot.

Laüe diagrams performed on $\langle 111 \rangle$ and $\langle 100 \rangle$ orientated crystals have shown that the misorientations in the ingot (30 x 60 mm) are less than 1°. No significative misorientations can be observed on different spots of the wafer.

Etch pits have been revealed on $\langle 111 \rangle$ A face using chemical etching (10). Depending on the analysed wafer the density is approximatively in the order of $10^5 - 10^6$ cm^{-2}. This density seems to be reduced by using isoelectronic doping elements (Zn or Se).

The microhardness of MCT versus composition x is shown in fig. 5. A peaked form is observed similar to many binary or pseudobinary solid solution. This variation can be described by the empirical relation : $H_V = (H_1 - H_0) x + H_0 + Kx (1-x)$ where H_1 and H_0 are respectively the microhardness of CdTe and HgTe. Maximum value of H_V is obtained for x = 0.75. Small differences between the hardness of $\langle 111 \rangle$ A (metallic) face and $\langle 111 \rangle$ B face are observed similar to III V compound[11] : the A face is harder than B face with $\frac{\Delta H_V}{H_V} \simeq$ 5 - 10 %. No difference in hardness is observed for N and P type.

. Some specific properties of high x material.

The bandgap energy is related to x by the following empirical relation $Eg = -0.115 + 1.316 x + 0.327 x^3$ (12) at 300°K. Concerning the Γ point spin orbit splitting Δ_o few experimental results have been published except HgTe with $\Delta_o = 1.15$ ev[13] and CdTe $\Delta_o = 0.91$ ev[14]. By using an electrolyte electroreflectance method the Δ_o values have been

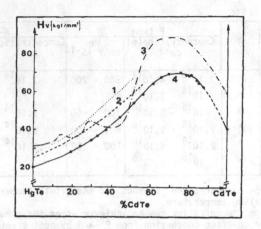

Figure 5 : Microhardness vs composition for $Cd_xHg_{1-x}Te$ after :
1 COLE[22]   2 SHARMA[23]
3 BALAGUROVA[24]   4 this work

measured in the range of x 0.6 - 1 using THM sample. Experimental values
of Δ_o = 0.8 ev[15] are measured for x = 0.55 to 0.7. Therefore $Eg = \Delta_o$
is occuring in the x $\simeq$ 0.6 region leading to a bandgap spin orbit splitting
resonance. This specific bandgap structure at the composition x $\simeq$ 0.6 has
been utilized to fabricate low excess noise factor APD[16].

The absorption below the bandgap of MCT high gap is also investigated.
It is found that this absorption is coming from two contributions[12].
. intervalence band absorption for p type
. free carriers absorption for n or p type
By using k.p approximation, the computed transmission curve is shown
compared to experimental one.

$Eg = 0,89\ eV\ (\lambda c = 1,39\ \mu m)$
After Annealing Figure 6

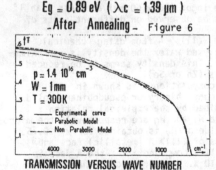

$p = 1.4\ 10^{16}\ cm^{-3}$
$W = 1\ mm$
$T = 300\ K$

—— Experimental curve
---- Parabolic Model
—·— Non Parabolic Model

TRANSMISSION VERSUS WAVE NUMBER

Figure 7 CHART $Eg = 0,89\ eV$
$(W = 1\ mm , T = 300\ K)$

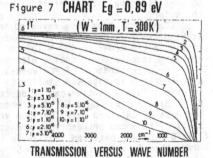

1 : $p = 1\ 10^{15}$
2 : $p = 3.10^{15}$
3 : $p = 5.10^{15}$ 8 : $p = 5.10^{16}$
4 : $p = 7.10^{15}$ 9 : $p = 7.10^{16}$
5 : $p = 1.10^{16}$ 10 : $p = 1.10^{17}$
6 : $p = 2.10^{16}$
7 : $p = 3.10^{16}$

TRANSMISSION VERSUS WAVE NUMBER

The fig.7 presents the influence of [p] on long wavelength absorption
for [p] ranging from 10^{15} to $10^{17}\ cm^{-3}$.
It can be pointed out that for low carrier concentration absorption
coefficient as low as 0.3 cm^{-1} is observed for [p] = 10^{15} and $10^{-2}\ cm^{-1}$
for[n]type $10^{14}\ cm^{-3}$.

The direct dependance between the absorption curve and p concentration has made it possible to use IR transmission for measuring p concentration. This optical method is a non contacting and non destructive method. The optical measurement is easier than Hall effect measurement for the high gap MCT material which presents the difficult problem of contacting on high resistivity p type material.

AVALANCHE PHOTODIODE TECHNOLOGY

The starting material is p type with concentration in the order of $1-2.10^{16} cm^{-3}$. The technology used for the MCT photodiode processing is a conventional planar photolithographic one[17]. The main steps of the process are : . Wafers chemo-mechanical polishing
. Insultating layers deposition by sputtering
. Guard ring and implantation window etching
. Al++ implantation of the N+ layer and post annealing
. Deposition of the passivation and AR layers
. Metallization of N and P contacts.
The structure of the APD is schematically presented by fig.8.

Figure 8 :
Avalanche
photodiode
structure

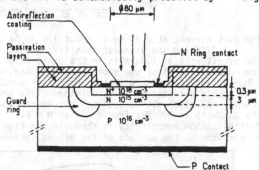

The N+-N-P structure was deduced from the following measurements.
. Capacitance and spectral responses vs bias voltage
. Electron beam induced current
The capacitance versus bias voltage shows a $V^{-1/3}$ dependance at low voltage (up to 10 v) then a $V^{-1/2}$ dependance at high voltage. These dependances can be explained by the fact that the depletion region is spread into the N side of the N/P junction reaching the N+ region at approximatively -10 V. For higher bias voltage the capacitance decreases slowly with the bias; The extension of the depletion region toward the surface is also supported by EBIC measurements on a cleaved photodiode, which also indicate a junction depth of about 3.5 µm at zero bias.

As it is pointed out the main physical characteristic of the MCT Avalanche Photodiode is originated at the specific bandgap structure with x composition close to 0.6. For this particular x value the bandgap energy Eg is equal to the spin orbit splitting energy Δ_o resulting in a resonant effect. Hildebrand et Al[16] have demonstrated that the threshold of hole ionization is reduced to a minimum value when $\Delta_o = Eg$ therefore the hole to electron ionization coefficient ratio β/α reaches a maximum value leading to low noise MCT avalanche photodiode having $x \simeq 0.6$[18] corresponding .. to 1.55 µm and 1.3 µm wavelength, the N side of the APD being illuminated.

The typical characteristics of the 1.3 µm APD are the following (17).

The breakdown voltage at 300°K is about 100 V and it decreases with the temperature as shown by the figure. This linear variation with a temperature coefficient $\gamma = \frac{\Delta V_b}{V_b} = 10^{-3} °C^{-1}$ indicates that the avalanche phenomenon is the prevailing one and no Zener effect contribution is observed even at high temperature.

Figure 9 :

Log Io versus V
with different
temperatures

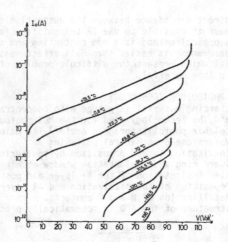

- The dark current at 0.95 V_B is about 1.5 10^{-7}A at room temperature
and 10^{-12} A at 77°K (80 μm diameter sensitive area).
- The sensitivity at 1.3 μm is in the range of 0.7 - 0.8 A/W.
- The multiplication of the photocurrent is measured under low light
level ($I_{p_0} \simeq 50.10^{-9}$ A) using a lock in amplifier synchronized with the
illuminated source : a gain M= 30 is obtained. This gain remained constant
under dynamic operation with a 1.3 μm laser source modulated at 500 MHz.
- The bandwith is 800 MHz from M = 1 to M = 20.
- Multiplication homogeneity is shown in the figure using a 10 μm
diameter spot scan on the surface of the APD at various reverse bias.

Figure 10

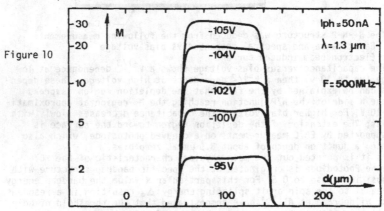

PHOTORESPONSE VERSUS SPOT POSITION ACROSS A PHOTO
① DIODE DIAMETER WITH DIFFERENT BIAS VOLTAGES

- $\beta/_\alpha$ ratio is in the order of 10 leading to a low excess noise factor
$F(M) = M^a$ with a = 0.4 (to be compared with a = 0.9 for germanium APD).
APD photodiodes have been produced with different wavelengths typically 1.3 μm - 1.55 μm and also 2.55 μm.
For the 2.5 μm photodiode $\Delta \geqslant Eg$ the hole ionization process due to
the spin orbit splitting is no more involved. In this composition range

$\frac{\Delta_\circ}{Eg} > 1$, the electron ionization[19] seems to be prevailing compare to hole ionization : a high energy electron losses its energy which generates an electron transition from the valence to the conduction band resulting in the final state two electrons and a (light or heavy) hole. Therefore the multiplication is occuring in the P type region when $\Delta_\circ > Eg$. Experimentally 2.5 μm APD photodiode have been measured : a multiplication factor M = 30 is observed by illuminating P region with 30 V reverse biais. The electron to hole ionization coefficient ratio α/β is in the order of 10 at room temperature, the contribution of tunnel current is found to be negligible.

The main characteristics of MCT APD 80 μm diameter operating at room temperature are summerized as follows..

Wavelength (μm)	1.3	1.55	2.5
V_B (volt)	100	80	30
Dark current at 0.95 V_B	50 nA	300 nA	10 μA
Responsivity at M = 1(A/W)	0.8	1	1.5
Capacitance pF	0.5	0.5	0.5
$k = \beta/\alpha$	10	20	1/10
Multiplication factor M	10	20	30
Excess noise a($F = M^a$)	0.4	0.3	0.3

Current PIN HgCdTe photodiodes[20] are also produced at different wavelengths with various sensitive area.The typical characteristics at room temperature are shown below.

Wavelength μm	1.3			1.55			2.5
Sensitive Area diameter	80 μm	1 mm	5 mm	80 μm	1 mm	5 mm	80 μm
Responsivity A/W	0.8	0.8	0.8	1	1	1	1.5
Dark current nA	0.5 at-10V	20 -5V	300 -5V	1 -10V	100 -5V	1500 -5V	500 -2V
Capacitance pF	0.8	60	1400	0.8	60	1400	0.7

CONCLUSION

Travelling heater method (THM)[21] has been successfully used : this method is characterized by :

. low temperature and low pressure process
. single crystal with defined orientation
. large dimensions ingot
. suitability for any range of composition.

The electrical properties as well as the optical properties have shown a high purity material.

Components for fiber optic communication (1.3 μm - 1.55 μm)and also 2.5 μm have been produced operating at room temperature either in Avalanche or PIN photodiode. Depending on the x composition, 1.3 and 1.5 μm APD present a low excess noise when N side is illuminated due to $\Delta_o \simeq Eg$. For longer wavelength for example 2.5 μm, $\Delta_o > Eg$, electron ionization becomes prevailing leading to low excess noise when P side of the APD is illuminated.

A wide range of wavelengths from 1.2 μm up to 16 μm optronic components are therefore easily obtainable by using MCT grown by the travelling heater method.

REFERENCES

1   M.J Brau U.S Patent N° 3 656 944 (1972)
2   J.L Schmit J. Cryst. Growth 65 249 (1983)
3   C.C Hein U.S Patent N° 2 747 271
4   P.H Stello U.S Patent N° 2 829 994
5   J.D Broder, G.A Wolff J. Electrochem. Soc. 110 (11) 1963, 1150
6   R. Triboulet Y. Marfaing J. Cryst. Growth 51 89 (1981)
7   R. Triboulet D. Triboulet, G. Didier J. Cryst Growth 38 82 (1977)
8   R. Triboulet, I.N Guyen duy, A. Durand J. VAC. SCI.Tech. A3(1) 95 1985
9   T.C Harman J. Electron. Mater. 2(1) 230 (1972)
10 P.F Fewster, S. Cole, A.I.W Willoughly, M. Brown J.Appl. Phys. 52, 7, 4568 (1981)
11 P.B Hirsh, P. Pirouz, S.G. Roberts, P.D Warren, Phil. mag. (1985)
12 T.Brossat, F. Raymond J. Cryst. Growth 72 280 (1985)
13 P.M Amirtharaj, F.H Pollak, J.K Furdyna Solid. St. COM. 39 35 (1981)
14 M. Cardona, K.L. Shaklee, F.H Pollak phys. Rev. 154, 696 (1967)
15 C. Nguyen Van Huong et Al J.Cryst. Growth 72 419 (1985)
16 O. Hildebrand, W. Kuebart, M.H Pilkuln Appl. Phys. Letters 37 801 (1980)
17 J. Meslage, G. Pichard et Al SPIE Proc. 18-22 April Geneva (1983)
18 C. Verié et Al J. Cryst. Growth 59 342 (1982)
19 B. Orsal, R. Alabedra et Al Opto 86 to be published
20 J.C Flachet, M. Royer, Y. Carpentier, G. Pichard SPIE Proc. Cannes 1985
21 A. Durand, J.L Dessus, T. Nguyen duy SPIE Proc Innsbruck 14-18 April 1986
22 S.Cole, A.F.W. Willoughby, M. Brown. J. Cryst. Growth 59, 370, (1982)
23 B.B Sharma, S.K. Metha. Phys. Stat Sol a, 60 k, 105 (1980)
24 Balagurova, Khabarov. Journ. Soviet Phys. (1976)

STRUCTURE-PROPERTY RELATIONSHIPS
IN SEMICONDUCTOR ALLOYS

A. SHER,* M.A. BERDING,* S. KRISHNAMURTHY,*
M. VAN SCHILFGAARDE,* A.-B. CHEN,** AND W. CHEN**
* SRI International, Menlo Park, CA 94025
** Auburn University, Auburn, AL 36849

ABSTRACT

We have demonstrated that the atomic distribution of constituents in semiconductor alloys *is never truly random*. There are always interactions causing correlations; the degree and nature of the correlations depend on which interactions dominate and on the growth conditions. While we have identified most of the interactions which are expected to cause correlations, not all of them have been treated completely to date. Therefore, some details remain unclear, but the principal effects can now be appreciated in broad terms.

TECHNICAL DISCUSSION

In the formalism reported here, we start by focusing on small clusters of atoms that are called microclusters. [1-4] Once the microcluster size is selected, then the total energy of the solid is expressed as a sum of cluster energies; and the number of configurations of the solid corresponding to a given total energy is calculated. Microcluster-microcluster interactions are neglected and there are approximations in the microcluster energy calculations, but once the approximations are made, no appreciable additional inaccuracy is introduced in the statistical mechanics arguments leading to microcluster population distributions. The accuracy of the final result for a given physical property, e.g. critical order-disorder transition temperature, differs for different properties, but in general improves as cluster size increases. Two-atom clusters give most trends properly, but details differ from the answers found for the five-atom, sixteen-bond clusters that are the basis for most of the numerical results we present here. We have not attempted to extend the numerical results for larger clusters.

We have demonstrated [4] that for an n-atom microcluster in state j, represented schematically as $A_{n-n_j(B)}B_{n_j(B)}$, corresponding to a given number $n_j(B)$ of B atoms if the degeneracy $g_j = [^n_{n_j(B)}]$ of a given energy state ε_j is not split, and if ε_j depends linearly on $n_j(B)$, then the average population distribution $\bar{x}_j$ is always that of a random alloy x_j^0. Therefore, only interactions that split the degeneracy or cause a nonlinear variation of ε_j on $n_j(B)$ drive correlations. To be precise, as can be seen from the detailed analysis, the energies $\varepsilon_j - n_j(B)\mu(B)$, where $\mu(B)$ is the B atom chemical potential in the grand partition function formalism, are responsible for populations of state j.

We have identified three mechanisms that cause appropriate nonlinear variations of ε_j. The first is strains resulting from bond length mismatches between the constituents. This is illustrated for $Hg_{1-x}Zn_xTe$ in Fig. 1 where

$$\Delta_j = \varepsilon_j - \frac{(n - n_j(B))}{n}\varepsilon_0^0 - \frac{n_j(B)}{n}\varepsilon_n^0$$

and ε_0^0 and ε_n^0 are the cluster energies in the pure materials. The second, referred to as chemical interactions, is based on potential differences between the constituents responsible for charge shifts among the atoms. In tight-binding terminology, these are the ionic, covalent, and metallization contributions to bond energies. The values of Δ_j for $Hg_{1-x}Cd_xTe$ are presented in Fig. 2. The third source of nonlinearity is electron-electron Coulomb interactions as modified by long-ranged Madelung sums. [5]   As a rule, the bond-length mismatch terms dominate, but the other terms can introduce substantial corrections. In the exceptional cases, such as $Hg_{1-x}Cd_xTe$, where there is a near-bond-length match, the other terms are all that remain.

Until recently, cluster energies were thought to be nearly independent of composition. [6]   In consequence, if the average of the AA and BB interaction energies exceeded the AB energy, then compound formation was thought to be favored; if not, spinodal decomposition was favored. The entropy terms always favor the intermediate random distribution. Now we know that this picture is flawed and the cluster excess energies are in fact highly composition dependent, as inspection of Fig. 1 shows. In the strain terms, the cluster whose volume most closely matches the average volume per cluster for the alloy will have the lowest energy. As a consequence, certain alloy compositions where simple stoichiometric compounds with long-range order could exist, e.g. x = 0.25, 0.5, 0.75, have comparatively low excess free energies. It is therefore possible to have a positive mixing entropy parameter defined by $\Omega \equiv \dfrac{-\Delta F}{x(1-x)}$ and still have compound formation favored for some special compositions x. These tendencies are quite weak in most alloys. For example, in Fig. 3, the excess enthalpy ΔE for $Hg_{1-x}Zn_xTe$ has only a small downward shift at 300 K and x = 0.5. This feature is so weak that it is wiped out in ΔF as seen in Fig. 4. Hence, HgZnTe should show normal spinodal decomposition with a low critical temperature. However, for materials such as $GaP_{1-x}Sb_x$, the shape of the ΔF versus x curve at 300 K resembles that in Fig. 4 but superimposed with a curve having a sharp minimum at x = 0.5. Therefore, $GaP_{1-x}Sb_x$ alloys with, for example, compositions in the range 0.05 < x < 0.5, grown at this low temperature will, to minimize their free energy, tend to exhibit spinodal decomposition into a nearly ordered Ga_jPSb compound and a $GaP_{0.95}Sb_{0.05}$ nearly random alloy. The ΔF curves for HgCdTe are even smoother. However, we have resisted the temptation to present phase diagrams, i.e. critical temperature versus composition, because they will be modified substantially by Coulomb interactions that are not yet incorporated completely into the formalism.

The chemical interactions modify the strain effects in $Hg_{1-x}Zn_xTe$ only slightly. They tend to cause a slight asymmetry in the excess-enthalpy variation with x (at about x = 0.5) and to shift the overall curves. For $Hg_{1-x}Zn_xTe$, the asymmetry causes features on the low-x side to have lower energies than corresponding features on the high-x side. We have demonstrated that, in general, the absolute shifts of the chemical excess energies are positive for the anion-substituted alloys and negative for the cation-substituted alloys. [2]   Hence, for the same lattice constant mismatch, a cation-substituted alloy will have a smaller mixing enthalpy than an anion-substituted alloy.

While the critical temperatures for spinodal decomposition have been demonstrated to be quite low for $Hg_{1-x}Cd_xTe$, $Hg_{1-x}Zn_xTe$, and $Cd_{1-x}Zn_xTe$, cluster populations deviate substantially from random distribution. To illustrate this point, Figs. 5 and 6 show $\overline{x}_j - x_j^0$, the deviations of the various cluster populations from that of a random alloy, for $Hg_{1-x}Zn_xTe$ and $Hg_{1-x}Cd_xTe$ for materials that we suppose to be grown at 300 K by, for example, molecular beam epitaxy. Generally, the population deviations are larger in $Hg_{1-x}Zn_xTe$ than in $Hg_{1-x}Cd_xTe$, but in both cases they can be several percent of a majority species population. In the region around x = 0.25, which is typical of materials used in narrow band-gap infrared devices, we find the

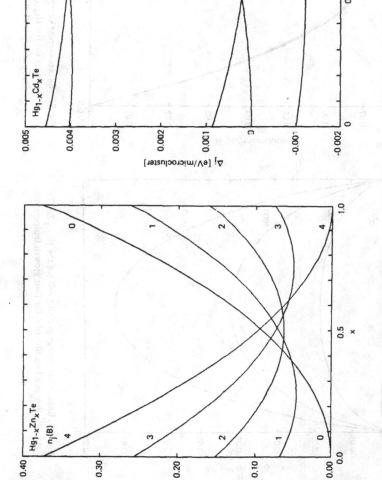

Fig. 1. Excess microcluster energies for the $Hg_{4-j}Zn_jTe$ [n_j (B) = j] 5-atom 16-bond clusters of the alloy $Hg_{1-j}Zn_xTe$ as a function of x.

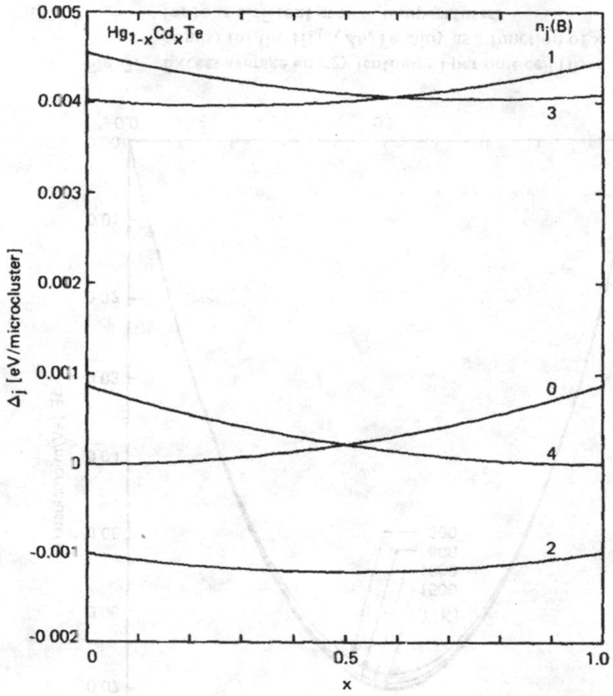

Fig. 2. Excess microcluster energies of the $Hg_{4-j}Cd_jTe$ [n_j (B) = j] 5-atom 16-bond clusters for the alloy $Hg_{1-x}Cd_xTe$ as a function of x.

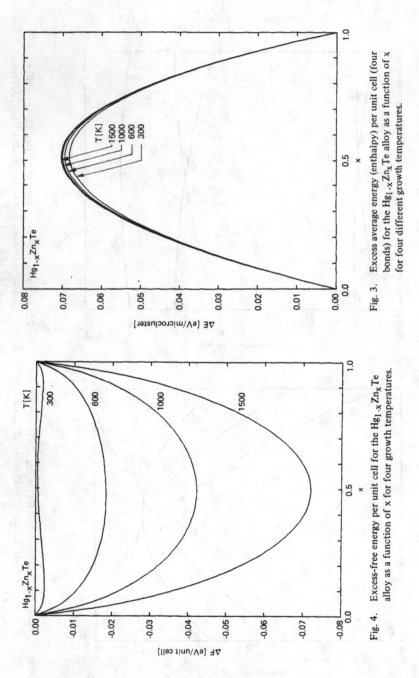

Fig. 3. Excess average energy (enthalpy) per unit cell (four bonds) for the $Hg_{1-x}Zn_xTe$ alloy as a function of x for four different growth temperatures.

Fig. 4. Excess-free energy per unit cell for the $Hg_{1-x}Zn_xTe$ alloy as a function of x for four growth temperatures.

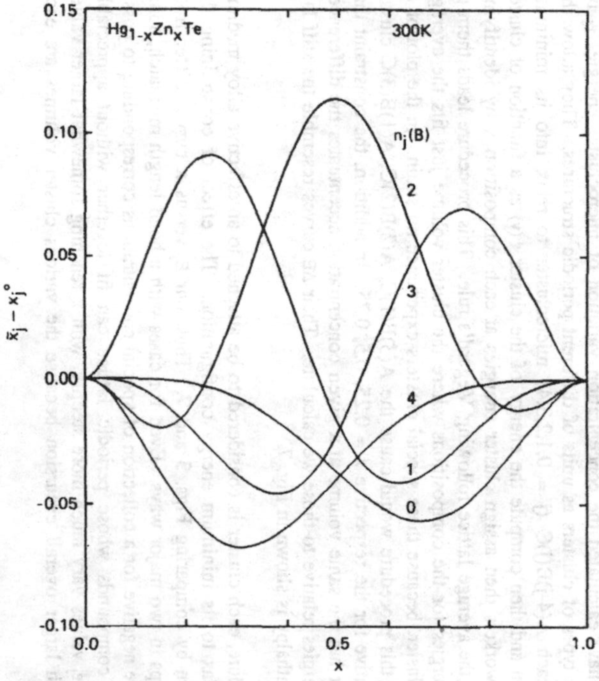

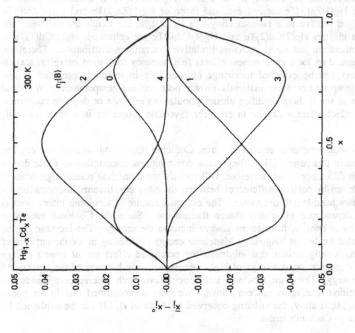

Fig. 5. Microcluster populations relative to those of a random alloy as a function of x for Hg$_{1-x}$Zn$_x$Te grown at 300 K.

Fig. 6. Microcluster populations relative to those of a random alloy as a function of x for Hg$_{1-x}$Cd$_x$Te grown at 300 K.

following: in HgZnTe, the population of the cluster Hg(3)Zn(1)Te is enhanced, those of Hg(0)Zn(4)Te and Hg(1)Zn(3)Te are reduced, and those of Hg(2)Zn(2)Te and Hg(4)Zn(0)Te are nearly unchanged relative to a random alloy. In $Hg_{1-x}Cd_xTe$, the trends are different; the populations of the clusters Hg(2)Cd(2)Te and Hg(4)Cd(0)Te are enhanced, Hg(3)Cd(1)Te is reduced, and the others are just about unchanged relative to a random distribution. Therefore, phenomena that depend on local environment effects (e.g. vacancy formation energies, scattering centers, doping), can be expected to change considerably in materials prepared at low effective growth temperatures from materials grown near melting temperatures. We must examine each case to see if these modified cluster populations enhance or degrade properties. In any case, this effect offers a chance to engineer favorable properties into such materials systems.

The configuration-dependent electron-electron Coulomb interactions are not included in the results reported in this paper. [5] They make contributions comparable to those driven by the bond-length differences, and, therefore, will modify the numerical results significantly. These terms are driven by polarity differences between the alloy constituents in contradistinction to the customary bond-length difference. The essential feature of Coulomb interactions is the configuration dependence of spatial charge fluctuations. Since the Coulomb energy is nonlinear in the charge density, fluctuations always increase the energy. The increase can be partially compensated for by the long-range Madelung energy originating in a coherent sum of alternating charges. Configurations that minimize the combined effect are of lowest energy while the random configuration is always of higher energy, mainly because of the weakening of the Madelung energy. Therefore, Coulomb interactions favor both ordered compounds and spinodal decomposition relative to random alloys. We have demonstrated that in the bond-length-matched $Ga_{1-x}Al_xAs$ alloy, the ordering observed by Kuan et al. [7] can be explained by these electron-electron Coulomb terms.

Others [8,9] have calculated the concentration variation of microcluster energies ε_j by treating the various types of clusters as units of different periodic structures. They allow the central cation in each A(4-j)B(j)C (j = 0,1,2,3,4) microcluster to relax into its minimum energy configuration and then compute the energy of the cluster $\varepsilon_j(v)$ as a function of cluster volume v. These workers then assign cluster energies at each composition x by identifying the v to be that of the average lattice following Vegard's rule. This procedure leads them to negative cluster energies for the compositions where the cluster volume just fits the average alloy volume per cluster, because these special clusters experience no strain. In the plot of Δ_j versus x in Fig. 1, this procedure would cause the A(3)B(1)C, A(2)B(2)C, A(1)B(3)C cluster energies to be negative for the respective x = 0.25, 0.5, 0.75. In addition, the constraint that each type of cluster has the same volume at a given concentration accentuates the differences between cluster energies relative to those we calculate. Their ΔE curves resemble the stiff lattice result whose enthalpy is shown in Fig. 7.

In our procedure, each cluster is considered to be attached to an effective alloy medium and allowed to relax to its minimum energy configuration. The effect of constraining the volume can be seen by comparing Figs. 3 and 7. Thus, our E_j versus x curves differ from those of other groups in two major ways. Even for cases with a bond-length mismatch, some of their ε_j values are negative for a collection of special concentrations corresponding to possible stoichiometric compounds whose periodic lattice can fit together without appreciable strain. Also, their ε_j values vary much more steeply, with x reaching somewhat larger values and having a much larger overall excursion because the various cluster volumes are each

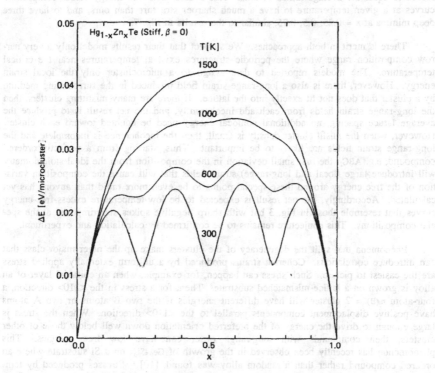

Fig. 7. Excess enthalpy per unit cell for the $Hg_{1-x}Zn_xTe$ alloy under the assumption of an excessively stiff lattice for four growth temperatures.

forced to equal the average lattice volume. This causes their free-energy versus composition curves at a given temperature to have a much sharper structure than ours, and to have three deep minima at x = 0.25, 0.5, 0.75, similar to the results in Fig. 7.

There is merit in both approaches. We suspect that their results model only a very narrow composition range where the periodic structures exist at temperatures near the critical temperature. The models reported to date assign to a microcluster only the local strain energy. However, there is also a long-range strain field produced in the surrounding medium by a cluster that does not fit exactly into the lattice. If there are many misfitting clusters, then the long-range strain fields from each add incoherently, and the net result is to produce the average lattice spacing; no additional nonlocal energy need be counted toward each cluster. However, when the misfit cluster density is small, then the incoherence is incomplete and the long-range strain fields are likely to be important. Thus, starting from a perfectly ordered compound, e.g. ABC_2, the first small deviation in the composition from the ideal stoichiometry will introduce large (local and long-range) strain fields that will cause the composition variation of the free energy around these special points to be even more rapid than anyone has yet calculated. Accordingly, the net result is expected to be low-temperature excess-free-energy curves that resemble those in Fig. 3 but with sharp negative spikes superimposed at the special compositions. This conjecture remains to be confirmed by calculation and experiment.

Phenomena that split the degeneracy of the clusters make up the other major class that can introduce correlations. Coherent strains produced by a uniform externally applied stress are the easiest to picture. Such stress can happen, for example, when an epitaxial layer of an alloy is grown on a lattice-mismatched substrate. Then, for a stress in the <110> direction, a four-atom $n_j(B) = 2$ cluster will have different energies if the two B atoms or two A atoms have positive displacement components parallel to the <110> direction. When the stress is large enough to drive the energy of the preferred orientation down well below those of other clusters, then compounds with long-range order can have low free energies. This phenomenon has recently been observed in the growth of $Ge_{0.5}Si_{0.5}$ on a Si substrate where an ordered compound rather than a random alloy was found. [10] Stresses produced by temperature gradients behind a growth front can cause similar effects. In this discussion, we have recognized the potential importance of applied stresses and temperature gradients in driving microcluster population distributions. However, a comprehensive quantitative theory must still be formulated.

A number of nonequilibrium growth processes are proving to be valuable additions to our materials preparation methods. Included in this category are all growth methods in which the substrate is held at a temperature well below the melting point of the growing material, e.g. molecular beam epitaxy (MBE), metal-organic chemical vapor deposition (MOCVD), and various energy assisted epitaxies (EAE). [11,12] The EAE methods are those in which some form of energy, e.g. laser light or ion bombardment, is supplied to the growing surface. Even without energy assistance, local bonding arrangements in the layers just beneath the growth surface can reorder to attain local minimum-free-energy configurations driven by the energy released when the new atoms arrive and bond, typically a few eV per atom. If one thinks in terms of an effective growth surface temperature that determines the nature of the order-disorder phase state of the material, then in normal MBE and MOCVD T_{eff} probably lies below the melting temperature T_m. For liquid-phase epitaxy, one has $T_{eff} \approx T_m$; for EAE, one has $T_{eff} > T_m$. This single T_{eff} parameter model is undoubtedly an oversimplification, but it serves to establish an order among the trends of a wide range of experimental results reported recently. When T_{eff} is small (MBE and OMCVD), then correlations are high; ordered crystals and crystals with ordered arrays of domains can occur. When $T_{eff} \approx T_m$, then correlations are smaller, and depending on the alloy and composition, more nearly random arrangements or normal spinodal decomposition are more likely. When $T_{eff} >> T_m$, then it is possible to grow materials in the form of random alloys that do not exist in equilibrium. While these materials are metastable, they may still be useful and open a whole new treasure trove to device science.

ACKNOWLEDGMENTS

This work is supported in part by Contracts NASA-NAS1-18232, ONR-0014-86-K-0448, and AFOSR-F4960-85-K-0023.

REFERENCES

1. A.-B. Chen and A. Sher, Mat. Res. Symp. Proc., *46*, 137 (1985).

2. A.-B. Chen and A. Sher, Phys. Rev., *B32*, 3695 (1985).

3. A. Sher, A.-B. Chen, and M. van Schilfgaarde, J. Vac. Sci. Technol., *A4*, 1965 (1986).

4. Detailed calculation will be published in J. Appl. Phys.

5. M. van Schilfgaarde, A.-B. Chen, A. Sher, Phys. Rev. Lett., *57*, 1149 (1986).

6. E.A. Guggenheim, *Mixtures*, (Oxford at the Clarendon Press, London, 1952).

7. T.S. Kuan, T.F. Kuech, W.I. Wang, and E.L. Wilkie, Phys. Rev. Lett., *54*, 201 (1985).

8. M.T. Czyzyk, M. Podgorny, A. Balzarotti, P. Lelardi, N. Motta, A. Kisiel, and M. Zemnal-Slarnawska, Z. Phys. B-Cond. Matter, *62*, 153 (1986).

9. G.P. Srivastava, J.L. Martins, and A. Zunger, Phys. Rev. B, *31*, 2561 (1985).

10. J. Bevk, J.P. Mannaerts, L.C. Feldman, and B.A. Davidson, Appl. Phys. Lett., *49*, 286 (1986).

11. J.E. Green, J. Vac. Sci. Technol. B,*1*, 229 (1983).

12. R.N. Bicknell, N.C. Giles, J.F. Schetzina, Appl. Phys. Lett., *49*, 1095 (1986).

PART IV

Bulk Growth Techniques and Structure—Property Relations

GROWTH AND CHARACTERIZATION OF HIGH QUALITY, LOW DEFECT, SUBGRAIN FREE CADMIUM TELLURIDE BY A MODIFIED HORIZONTAL BRIDGMAN TECHNIQUE

W. P. ALLRED[*], A. A. KHAN[**] C. J. JOHNSON[+], N. C. GILES[++] AND J. F. SCHETZINA[++]
* Galtech Inc., North State Street, Mt. Pleasant, UT 84647
** Washington State University, Dept. of ECE, Pullman, WA 99164-2752
+ II-VI Inc., Saxonburg Boulevard, Saxonburg, PA 16506
++ North Carolina State University, Dept. of Physics, Box 8202, Raleigh, NC 27695-8202

ABSTRACT

A low stress modified horizontal Bridgman technique has been developed and used to grow low defect, large area, subgrain free CdTe crystals for use as substrates in the epitaxial growth of HgCdTe and related IR detector materials. CdTe wafers cut from horizontal Bridgman grown boules exhibit resistivities in the 10^7 ohm-cm range. Etch pit counts are in the $10^4 cm^{-2}$ range. Etch pit patterns as well as x-ray topographs indicate the absence of low-angle grain boundaries. Double crystal x-ray rocking curves are single peaked and very narrow with FWHM(333) as low as 9 arc-sec. Rocking curves of FWHM(333) = 9 to 15 arc-sec, measured at several different laboratories, have been obtained for CdTe wafers cut from several boules. This is in contrast to standard vertical Bridgman grown CdTe samples, which generally show broader x-ray rocking curves sometimes with multiple peaks as a result of subgrain structure. Low temperature (1.6-4.5 K) photoluminescence (PL) measurements on these low defect samples reveal bright edge emission lines which are the main feature of the spectrum. Additional bound exciton lines and other sharp features associated with donor and acceptor impurities are also present. The very weak defect band luminescence (1.40-1.46 eV) provides additional evidence of sample quality.

INTRODUCTION

Cadmium Telluride is a technologically important member of the II-VI family of compound semiconductors. It is used in the fabrication of electro-optic modulators, gamma ray detectors, laser optics and as epitaxial substrate for the growth of epitaxial mercury cadmium telluride. Cadmium telluride has been successful in these applications because of a unique combination of characteristics such as large mean atomic number (50) and reasonably high band gap (1.44) eV) [1]. It also has other characteristics such as low thermal conductivity, large bond lengths, low hardness (2.75 on Moh's scale [2]), which makes its bulk processing a very delicate task. The bulk crystals are normally polycrystalline and the cutting and polishing damage extends several microns beneath the polished surface. The single crystal substrates exhibit intensive dislocation networks and subgrain structure. The etch pit counts of commercially available CdTe is in the low $10^5 cm^{-2}$ range. This material is generally grown by the vertical Bridgman (VB) method.

We have investigated the growth of single crystal CdTe by the horizontal Bridgman (HB) technique. The early results of these growth efforts have been published elsewhere [3]. These preliminary results clearly indicated that it is possible to grow high quality, sub-grain free, single crystal CdTe with etch pit counts as low as $10^4 cm^{-2}$ and x-ray rocking curve FWHM in the 9-15 arc-sec range. The purpose of this paper is to report the results of more extensive characterization of HB grown CdTe single crystals.

Mat. Res. Soc. Symp. Proc. Vol. 90. ' 1987 Materials Research Society

EXPERIMENTAL RESULTS

The growth procedure has been described earlier [3]. In successful runs 60 to 80 percent of the ingot volume is single crystal and oriented in the seed direction. Samples are cut with the (111) orientation for characteriza- tion purposes. Resistivities, etch pit densities, x-ray rocking curves, x-ray reflection topographs and photoluminescence spectra have been measured.

The resistivities are in the 10^7 ohm-cm range. The etch pit density pro- file varies along the length of the ingot from as low as 10^4 cm^{-2} in the middle of the ingot to low 10^5 cm^{-2} at the extremes. Infrared microscopy has revealed the presence of 10 micron and smaller size inclusions. In vertical Bridgman CdTe samples, we have identified such inclusions to be tellurium precipitates. Furthermore our experience shows that such small inclusions can be dissolved by whole ingot or wafer annealing procedures. However, no annealing experiments have been performed on HB grown material. We have noted that in HB grown ingots, the density of inclusions increases from the top (very clear) to the bottom (inclusion networks) of the D-shaped boule. Figure-1 shows an interesting crystallographic distribution of these inclusions.

Double crystal x-ray rocking curves have been measured using (333) reflections off of (111) oriented CdTe surfaces, using a (111) oriented InSb wafer as a reference crystal. The incident beam of CuKα radiation had nominal 1 mm diameter. Figure-2 illustrates the results of a two dimensional mapping of x-ray rocking curve FWHM on a typical HB CdTe wafer. Clearly most of the wafer shows excellent crystalline quality. A subsequent x-ray reflection topograph confirmed that the crystal had either structural or polishing surface damage where multiple peaks are encountered. Note, however, that the FWHM values at these multiple peak locations are still good in comparison to the values normally seen in commercial grade CdTe. The best value seen for HB grown CdTe is around 9 arc-sec. Nominally samples show FWHM in the 10-15 arc sec range. The best veritical Bridgman sample was found to have FWHM in the 20 are sec range. Most vertical Bridgman samples were found to range from 30-50 arc-sec. All of the samples were chemically polished to reduce the effects of surface damage normally found in conventionally polished compound semiconductors. Figure 3 shows a reflection topograph of a large area HB CdTe wafer. Features due to slip, lattice distortions and sub-grain boundaries are noticeably absent.

Infrared transmission has been measured in the 2.5 to 20 micron wave- length range. In all samples measured to date, the IR transmission is slightly lower than the best vertical Bridgman samples. We initially suspected that impurities may be a cause of this lower value. Recent photo- luminescence results tend to support this view.

For photoluminescence measurements, the samples were chemo-mechanically polished. The samples were then immersed in liquid helium, pumped below the lambda point, or were suspended in flowing helium vapor to achieve sample temperatures in the range from 1.6 to 4.5 K. Representative spectra are shown in Figures 4 through 7.

Fig. 1 Infrared photograph of a HB CdTe sample showing
crystallographic distribution of micro-inclusions.

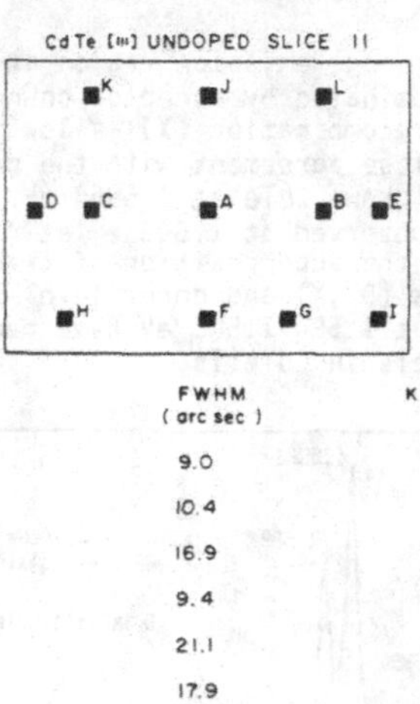

CdTe [III] UNDOPED SLICE II

POSITION	FWHM (arc sec)	K counts/sec
A	9.0	3.39
B	10.4	3.12
C	16.9	2.05
D	9.4	3.33
E	21.1	1.80
F	17.9	2.25
G	{10.0 / 19.9}	{0.70 / 1.73}
H	11.6	3.27
I	[at least 4 unsolved peaks in $\Delta\theta$ of 100 arc sec]	
J	8.9 [low intensity shoulder on high θ side of peak]	3.98
K	9.1	3.40
L	9.6	3.65

Fig. 2 A two dimensional map of double crystal
x-ray rocking curve FWHM values for one
of the HB CdTe substrates.

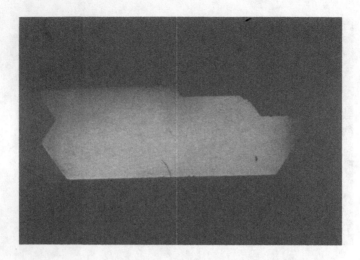

Fig. 3 X-ray reflection (333) topograph of
a (111) oriented HB CdTe sample.

Figure-4 shows the near edge emission region at 1.6 K from one of the
samples. This spectrum is dominated by acceptor-bound excitonic transitions
at 1.5894 eV. Free exciton recombination (X) is low in amplitude, but observed at about 1.5966 eV, in close agreement with the position of the X transition found in thin film MBE grown CdTe at 1.5964 eV. In bulk CdTe free exciton recombination has been observed at 1.596 eV at 4.2 K [4]. The peak at
1.5925 eV is believed to be the superposition of transitions involving
neutral donor-bound excitons (D^o,X) and donor level-valence band recombinations (D-h). The emissions at 1.584-1.585 eV have been related to the presence of donor impurity levels in CdTe:In.

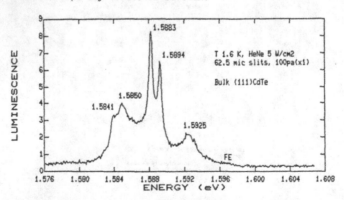

Fig. 4 PL spectrum of a typical HB CdTe sample.

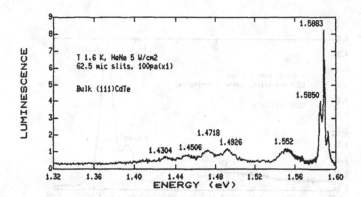

Fig. 5   PL spectrum of the same HB CdTe
samle as in Fig. 4

Figure-5 shows the lower energy transitions seen in the PL spectra
from the same sample whose spectrum is shown in Figure-4. The emission at
1.552 eV occures in the region where electron-acceptor level recombination
and donor-acceptor pair recombination has been reported for CdTe. The emission peaks below 1.5 eV are separated by about one LO phonon energy, which
for CdTe is about 21.3 meV. The zero phonon transition which is observed at
1.4926 eV has been seen in self-compensated CdTe:In and has been related to
complexes involving donors.

Although several of the bulk CdTe samples exhibited PL spectra similar
to those shown in Figures 4 and 5, quite different spectra were obtained
for two of the slices. Figures 6 and 7 show the PL emissions from one of
these samples.

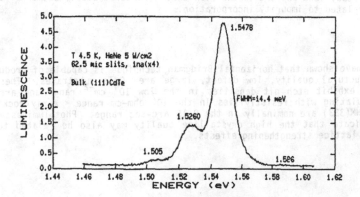

Fig. 6   PL spectra from an unusual HB CdTe
sample.

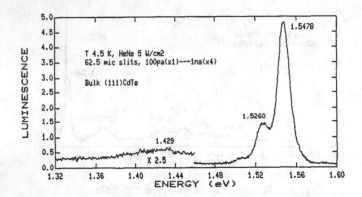

Fig. 7 PL spectra from the same HB CdTe sample
as in Fig. 6.

These spectra show that the near edge recombination involving excitons is absent from these unusual HB CdTe samples. This is true for both free and bound excitonic transitions. The main bright feature in these spectra occurs at 1.5478 eV accompanied by phonon replicas at 1.5260 and 1.505 eV. PL emission in the 1.4 eV defect region is present but is low in amplitude.

Our PL data are in marked contrast to the usually accepted view that high quality CdTe will exhibit sharp, bright excitonic emission lines in the near edge energy region at low temperatures. Although we have not been able to identify impurities from PL emissions, both types of PL spectra recorded in this study lend support to the view that structural stabilization of CdTe may be related to impurity incorporation.

SUMMARY

We have shown that horizontal Bridgman technique is capable of producing high structural quality, low defect, large area CdTe substrates. These crystals exhibit etch pit densities in the low 10^4 cm^{-2} range and are semi-insulating with resistivities in the 10^7 ohm-cm range. X-ray rocking curve FWHM(333) are nominally in the 9-15 arc-sec range. Photoluminescence data indicate that the high crystalline quality may also be related to impurity lattice strengthening effects.

REFERENCES

1. A. J. Strauss, Revue De Physique Appliquee, 12, 168(1977).
2. D. F. Weirauch, J. Electrochem. Soc., 132, 250(1983).
3. A. A. Khan, W. P. Allred, B. Dean, S. Hooper, J. E. Hawkey and
   C. J. Johnson, J. Electronic Materials, 15, 181(1986).
4. R. Triboulet and Y. Marfaing, J. Electrochem Soc., 120, 1260(1973).

HIGH QUALITY CdTe GROWTH BY GRADIENT FREEZE METHOD

A. Tanaka, Y. Masa, S. Seto and T. Kawasaki
Electronics Materials Laboratory, Sumitomo Metal Mining Co., Ltd.
Suehiro-cho, Ohme-shi, Tokyo 198, Japan

ABSTRACT

The vertical Bridgman method, a gradient freeze technique, is feasible for the growth of high quality, large CdTe crystals. The equi-composition contour of Zn in doped crystals has been used to reveal the solid-liquid interface shape in the growth process. A slow cooling rate is necessary to obtain a convex interface shape. A low temperature gradient at the solid-liquid interface, down to 2 $^{\circ}$C/cm, and Zn doping are found to be effective to reduce the etch pit density to 5×10^3 /cm^2 (minimum) to 10^4/cm^2 (average). The crystallographic quality has been evaluated by means of X-ray diffraction, X-ray topography, etch pit delineation with the Nakagawa etchant, infrared measurements, photoluminescence and Hall effect measurements. CdTe crystals are found to be free from subgrain structure, Te precipitates and deep levels, and have high electron mobility.

INTRODUCTION

Cadmium telluride is one of the ideal substrates for HgCdTe (MCT) epitaxial layers because of the good lattice matching and wide band gap of CdTe. However, it is reported that CdTe crystals usually have a low angle subgrain structure, and a dislocation density of 10^5-10^6/cm^2 [1-2]. The minority carrier lifetime in MCT epitaxial layers decreases with increasing dislocation density of substrate when the density is above 10^5/cm^2 [3]. Low dislocation density, subgrain-free, large CdTe substrates are therefore necessary to obtain devices of MCT epitaxial layers [4].

The vertical Bridgman (VB) method has been found to be most suitable to grow high-quality, large CdTe crystals because: 1) a crystal can be grown using a low temperature gradient near the growth interface; 2) stoichiometry control of the melt and the crystal are possible using a Cd or Te reservoir; 3) the growth rate is fast; 4) large crystals can be grown; and 5) the growth system is simple. The first reason is particularly important because a low temperature gradient is one of the key factors in reducing dislocation density.

In this paper, we report on the crystal growth of CdTe by the gradient freeze (GF) technique, a method to measure the shape of the growth interface, the factors influencing the dislocation density, and the properties of the crystals.

CRYSTAL GROWTH OF CdTe

A schematic diagram of the growth system is shown in Fig. 1. The starting materials, Cd and Te, were refined by sublimation and zone melting. The nominal purities of these materials were over 99.9999%. Charge materials were sealed in an evacuated quartz ampoule under a pressure of 10^{-6} torr. The inside wall of the ampoule was coated with pyrolytic carbon to avoid reaction of residual Cd-oxide with the quartz. CdTe poly crystals were synthesized from metallic Cd and Te in the ampoule. Growth conditions used in this investigation are summarized in Table I. The average growth rate was kept in the range of 0.6-5 mm/hr.

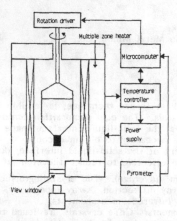

Fig. 1. Schematic block diagram of a VB growth system.

Table 1. Growth conditions.

Growth parameters	Conditions	
Diameter of growth ampoule	30-60	(mm)
Weight of source materials	500-900	(gms)
Stoichiometry of melt	0.999-1.001	(Cd/Te)
Temperature gradient	1.5-25	(^{o}C/cm)
Cooling rate	0.1-3	(^{o}C/hr)

It is well known that a small convex or flat shape of the growth interface (away from the liquid) is necessary to grow a good crystal. A number of experiments [5-7] and model calculations [8] have been carried out to define the shape of growth interface of CdTe grown by the VB method. However, the relationship between the interface shape and the growth conditions has not been well understood as yet. We utilized the equi-composition contour of dopants to preserve the growth interface shape. Zn was chosen as a dopant because it is an isoelectronic element in CdTe and has a favorable segregation coefficient of 1.3-1.4. The spatial distribution of Zn concentration in a crystal was measured by atomic absorption spectrometry.

The contour maps of Zn concentration in central vertical sections of doped crystals grown at two different cooling rates are shown in Fig. 2. The increment in Zn concentration between the neighbouring contour lines is 0.2 at.%. These crystals of 50 mm-diameter were grown from a $Zn_{0.015}Cd_{0.985}Te$ melt under the same temperature gradient of 2 ^{o}C/cm. When the cooling rate was 0.6 ^{o}C/hr, the shape of the growth interface changed drastically from flat to concave during the early stages of growth as shown in Fig. 2(a). For the cooling rate of 0.1 ^{o}C/hr, the shape remained convex or flat during the first half of the growth, and became concave at the later part of the growth as shown in Fig. 2(b). This is the first evidence that a convex growth interface was obtained in 50 mm-diameter CdTe crystals grown by the Bridgman method. This dependence of the interface shape on the cooling rate and the solidified fraction suggested that latent heat liberated on solidification could not escape downward effectively

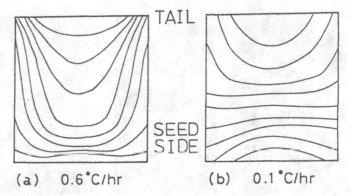

Fig. 2. Contour maps of 1.5 at.% Zn-doped CdTe crystals grown with different cooling rates, taken from central vertical sections of 50 mm-diameter crystals.

through the solidified CdTe. This can be attributed to the lower thermal conductivity of solid CdTe as compared to quartz and liquid CdTe, and to a positive radial temperature gradient at the bottom of the growth ampoule which gradually reverted to a negative gradient during the growth. Continuous grain growth from seed to tail was attained in the CdTe crystal grown at a slow cooling rate and low temperature gradient.

Dislocations in CdTe have usually been evaluated by a chemical etching technique with two typical etchants, E-Ag1 [9] and the Nakagawa [10] etchant. We had found that the Nakagawa etchant was suitable for investigating the dislocation structure in CdTe. The etch-pit patterns delineated on the (111) Cd surface [11] are compared with the images obtained by X-ray Lang topography in Fig. 3. Characteristic lineage structures which are composed of linearlly arranged dislocations can be found in both photographs. In contrast, the E-Ag1 etchant could not provide such a correspondence.

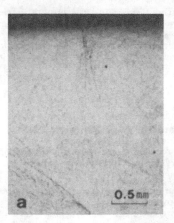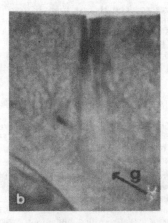

Fig. 3. Comparison between an etch-pit pattern and an X-ray diffraction image of the same crystal: (a) etch-pit pattern; (b) Lang topograph, (220) reflection of AgKa line from a (111) 0.1mm thick wafer.

114

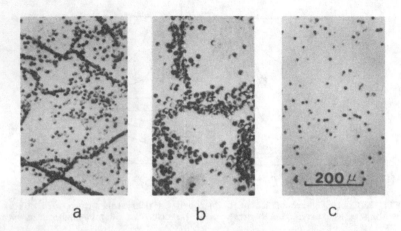

Fig. 4. Dislocation pit patterns in crystals grown under various temperature gradients; (a) 25 °C/cm , (b) 15 °C/cm and (c) 7.5 °C/cm .

The temperature gradient was expected to have a large influence on the dislocation density in CdTe crystals. Typical dislocation-pit patterns of the crystals grown with various axial temperature gradients in the range from 7.5 to 25 °C/cm are shown in Fig. 4. These crystals were grown using ampoules of 30 mm diameter and the same growth rate of 2.5mm/hr. For temperature gradient larger than 20 °C/cm, the crystals had high etch-pit density (EPD), e.g., over $10^6/cm^2$ and exhibited subgrain structure. By decreasing the temperature gradient, the number of dislocations constituting the subgrain boundaries were reduced. When the temperature gradient was decreased below 10 °C/cm, the EPD was reduced to less than $10^5/cm^2$ without subgrain-structure. This result indicated that the temperature gradient was one of the main factors influencing the dislocation density in CdTe.

Zinc doping was also effective in decreasing the EPD. The distribution of EPD in a 1.5 at.% Zn doped crystal is shown in Fig. 5. The EPD of this crystal was as low as $5\times10^3/cm^2$.

EVALUATION OF CRYSTAL PROPERTIES

Crystallographic qualities were evaluated by X-ray diffraction, IR measurements, photoluminescence and Hall effect measurements.

The double crystal X-ray diffraction rocking curves obtained for undoped crystals are shown in Fig. 6. The first crystal was (111) CdTe, without subgrain structure, and with a dislocation density of less than $1\times10^5/cm^2$. The incident beam was CuKa radiation. The rocking curves obtained for the crystal free from subgrain structure had a single peak as shown in Fig.7(a). The value of the full width at half maximum (FWHM) was 9.1 arc seconds. The rocking curve obtained for a poor crystal which had clear subgrain structure is shown in Fig. 6(b). The degree of misorientation between subgrains in this crystal was up to several hundred arc seconds.

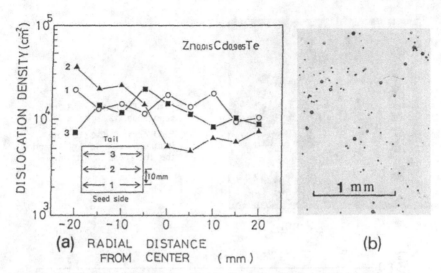

Fig. 5. Distribution of dislocation density in a Zn-doped crystal (a). Etch-pit pattern at the lowest density ($5 \times 10^3 / cm^2$) (b).

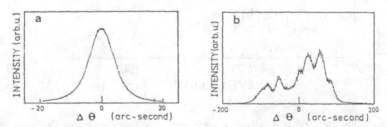

Fig. 6. Double crystal X-ray rocking curves (a) low EPD crystal ($1 \times 10^4 / cm^2$), (b) crystal with cell structure. X-ray reflection is (333). The diameter of the incident beam is 1 mm.

A Lang X-ray topograph obtained for a crystal with a low EPD ($2-3 \times 10^4$ /cm^2) is shown in Fig. 7. Note that the images of the individual dislocation lines can be separated since the density of dislocations is low. It is also of interest that the dislocations are 60°-edge type and are not responsible for subgrain formation. However lineage structure which sometimes leads to polygonization can be recognized at the lower half of this photograph. The reduction of the lineage will be the next important problem on the growth of low EPD CdTe crystals.

It has been known that metallic tellurium tends to precipitate in CdTe and decreases its optical transmittance [12]. The transmittance of an undoped crystal in the far-infrared region at room temperature is shown in Fig. 8. The sample was 1 mm thick and was chemically polished using a 2% Br-methanol solution. The measured transmittance exceeded 64 % at 400 cm^{-1}, suggesting that Te precipitates were not present.

116

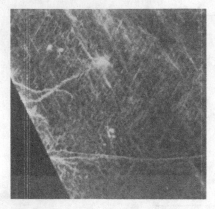

Fig. 7 Transmission Lang
topograph, (220) of a low
EPD crystal ($2\text{-}3\times10^4/cm^2$).
Incident beam is AgKa, and
the thickness is 0.1mm.

,2mm,

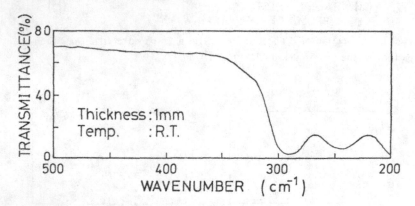

Fig. 8. Far-infrared transmittance spectrum of undoped as-grown CdTe.

A photoluminescence spectrum measured at liquid helium temperature of an undoped p-type crystal grown from a Te rich (Cd/Te=0.999) melt is shown in Fig. 9. The line Ao-X at 1.5906eV was the most intense line and resulted from a recombination of an exciton bound to a shallow acceptor. The first and second L.O. phonon replicas of line Ao-X were clearly seen at 1.5685 and 1.549eV respectively. The luminescence lines from Do-X (1.593eV) and the free exciton (1.596) were visible at the higher energy side of the Ao-X line. In the mid gap range there was no D-A pair recombination line. This spectrum is similar to that obtained for a crystal grown from carefully purified source materials by the traveling heater method (THM) [13].

The electrical properties of GF crystals depended strongly on the compositional deviation from stoichiometry of a melt. As is summarized in Table II, in the Te-rich side the crystal shows p-type conductivity, while in the Cd-rich side, it show n-type conductivity. At liquid nitrogen temperature the resistivity of the p-type crystal in Table II was over 10^6 ohm-cm, allowing for its usage as a semi-insulating substrate for MCT.

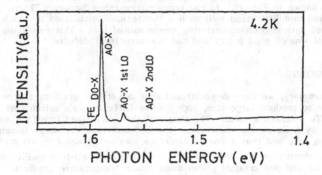

Fig. 9. Photoluminescence spectrum of undoped p-type crystal.

Table II. Electrical properties of undoped as-grown CdTe crystals
at room temperature.

No.	Melt Comp. (Cd/Te)	Cond. type	Carrier Conc. (cm^{-3})	Carrier Mob. (cm^2/V-sec)	Resistivity (ohm-cm)
1	0.999	p	$<3 \times 10^{15}$	60-90	35-5000
2	1.001	n	$<4.5 \times 10^{14}$	700-1100	20-50

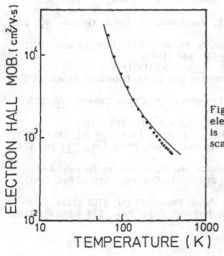

Fig. 10. Temperature dependence of electron Hall mobility. The solid line is a theoretical calculation assuming scattering by LO phonons.

118

The temperature dependence of the electron Hall mobility of an n-type crystal is shown in Fig. 10. In the temperature range between 77 and 300 K, the electron mobility fitted well with a theoretical calculation in which only longitudinal optical phonon scattering was assumed [14]. This result suggests that the crystal was of good purity and had low structural defects.

CONCLUSIONS

In summary, we have demonstrated that a vertical gradient freeze method is feasible to produce large size, high quality CdTe crystals suitable as substrates for HgCdTe epitaxial growth. The shape of the solid-liquid interface was evaluated by a Zn doping technique. It was found that the shape is sensitive to the cooling rate and that a convex interface can be obtained in VB growth. Low dislocation density on the order of $10^4/$ cm^2 and subgrain-free quality are obtained for undoped crystals grown using a low temperature gradient. Zn doping is also effective to reduce dislocation density. Dislocation density in a 1.5 at.% Zn doped crystal is as low as $5\times10^3/$cm^2. Crystallographic properties were evaluated by X-ray diffraction, IR measurements, photoluminescence, and Hall effect measurements. Good CdTe crystals which are free from subgrain structure, Te precipitates and deep level defects, and have high electron mobility were obtained.

ACKNOWLEDGMENTS

We would like to thank A. Nakamura and M. Kawashima for stimulating discussions. We also thank T. Kazuno for electrical and optical measurements, and S. Dairaku, M. Hino, R. Taniguchi, S. Yamauchi and Y. Maekawa for helpful assistance.

REFERENCES

1. L. O. Bubulac, W. E. Tennant, D. D. Edwall, E. R. Gertner and J. C. Robinson, J. Vac. Sci. Technol. A3 (1), 163 (1985)
2. Y. C. Lu, R. S. Feigelson, R. K. Route and Z. U. Rek, J. Vac. Sci. Technol. A4 (4), 2190 (1986)
3. T. Yamamoto, Y. Miyamoto and K. Tanikawa, J. Cryst. Growth 72, 270 (1985)
4. A. A. Khan, W. P. Allred, B. Dean, S. Hooper, J. E. Hawkey and C. J. Johnson, J. Electronic Mater. 15 (3), 181 (1986)
5. N. R. Kyle, J. Electrochem. Soc. 118, 1790 (1971)
6. L. Wood, E. R. Gertner, W. E. Tennant and L. O. Bubulac, Procc. SPIE 350, 30 (1982)
7. R. K. Route, M. Wolf and R. S. Feigelson, J. Cryst. Growth 70, 379 (1984)
8. T. Jasinski and A. F. Witt, J. Cryst. Growth 71, 295 (1985)
9. M. Inoue, I. Teramoto, S. Takayanagi, J. Appl. Phys. 33 (8), 2578 (1962)
10. K. Nakagawa, K. Maeda and S. Takeuchi, Appl. Phys. Lett. 34 (9), 574 (1979)
11. T. Haga, H. Suzuki, H. Abe, A.Tanaka and K. Suzuki, to be published
12. S. H. Shin, J. Bajaj, L. A. Moudy and D. T. Cheung, Appl. Phys. Lett. 43 (1), 68 (1983)
13. R. Triboulet and Y. Marfaing, J. Appl. Phys. 45 (6), 2759 (1974)
14. B. Segall, M. R. Lorenz and R. E. Halsted, Phys. Rev., 129 (6), 2471 (1963)

GROWTH AND CHARACTERIZATION OF Tl_3PSe_4 SINGLE CRYSTALS

N. B. Singh, R. H. Hopkins, R. Mazelsky and M. Gottlieb
Westinghouse R&D Center, 1310 Beulah Rd., Pittsburgh, PA 15235

ABSTRACT

Good quality single crystals of Tl_3PSe_4 were grown from the melt by the Bridgman technique following improvements in the method of purifying the parent components, and optimization of growth parameters. Crack-free crystals 8 cm in length and 17 mm in diameter were produced.

The quality of the crystals was evaluated by optical transmittance and metallographic techniques. In the range 0.7 to 14 μm the optical transmittance shows elimination of absorption bands exhibited in crystals grown without special purification steps. Etchpit studies showed that the crystals were free from inclusions and lamellar twins and that they show a uniform cross sectional etch pit density.

INTRODUCTION

Increasing demands for better infrared laser modulators, Q-switches, deflectors, signal processors and optical filters have driven[1-3] research to improve the optical materials from which these devices are made. Since the mid-nineteen seventies we have been exploring[3-5] the growth and properties of ternary sulfosalt compounds and significant improvements have been made in their growth and device fabrication. Among the sulfosalts, Tl_3AsSe_3, Tl_3VS_4, Tl_3AsS_4, and Tl_3PS_4 were the subjects of extensive studies which showed that these materials have very favorable acousto-optic properties.

The acousto-optic figure of merit, $M_2 = \eta^8 p^2 / \rho v^3$, is a measure of the inherent material diffraction efficiency independent of device geometry. This parameter was observed to be extremely high for sulfosalt materials. For example, Figure 1 shows the variation in M_2 with acoustic attenuation for a number of important materials. Here we have computed the figure of merit relative to quartz at $\lambda = 0.6328$ μm the for photoelastic coefficient p.31. For devices where the diffraction efficiency is not important, other relevant figures of merit can be defined for such materials.

Thallium phosphorous selenide is an orthorhombic material with acentric crystal structure. In earlier studies of Tl_3PSe_4 (TPS) crystal growth, TPS crystals exhibited inclusion formation, compositional inhomogeneity and a tendency to cleave preferentially along (010).

In a continuation of our efforts to improve the optical quality of the chalcogenides, a systematic study was made of Tl_3PSe_4 synthesis and crystal growth. A significant improvement in the crystal size and quality was achieved, and the results are reported in the present article.

2. EXPERIMENTAL PROCEDURE

2.1 Materials and Purification

The Tl, P, and Se used in the present study were identified by their manufacturers as 5N purity material. Further purification of Tl and Se was carried out to reduce the residual impurities. Reactant mixtures for crystal growth were prepared by weighing the starting materials in stoichiometric proportions of 3:1:4. The mixture was homogenized by raising the temperature above 800°C and keeping it molten liquid for several hours. Following homogenization the reactant material was subjected to directional freezing.

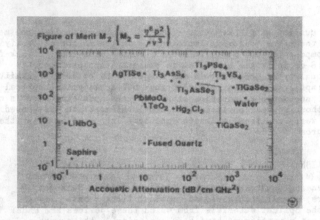

Figure 1 - Acoustooptic merit of various compounds

2.2 Crystal Growth

The homogenized and zone refined mixture of Tl_3PSe_4 was loaded into a well cleaned quartz growth tube (17 mm outside diameter and 20 cm length) and evacuated several hours until a pressure of 10^{-7} torr was achieved. The sample was heated from time to time while under evacuation to drive off residual moisture. The growth tube was then sealed and placed in a two zone Bridgman furnace. By adjusting the furnace temperature the thermal gradient used to grow Tl_3PSe_4 crystals was set at values between 60 and 80°C/cm; the growth speed was fixed at specific values between 0.50 to 0.70 mm/hr.

2.3 Characterization

2.3.1 Bulk Transparency: The overall quality of the crystals was measured qualitatively by observing the crystal through parallel polished inspection faces with an infrared television system. A 120 mesh screen placed behind the crystal was used to define a reference image for optical homogeniety. No significant distortion of the image grid was observed.

2.3.2 Homogeneity: The microscale optical homogeneity of the crystals was characterized by IR photomicrography. Opposite faces of the crystals were polished and the internal microstructure was observed and recorded with a microscope and attached polaroid camera fixture.

2.3.3 Optical transmission measurements: Optical transmission spectra were obtained over the 0.5 to 14 μm range by means of Beckman spectrophotometer. The thickness of the samples for these measurements was 8 mm.

2.3.4 Crystal Etching: Preferential etching is a convenient and relatively rapid method for defect analysis. Samples for the etchpit study were prepared in three stages, sectioning, grinding, and mechanical polishing as described in Reference 6. Final polishing was done with Br-methanol solution to produce a damage-free surface. A solution of NH_4OH and H_2O_2 was then used to delineate etchpits on either the chemically polished or cleaved surfaces. The time period of etching varied from 1 to 4 minutes depending on the concentration of the etching solution.

Figure 2 - Single crystal of Tl_3PSe_4

RESULTS AND DISCUSSION

Figure 2 illustrates a typical crystal grown as part of the present study. Laue x-ray diffraction was employed to verify the orientation of the unseeded crystals. A Laue pattern for the [010] direction is shown in Figure 3. Tl_3PSe_4 exhibits a strong cleavage along (010); cracking was observed in some cases due to this cleavage. The values of the lattice parameters on material grown from the purified reactants are given in Table 1. The values are in good agreement with those previously published.[4]

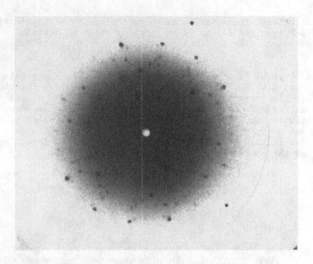

Figure - 3   Laue Pattern for (010) Plane

The crystal quality was evaluated for a piece (8 mm thickness) cut from the as-grown crystal. The crystal was very uniform in color and free from twins or cracks. The transparency of the crystal is shown in Figure 4 with and without a background grid to characterize image distortion. We found no significant optical distortion in any of the samples tested in this way.

In order to identify any inclusions or bubbles in the bulk of crystal, detailed infrared microscopic studies were carried out. A typical micrograph is shown in Figure 5. In general the photographs were feature free. At one or two spots, dark phases 3 to 5 μm in size were present. The exact composition of these phases has not been measured. Neither is it known whether these phases are the result of impurities derived from the parent components are due to fluctuations of the furnace temperature or growth speed.

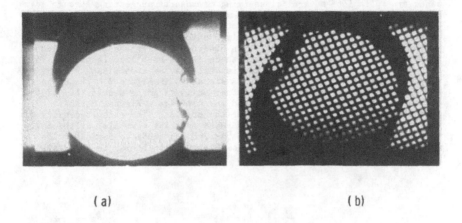

(a)   (b)

Figure 4 - Transparency of 8 mm thick crystal a) without mesh b) with mesh

Figure 5 - Infrared Micrograph   200 μm

The optical transmission of TPS, Figure 6, exhibits a number of strong absorption bands located near 7.5, 9.0, and 11-12.3 μm. These absorption bands are split into several close-lying components suggesting strong coupling between optical and acoustic waves. The short wavelength cutoff is 0.76 μm. Compared to transmission spectra from earlier TPS crystals[4], these spectra show fewer features, apparently due to improved purification.

Since cracking and voids have been observed in crystals of this class of materials, chemical etching was used to delineate such imperfections. To test the etchant for sensitivity to dislocations, we etched each half of a matched cleavage surface. The results are shown in Figure 7. The etched features are mirror images of one another suggesting the sensitivity of the etchant to linear defects which thread the cleavage. Figures 8 shows etchpits on the two different faces of the crystal. The etchpit density was measured at 6.2×10^2 per cm^2. At some spots near the crystals' cylindrical surface a higher etchpit density was observed. A study to correlate the growth conditions with the etchpit density and distribution is underway.

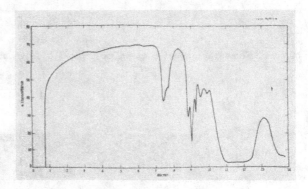

Figure 6 - Transmission Through a Crystal of 8 mm Thickness

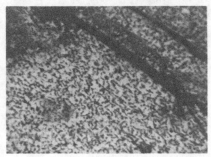

200 μm

Figure 7 - Etchpits on a Cleaved Surface

4. SUMMARY

• Good optical quality crystals of Tl$_3$PSe$_4$ have been grown without the extensive cracking previously reported.

• The bulk transparency and infrared transmission micrography examination showed no gross optical distortion or inclusions in the crystal.

• Several bands in the absorption spectrum seem to be split into close-lying components implying coupling between optical and acoustic waves. Improved purification reduces the absorption level.

• An NH$_4$OH-H$_2$O$_2$ etching technique was developed for this material which revealed well defined etchpits. The etchpit density was a few hundred per cm^2. It is expected that further improvements in purification of the parent components and optimization of crystal growth conditions will reduce the etchpit density and enhance the transmittivity further.

Table 1

Lattice Parameters for Tl$_3$PSe$_4$ (orthorhombic)

a(Å)	b(Å)	c(Å)
9.299	11.073	9.063
9.276[*]	11.036[*]	9.058[*]

[*]Reference 4

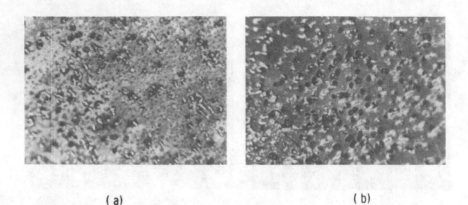

(a)   (b)

$\lfloor$100 µm$\rfloor$

Figure 8 - Etchpits on the Different Surfaces of the Crystals

5. ACKNOWLEDGEMENTS

Authors are thankful to Ms. Debbie Todd for preparing the manuscript.
Technical assistance from time to time was provided by H. Dorman and W.
E. Gaida.

6. REFERENCES

1. J. L. Shay and J. H. Wernick, "Ternary chalcopyrite semiconductors,
    growth, electronic properties and applications" (Pergamon Press, NY,
    1975).

2. R. S. Feigelson and R. K. Route, J. Crystal Growth 49 p. 261,
    (1980).

3. N. B. Singh, R. H. Hopkins, R. Mazelsky, W.Gaida, D. H. Lemmon, J.
    D. Feichtner and R. Mazelsky, Mater. Letter 4 p. 21 (1985).

4. M. Gottlieb, T. J. Isaacs, J. D. Feichtner and G. W. Roland, J.
    Appl. Physics 45 p. 5145 (1974).

5. N. B. Singh, R.H. Hopkins and R. Mazelsky, J. Crystal Growth (to be
    communicated).

6. N. B. Singh, T. Gould and R. H. Hopkins, J. Crystal Growth 78 p. 43
    (1986).

7. I. Fritz, T. J. Isaacs, M. Gottlieb, and B. Morosin, Solid State
    Communications 27 p. 535 (1978).

HgCdTe VERSUS HgZnTe: ELECTRONIC PROPERTIES AND VACANCY FORMATION ENERGIES

M.A. BERDING*, A.-B. CHEN**, AND A. SHER*
* SRI International, Menlo Park, CA 94025
**Auburn University, Auburn, AL 36849

ABSTRACT

The alloy variation of the band gap and the electron and hole effective masses have been calculated for HgCdTe and HgZnTe. Band-gap bowing is larger in HgZnTe than in HgCdTe because of the larger bond length mismatch of HgTe and ZnTe; electron and hole effective masses are found to be comparable for the two alloys for a given band gap. We have calculated the electron mobility in both alloys with contributions from phonon, impurity, and alloy scattering. Contributions to the E_1 line width due to alloy and impurity scattering in $Hg_{0.7}Cd_{0.3}Te$ have been calculated. Results of calculations of the vacancy formation energies in HgTe, ZnTe, and CdTe are discussed.

INTRODUCTION

HgCdTe, the most popular material for current narrow-gap device applications, is plagued by poor structural properties; it is subject to Hg loss at surfaces, contains large as-grown dislocation densities, and is generally fragile. HgZnTe has recently been proposed [1] as an alternative to HgCdTe based on extended bond-orbital calculations. Alloying HgTe with ZnTe instead of CdTe should produce a structurally superior material because the shorter bond length of ZnTe should result in solid solution (alloy) hardening of the HgTe. Alloy hardening in the nearly lattice-matched HgCdTe system would be expected to be less. Recent Knoop hardness measurements on $Hg_{0.89}Zn_{0.11}Te$ and $Hg_{0.80}Cd_{0.20}Te$, both corresponding to near-zero band gaps, found the HgZnTe to be the harder material [2]. Preliminary results of bulk and epitaxially grown HgZnTe showed dislocation densities comparable to, or less than, those of HgCdTe for compositions of technological interest [2,3]. HgZnTe has yet to prove itself as a replacement for HgCdTe, although indications to date are that it is fulfilling its promise as an electronically equivalent, structurally superior replacement for HgCdTe in narrow-gap device applications.

This paper addresses both electronic and structural issues which are important in comparing the relative properties and merits of the two alloys under consideration here. We present recent results on the electronic properties of HgZnTe and compare them to those of HgCdTe, using the coherent potential approximation (CPA) [4] to compute alloy band structures. The alloy variation of the band gap and the electron and hole effective masses for HgZnTe and HgCdTe are calculated and compared. Electron mobilities for HgCdTe and HgZnTe are calculated and the contributions to the E_1 electroreflectance line width from impurity and alloy scattering are estimated for HgCdTe. We also discuss the results of our calculation of the vacancy formation energies in the II-VI compounds HgTe, CdTe, and ZnTe, and their alloys. The vacancy formation energy is related to diffusion activation energy and provides one estimate of the relative stability of the two alloys, HgZnTe and HgCdTe.

ELECTRONIC PROPERTIES

We calculated compound band structures as discussed by Berding et al. [5] using empirical local pseudopotentials [6] plus a tight-binding Hamiltonian to adjust important gaps to the experimentally measured values where available. Spin-orbit splitting was also included. The compound band structures were then used to calculate the alloy band structures using CPA.

128

The CPA calculation includes both diagonal disorder, arising from differences in local bond energies, and off-diagonal disorder, arising from covalent energy differences resulting from bond length differences. In both HgZnTe and HgCdTe, the cation p-state energy difference is small, resulting in little alloy diagonal disorder at the top of the Γ_8 valence-band edge. The cation s-state disorder present in HgZnTe differs from that in HgCdTe. The Hg-Cd s-state energy difference is 1.4 eV, compared with 1.0 eV for the Hg-Zn s-state energy difference. The s-state disorder will be important at the bottom of the conduction band edge of Γ_6 symmetry. Off-diagonal disorder is small for HgCdTe because of the near lattice match of the constituents; such disorder is substantial for HgZnTe, with its lattice mismatch of 6 percent, and contributes significantly to band-edge properties.

The results of CPA calculations for the alloy variation of the band gap as a function of concentration, x, for $Hg_{1-x}Zn_xTe$ and $Hg_{1-x}Cd_xTe$ are shown in Fig. 1. The bowing parame-

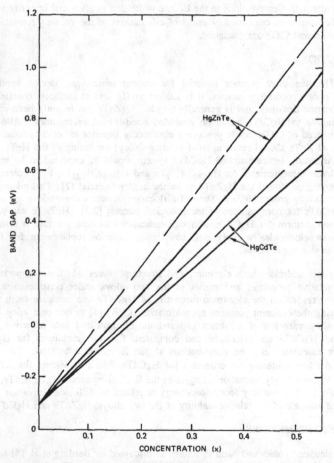

Fig. 1. Band gap versus concentration x for HgCdTe and HgZnTe at 0 K. The broken line represents a straight-line average, and the solid line represents the results of a CPA calculation.

ter, b, defined by $E_g = E_g(\text{linear}) - bx(1-x)$ where $E_g(\text{linear})$ is the linear average of the constituent compound band gaps, is a measure of the band-gap deviation from linearity. From Fig. 1, it is seen that the HgZnTe bowing is larger than that of HgCdTe, with bowing parameters $b(\text{HgZnTe}) = 0.79$ and $b(\text{HgCdTe}) = 0.36$. The larger bowing in HgZnTe arises from the bond-length mismatch of HgTe and ZnTe. The bowing in HgZnTe results in a reduction of the slope of E_g versus x. For a band gap of 0.1 eV we find $dE_g/dx = 2.1$ eV for HgZnTe and $dE_g/dx = 1.9$ eV for HgCdTe. The comparable slopes for HgZnTe and HgCdTe will require comparable control on the concentration of Zn and Cd during alloy growth. The concentrations of Zn and Cd necessary to achieve several specific band gaps are listed in Table I.

Table I. Concentration, x, electron effective mass, m_e^*, and heavy-hole effective mass, m_{hh}^*, for several band gaps. Effective masses are in units of m_0, the free-electron rest mass.

	x		m_e^*	
E_g - eV	HgZnTe	HgCdTe	HgZnTe	HgCdTe
0	0.14	0.18	--	--
0.1	0.18	0.24	0.015	0.013
0.2	0.23	0.30	0.023	0.021
0.3	0.28	0.35	0.030	0.027

$m_{hh}^*(100) = 0.6$
$m_{hh}^*(111) = 1.0$

The alloy variation of the electron and hole effective mass was computed for both alloy systems. The results of alloy disorder on the effective masses were found to be small. The electron effective mass (defined as positive) decreases with increasing concentration, x, until the semimetal to semiconductor transition is reached. At that point, the effective mass begins to increase with concentration. The minimum in the effective mass corresponds to the semimetal to semiconductor transition when the heavy-hole and light-hole Γ_8 and the Γ_6 bands are all degenerate at the Γ-point. The heavy-hole effective mass was found to be independent of concentration for the band gaps of interest and comparable for HgCdTe and HgZnTe. Table I lists results for the electron and heavy-hole effective masses for several band gaps.

Impurity, alloy, and phonon contributions to electron mobility were calculated for $Hg_{0.77}Zn_{0.23}Te$ and $Hg_{0.7}Cd_{0.3}Te$, corresponding to a band gap of 0.2 eV for each alloy. With singly charged impurity densities of $2.7 \times 10^{17} cm^{-3}$, we find mobilities of $1.3 \times 10^4 cm^2/V$-s and $1.4 \times 10^4 cm^2/V$-s for HgZnTe and HgCdTe, respectively, with phonon and impurity scattering limiting the mobility. Our result for HgCdTe is in good agreement with the recently reported electron mobility of $1.22 \times 10^4 cm^2/V$-s for epitaxial MOCVD-grown material [7]. For both HgCdTe and HgZnTe, the alloy contribution to the mobility is an order of magnitude higher than the measured mobilities and, therefore, can be neglected.

The alloy contribution to the various electroreflectance line widths can be computed from the CPA self-energies, which we have calculated for the determination of the alloy band structures. For $Hg_{0.7}Cd_{0.3}Te$ we have calculated the contribution to the E_1 line width due to alloy scattering and charged-impurity scattering in the valence and conduction bands. Complete alloy-band structures were used in the impurity scattering calculation and an upper limit of 10^{18} impurities was assumed. The contribution to the E_1 line width due to alloy scattering is 20.4 meV; we find the contribution from impurity scattering to be negligible (less than 0.1 meV) even for the high impurity concentrations assumed. Calculated line widths from impurity and alloy scattering are much smaller than experimentally measured widths [8-10].

VACANCY FORMATION ENERGIES

The presence of Hg vacancies in HgCdTe is in part responsible for its poor structural properties and also contributes to its electrical characteristics. It is, therefore, crucial to understand the reasons for the high vacancy concentrations in HgCdTe and to compare them with the expected vacancy concentrations in HgZnTe for an analysis of the relative properties of the two alloys. The calculation of vacancy formation energies is difficult; even for spatially localized potential perturbations in the lattice that result from the presence of a vacancy, vacancy wave functions have been found to extend well away from the vacancy site. The interpretation of experimental data relating to vacancy formation energies is often complicated by the existence of several charged states of the vacancy, interstitials, and other mechanisms that all contribute to, for example, the self-diffusion activation energy. In addition, diffusion activation energies have contributions from migration barrier energies.

An additional complication to the vacancy formation energy problem is the final state. As we show below, the vacancy formation energies depend strongly on the final state of the cation or anion removed from the lattice. This can be seen by considering the formation of a vacancy in steps. First, the four bonds are broken when a neutral atom is removed to a free-atomic final state, costing four times the cohesive energy per bond. Some energy is regained by the bonds adjacent to the vacancy through modifications in the metallization energies. In addition, the energy to remove the neutral atom is modified by rehybridization of the dangling hybrid orbitals at the vacancy and Coulomb energies arising from charge shifts about the vacancy site. Finally, some energy can be regained by the remaking of bonds at a surface or in the formation of some molecular state (e.g. gaseous Hg_2). The energy regained by an atom on the surface will vary, depending on the surface to which it migrates.

Consider, for example, ideal unreconstructed surfaces. An anion or cation can remake two bonds going to a (110) or (100) surface. In contrast, a cation can remake one bond on the (111)B surface, or three bonds on the (111)A surface. Similarly, an anion can remake one bond on the (111)A surface, or three bonds on the (111)B surface. The making of three bonds by the cation or anion on the cation or anion (111) surface depends on the number of such surface sites unoccupied. The making of bonds thus depends on the migration of the opposite species to the same surface, because the building up of the surface requires alternating cation and anion layers. The problem is further complicated by surface reconstruction, which is present in any real material.

We have estimated the vacancy formation energies for two specific final states: a free-atomic final state and a (111) A or B surface for a cation or anion, respectively. We have used a cluster containing 100 hybrid orbitals to model the vacancy and its near surroundings. Harrison's tight-binding model [11] is used to calculate the hybrid matrix elements. The vacancy formation energy is determined from the difference between the cluster energy with the vacancy atom in the cluster (initial state) and with it removed (final state). Energy shifts of bonds outside of the cluster caused by vacancy formation are included through the coupling of the cluster states to bond-orbitals outside the cluster in second-order perturbation theory. Contributions to the vacancy formation energy from rehybridization of the dangling hybrids at the vacancy site and Coulomb energies from charge shifts about the vacancy have also been included. For a complete discussion see Berding et al. [12].

There are other small corrections that are due to lattice relaxation. When four electrons are in the six-fold degenerate T_2 state, the Jahn-Teller distortion will lift the degeneracy and drive the electrons to the state that has lower energy and lower symmetry. However, these Jahn-Teller distortions have to compete with the Coulomb energy. Estimates from more detailed calculation for Si show that the net lowering of the energy is about 0.1 eV [13].

Finally, we have estimated the modifications to the total energy from a breathing mode relaxation about the vacancy site. We found approximately a 3 percent reduction in the total energy from these effects.

Calculated vacancy formation energies for cations and anions in HgTe, CdTe, and ZnTe are listed in Table II. We have considered neutral free-atomic final states and (111) surface final states where three bonds can be remade. The vacancy formation energies shown in Table II include corrections from Coulomb energies and rehybridization at the vacancy site. For comparison, we show the experimental cohesive energies per bond.

Table II. Experimental cohesive energies per bond, E_{ch}, and vacancy formation energies for a free-atomic final state, ΔE_v (∞), and a (111) surface where three bonds are remade, ΔE_v (111). All energies are in units of eV.

Compound	E_{ch}	ΔE_v (∞)		ΔE_v (111)	
		Cation	Anion	Cation	Anion
ZnTe	1.20	4.47	7.35	1.20	1.12
CdTe	1.10	4.10	6.71	1.12	1.15
HgTe	0.82	2.84	5.56	0.82	0.65

For all three compounds, the energy to remove a cation from the lattice to a free-atomic final state is roughly equal to four times the cohesive energy per bond. The energy to remove the cation to a (111) A surface is comparable to the cohesive energy. These energies correspond to the net number of bonds broken when the cation is removed from the bulk. For $\Delta E_v(\infty)$, four bonds are broken and for $\Delta E_v(111)$ one bond is broken (four bonds are broken to remove the atom from the bulk, but three bonds are remade by the rebonding at the surface). The idea of bond breaking should not be taken too far, as can be seen by the large difference between anion and cation removal energies. The vacancy formation energies for both the cation and the anion are found to be largest in ZnTe and smallest in HgTe. Note the anomalously low value of $\Delta E_v(111)$ for Te in HgTe. This results ultimately from the same relativistic s-state term value shift that is responsible for making HgTe a semimetal. The low Te vacancy formation energy helps to account for the presence of Te inclusions often found in HgTe and its alloys.

We have calculated the vacancy formation energy for dilute alloys. Results are listed in Table III. The Te vacancy formation energy is seen to be lower in HgTe than in either CdTe or ZnTe. Similarly, the Hg and Zn vacancy formation energies are higher in ZnTe. The cation vacancy formation energies in HgCdTe are approximately the same for the dilute alloy as for the pure compound.

Table III. Vacancy formation energies for dilute alloys. All energies are in units of eV.

Vacancy	Impurity	Host	ΔE_v (∞)	ΔE_v (111)
Hg	Hg	ZnTe	3.51	1.08
Zn	Zn	HgTe	3.71	1.02
Hg	Hg	CdTe	3.05	0.99
Cd	Cd	HgTe	3.90	0.95

DISCUSSION AND CONCLUSIONS

We have computed the alloy variation of the band gap for HgZnTe and HgCdTe. HgZnTe shows a larger band gap bowing than HgCdTe, resulting from disorder due to the bond-length mismatch of HgTe and ZnTe. Electron and heavy-hole effective masses are comparable for the two alloys for a given band gap. HgCdTe and HgZnTe are found to have different vacancy formation energies; the Zn, Hg, and Te vacancy formation energies are larger in HgZnTe. Thus, although we expect HgZnTe and HgCdTe to have comparable electronic properties, we expect HgZnTe to be more resistant to vacancy formation, resulting in a more stable material.

ACKNOWLEDGMENTS

Srinivasan Krishnamurthy contributed substantially to the electronic properties calculations, and Mark van Shilfgaarde gave us valuable insights on the vacancy formation energy calculation. This work was supported in part by NASA and NV&EOC Contract NAS1-12232, ONR Contract N00014-85-K-0448, and AFOSR Contract F4920-85-K-0023.

REFERENCES

1. A. Sher, A.-B. Chen, W.E. Spicer, and C.K. Shih, J. Vac. Sci. Technol. A 3, 105 (1985).

2. S. Sen, W.H. Konkel, R.C. Cole, T. Tung, J.B. James, E.J. Smith, V.H. Harper, and B.F. Zuck, presented at the 1986 MCT Workshop and to be published in J. Vac. Sci. Technol. A (1987).

3. T. Tung, E.J. Smith, S. Sen, W.H. Konkel, J.B. James, V.B. Harper, and B.F. Zuck, presented at the 1986 MCT Workshop and to be published in J. Vac. Sci. Technol. A (1987).

4. P. Soven, Phys. Rev. 156, 809 (1967).

5. M.A. Berding, S. Krishnamurthy, A. Sher, and A.-B. Chen, presented at the 1986 MCT Workshop and to be published in J. Vac. Sci. Technol. A (1987); references therein.

6. J.R. Chelikowsky and M.L. Cohen, Phys. Rev. B, 14, 556 (1976).

7. P.-Y. Lu, C.-H. Wang, L.M. Williams, S.N.G. Chu, and C.M.Stiles, to be published in Appl. Phys. Lett. (17 Nov. 1986).

8. P.M. Raccah, U. Lee, S. Ugur, D.Z. Xue, L.L. Abels, and J.W. Garland, J. Vac. Sci. Technol. A 3, 138 (1985).

9. P.M. Raccah, J.W. Garland, Z. Zhang, A.H.M. Chu, J. Reno, I.K. Suo, M. Boukerche, and V.P. Faurie, J. Vac. Sci. Technol. A 4 2077 (1986).

10. P.M. Amirtharaj, J.H. Dinan, J.J. Kennedy, P.R. Boyd, and O.J. Glembocki, J. Vac. Sci. Technol. A 4, 2028 (1986).

11. W.A. Harrison, Electronic Structure and the Properties of Solids (Freeman, San Francisco, 1980); Microscience (limited distribution report, SRI International, Menlo Park, CA), Vol. IV, 34 (1983).

12. M.A. Berding, A. Sher, and A.-B. Chen, presented at the 1986 MCT Workshop and to be published in J. Vac. Sci. Technol. A (1987); references therein.

13. G.A. Baraff, E.O. Kane, and M. Schluter, Phys. Rev. B 21, 5668 (1980).

STRUCTURAL STUDIES OF HYDROGENATED
AMORPHOUS CARBON INFRARED COATINGS

C. J. Robinson, M. G. Samant, J. Stöhr, V. S. Speriosu, C. R. Guarnieri* and J. J. Cuomo*
IBM Almaden Research Center, 650 Harry Road, San Jose, California 95120-6099
*IBM Thomas J. Watson Research Center, Yorktown Heights, New York 10598

ABSTRACT:

Hydrogenated amorphous carbon thin films are well known for their mechanical hardness and optical properties which make them useful for applications in infrared device coatings. In this work films have been prepared by plasma decomposition of methane using an RF diode reactor operating under conditions of high self bias potential ($V_b = 1$ KeV). The resulting ion bombardment during film growth leads to the formation of hard, insulating carbon coatings which have a band gap of ≈ 1.1 eV and are transparent in the IR. Nuclear reaction analysis has been used to quantify the atomic concentration of hydrogen incorporated in the films and extended x-ray absorption fine structure (EXAFS) has been used to determine local site geometry. Only first and second nearest neighbor bond lengths are observed with no evidence of further long range order or microcrystallinity. A model for atomic structure is proposed which includes both sp^2 and sp^3 bond configurations and direct comparisons are made with data obtained from sputtered carbon films, graphite and diamond.

INTRODUCTION

Research on hard carbon films has intensified in recent years as more applications are being found, particularly in the field of IR coatings [1,2]. A bibliography of papers and reports covering the historical development of these films has recently been published by Woollam et al. [3]. On studying the literature it is found that hard carbon films have been produced by a wide variety of techniques including plasma deposition [4] ion beam deposition [5], sputtering [6], reactive sputtering in hydrogen [7] microwave discharge processes [8] and other variations thereof. In each case, properties of the films produced are reported for the particular preparation technique employed. Recent work, for example, has reported the formation of polycrystalline diamond thin films [8]. Many films however, are produced from a hydrocarbon gas precursor (usually CH_4) or involve hydrogen in some way so as to result in the formation of C-H bonding in the deposited material. Such films are reported to contain both sp^2 and sp^3 hybridized bonds [9] and are often termed "diamond like" because of their hardness, high refractive index ($1.9 < n < 2.3$) and electronic insulating properties. The terminology is misleading since the films are invariably amorphous and the sp^3 bonding is not unambiguously that of C-C in the tetrahedral diamond configuration. What is unambiguous however, is the existence of C-H bonds as evidenced by specific absorption bands near 2900 and 1450 cm^{-1} in the IR [5,10]. The question as to the degree of C-H sp^2 and sp^3 bonding remains largely unanswered and is also highly process dependent. In producing IR coatings, a knowledge of bonding and structure in the material in addition to process dependencies is desired. In this work we have prepared films of hydrogenated amorphous carbon (a-C:H) by the somewhat conventional technique of plasma deposition. We describe the plasma conditions during film growth and characterize the films by conventional techniques. In particular we have measured the total hydrogen content in the films by nuclear reaction analysis and have used EXAFS to gain structural information on the material.

Film Preparation

Thin films of a-C:H were prepared by plasma deposition from methane (CH_4) in a conventional RF diode apparatus similar to that used by Holland et.al. [11,12] and more recently Bubenza et.al. [4]. We have used a grounded electrode of large area placed 16.0 cm distance from a water cooled driven cathode. The cathode and grounded electrode were circular of 10.7 cm and 56.0 cm diameter respectively and the positive glow of the plasma region was contained within the volume defined by the two electrodes. RF power at a frequency of 13.56 MHz was capacitively coupled to the cathode *via* a pi-type impedance matching network. A negative self bias potential V_b developed on the cathode and could be varied by changing the RF power, P_f, into the plasma. In this apparatus the maximum bias voltage obtained was 1 KV ($P_f \approx 1$ KW) and could be reduced to 200V ($P_f \approx 150$W) before plasma instabilities occurred. An estimate of plasma potential V_p was made by inserting a Langmuir probe into the positive glow region of the plasma. The probe potentials were positive with respect to ground and varied from a maximum of 16V to a minimum of 2V for the power and bias conditions mentioned above. True plasma potentials V_p are estimated to be several volts above the probe potentials measured [13]. The pressure, P, of CH_4 used was 2.7×10^{-1} Pa ($\approx 2.0 \times 10^{-3}$ torr) with a gas flow rate of ≈ 8.0 sccm. This pressure was chosen because it was close to the peak of the V_b *versus* P curve and the corresponding mean free path is large (≈ 2.5 cm) compared with the estimated width of the cathode dark space (≈ 0.3 cm for the plasma conditions described). Because of this we make the assumption that positive ions accelerated across the dark space from the edge of the positive glow arrive at the cathode with full accelerating potential ($V_p + V_b$).

Within the plasma, CH_4 molecules decompose into radicals and fragment to form a partially ionised plasma containing a variety of decomposition products. Ions are accelerated to the cathode surface where continued ion bombardment further reduces the a-C:H deposited film and hydrogen is desorbed. Films of a-C:H are deposited on both electrodes and elsewhere in the deposition chamber but only on the cathode surface under conditions of ion bombardment are the hard films of interest here produced. Deposition rates at the cathode were 0.6Å sec^{-1} ($V_b = 300$V) to 1.56Å sec^{-1}($V_b = 1000$V).

Film Characterization

The properties of these plasma deposited a-C:H films are influenced by the plasma conditions during growth. For the deposition conditions previously described the films have properties which are, in general terms, similar to those described extensively elsewhere in the literature [10-12,14-15]. That is, they are hard, electrically insulating (resistivity $\rho \approx 10^9$ohm cm), transparent in the IR, etc. However, to more fully understand and characterize the films, a detailed knowledge of the atomic structure and chemical bonding is required. Figure 1 shows a schematic diagram which illustrates some of the major bonding configurations possible in this material. Both sp^3 and sp^2 hybridized bonds can coexist in some ratio [9] and vary according to deposition conditions. It should be noted that sp^3 bonds are more likely to exist in the form of C-H bond configurations rather than purely tetrahedraly bonded C-C as in the diamond structure. The various forms of C-H bonding give rise to absorption bands within the IR [14] but as yet the existence of the diamond structure has not been established in films prepared by this technique. The high energy ion bombardment during growth of these films leads to cross-linking and degradation of any "polymer-like" material and desorption of hydrogen from the film [12,15]. For the hydrogen content we report in these films, it is unlikely that such polymeric regions exist. Long chain, aromatic or aliphatic hydrocarbon groups would be expected to decompose into smaller sub-groups by ion bombardment or else not form at all. The amorphous carbon is expected to contain voids by analogy with amorphous silicon (a-Si:H) prepared by the same technique [16]. The existence of voids leads to trapped hydrogen gas [15] and dangling bonds as detected by electron spin resonance [17] (ESR).

Band gap measurements have been made by a Tauc analysis and shown in Fig. 2. The dependence on substrate bias voltage V_b is small varying from $E_g \approx 1.4$eV($V_b = 300$V) to $E_g = 1.1$eV($V_b = 1000$V). Films deposited under low bias voltage conditions (for example on the grounded electrode) show a wider band gap of $E_g \approx 2.5$ eV and a correspondingly higher resistivity ($\rho \approx 10^{12}$ohm cm). These films however are softer and generally considered more polymeric in nature [15]. It should be noted that band gap measurements made by this technique are derived from an extrapolation to where the density of states in valence and conduction bands (assumed to be parabolic) reaches zero. The measured values of Eg are thus greater than the pseudo-gap and less than the mobility-gap as conventionally defined in amorphous semiconductors [18].

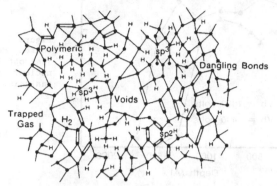

Figure 1.   Schematic diagram showing possible atomic configurations in a random network of hydrogenated amorphous carbon.

Films deposited under conditions of ion bombardment showed strong ESR signals with narrow line width varying with bias voltage from 0.60 Gauss (V_b = 1000V) to 5.50 Gauss (V_b = 300V). The calculated number of spins was within the range 0.9×10^{21} to 3.4×10^{21} spins/cm³ and showed no dependence on V_b. This number of spins is approximately 100 times higher than that reported elsewhere [17]. The existence of unpaired electron spins observed by ESR is consistent with the existence of dangling bonds within the material and is in agreement with analogous results obtained from plasma deposited amorphous silicon [16]. Localized energy levels within the band gap associated with dangling bonds results in charge transport by a hopping conduction mechanism [19]. From the standpoint of electronic properties our results from the a-C:H films produced here appear in good agreement with those reported elsewhere in the literature.

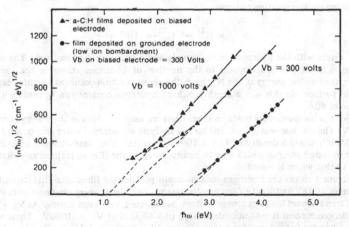

Figure 2.   Tauc plots derived from optical absorption data for a-C:H films. Where α is the absorption coefficient and $\hbar\omega$ is the photon energy.

136

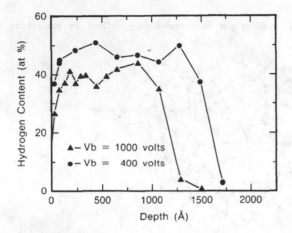

Figure 3. Hydrogen content depth profiles obtained from nuclear reaction analysis.

Surface contamination of the a-C:H films was investigated using x-ray photoelectron spectroscopy (XPS). The film surfaces were seen to have a high oxygen content as evidenced by a strong O_{1s} spectrum. However, the carbon C_{1s} spectra peaked at 284 eV binding energy with no spectra observed at the CO and CO_2 binding energies of 290.2 eV and 291.8 eV respectively. Thus the surface carbon was not in an oxidized state. Surface oxygen was attributed to adsorbed water vapor and was removed by a low energy Ar ion etch, leaving no trace above the detection limit of the instrument. No further contaminants were detected by XPS or Auger depth profiling of the sample.

Hydrogen Content

The hydrogen content in these films was measured by nuclear reaction analysis. The technique involves bombarding the sample with N^{15} ions at the resonance energy 6.38 MeV for the following nuclear reaction to occur:

$$N^{15} + H^1 \rightarrow C^{12} + He^4 + \gamma$$

The N^{15} reacts with the hydrogen in the sample and a gamma ray is emitted. The number of gamma rays is directly proportional to the number of hydrogen atoms in the sample. By increasing the incident energy of the N^{15} beam the above reaction occurs at a greater depth below the a-C:H surface. In this way a depth profile of hydrogen content can be made with a depth resolution of 40Å.

For the measurements made on these films we used N^{15} ions in the energy range 6.36 to 6.67 MeV. The ion dose was 0.6 μC and for the analysis we assumed a stopping power for carbon of 1.90 MeV/μm and a density of 11.4 $\times$ 10^{22} atoms/cm^3. The measurements of hydrogen are total and include hydrogen which may be incorporated in the film as trapped gas (dissolved) in addition to that which is bound.

Figure 3 shows the hydrogen content depth profiles for films of a-C:H deposited under conditions of 400 eV and 1000 eV ion bombardment. The higher energy ion bombardment shows a general trend toward lower hydrogen content, as expected, but is not strong. At $V_b = 400V$ the mean hydrogen content is $\approx$48 at% decreasing to $\approx$40 at% at $V_b = 1000V$. These values are lower by a factor of about two from those reported by Angus et al. [5] for films prepared under somewhat similar V_b conditions and analyzed by the same technique.

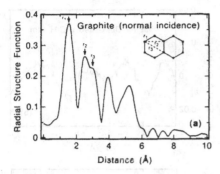 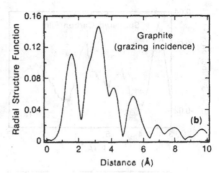

Figure 4.   Radial structure function obtained from graphite sample
(a) normal incidence data, (b) grazing incidence data.

It is possible that the hydrogen content in these films is primarily due to the thermal effects of ion bombardment rather than the absolute ion energy per-se. In our experiments the substrate (cathode) was water cooled resulting in only small variations of temperature ($< 40°C$) during deposition. Other work has shown a more significant dependence of hydrogen content on anneal temperature [20].

The lower hydrogen content at V_b = 1000V is consistent with the smaller band gap measured in Figure 2. As V_b is increased (and hence ion bombardment of the growing film), C-H bonds are broken and hydrogen is desorbed. The deposited films progressively become less "polymer like" ($V_b > 2.0eV$) and more typical of the films discussed here ($V_b \approx 1.0$ eV). Continued increase in ion bombardment further reduces the band gap and ultimately the films graphitize [15] through thermal heating. Similar results have been reported by Smith [21] showing the band gap reducing to zero as anneal temperature is increased to 750°C.

EXAFS Analysis

Structural information on these carbon films was obtained by measurement of first and second nearest neighbor bond lengths using the technique of Extended X-ray Absorption Fine Structure (EXAFS). Analysis was made on single crystals of diamond and graphite for calibration and thin films of amorphous carbon prepared by plasma deposition as described (V_b = 1000 Volts) and DC planar magnetron sputtering for comparison. Both thin film samples were deposited on silicon substrates and of 250Å thickness. The X-ray absorption spectroscopy (XAS) data were collected at the carbon K edge (284 eV) for all samples using detection of total Auger electron yield in an ultra high vacuum chamber described elsewhere [22]. The analysis of the spectra, based on well-documented procedures [23] yielded the structural information described here. The X-ray absorption edge was at 284 eV for all the samples, again indicating that the carbon was not in an oxidized state (also in agreement with the XPS data). The EXAFS function $\chi(k)$, (where k is the wave vector) for the thin film samples, is shown in Fig. 5a and 5b. The Fourier transform of the product $k.\chi(k)$ provides a radial structure function (RSF) as shown in Figs. 4a,b and Figs. 5c,d. Data for the single crystal graphite sample were obtained with the incident X-ray beam both normal and at grazing incidence to the lamina planes of the graphite structure. In normal incidence, information is only obtained from atomic distances within the plane of the hexagonal ring, $i.e.$, the first, second and third nearest neighbor distances (r_1, r_2 and r_3 in Fig. 4a). In grazing incidence a large contribution from the inter-planar spacing is observed which becomes difficult to deconvolve from the second and third neighbor peaks (Fig. 4b). The real C-C nearest neighbor distances were obtained from these data by using the phase shift derived from the diamond standard in which the sp³ C-C bond length of 1.545Å was assumed. Individual peaks in the RSF were reverse transformed and fitted to the $\chi(k)$ function in order to obtain specific values. Errors in atomic distances are ±0.03Å by this technique. From figure 4a the C-C nearest neighbor distance r_1 was measured to be 1.42Å and is in excellent agreement with

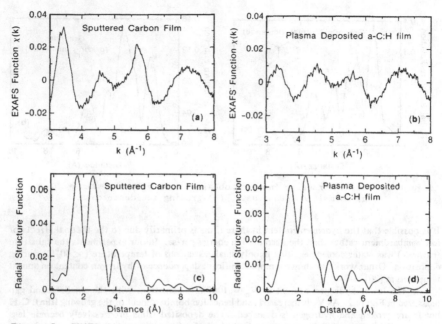

Figure 5.   EXAFS function and radial structure function for (a and c)
sputtered carbon film and (b and d) a-C:H film.

the theoretical sp^2 bond length for graphite (1.421Å). Thus by using diamond as a calibration standard and measuring the known bond length of graphite as a cross reference we demonstrate the validity of these data. The r_2 and r_3 peaks in figure 4a cannot be sufficiently deconvolved to permit measurement of 2nd and 3rd neighbor distances within the same error margin. However, it is nonetheless instructive to observe the relative amplitudes of these peaks for comparison with the thin film data.

The EXAFS data and radial structure functions obtained from the sputtered carbon and plasma deposited a-C:H thin film samples are shown in figure 5. In each case it can be seen that the films show a well defined first and second nearest neighbor distance. The third neighbor distance (corresponding to r_3 in fig. 4a) and inter-planar spacing (corresponding to the largest peak in fig. 4b) are missing or greatly reduced in amplitude so as to be within the noise of the data. Since the hexagonal graphite ring dimension r_3 is missing we conclude that complete rings are not an integral part of the atomic structure of the thin films. Thus there is no evidence of a complete unit cell or microcrystallinity (either graphite or diamond) in these samples and in this context they may be considered truly amorphous. The dominant first and second nearest neighbor distances suggest a microstructure based on a structural unit consisting of a three carbon atom group. Each group being bonded to the next but rotated so as to form a continuous three dimensional amorphous network. Within each group the bond angle is preserved and determined by the hybridization of the bond (ie. $120°$ for sp^2 bonds and $109°$ for sp^3 bonds). For the sputtered carbon film the first and second nearest neighbor peaks correspond well with those of graphite and since hydrogen is not a significant constituent it is most likely that the entire structure is sp^2 bonded. For the plasma deposited a-C:H film both sp^2 and sp^3 bonds are likely to exist in a ratio commensurate with the hydrogen content discussed earlier. However the data do not distinguish between sp^3 bonds which involve C-H groups and those which are purely C-C bonded since in each case the bond lengths are the same. The absence of definite structural information beyond the second neighbor distance again supports a model in which three atom sub-groups are bonded with random orientation so that long range order is not observed.

Conclusions

Thin films of hydrogenated amorphous carbon with applications as IR coatings, have been produced by conventional methods. The films have been characterized and shown to be similar to those generally reported for this preparation technique. Hydrogen content has been measured at $\simeq 40$ at% by nuclear reaction analysis and the use of EXAFS has proven a valuable technique in determining local structure and bonding information in these materials.The films show a well defined local atomic structure consisting of first and second nearest neighbor distances but no long range order beyond this or evidence of microcrystallinity is observed. To this extent the films may be described as truly amorphous.

ACKNOWLEDGMENTS

The authors wish to thank M. Le for technical support and R. A. Sigsbee for providing the samples of sputtered carbon films. Helpful discussion with J. Coburn and B. Street are also gratefully acknowledged.

REFERENCES

[1]   A. Bubenza, B. Dischler, G. Brandt and P. Koidl, *Opt. Eng.* 23, 153-6 (1984).
[2]   K. Enke, *Appl. Optics* 24, 508-512 (1985).
[3]   J. A. Woolam, H. Chuang and V. Natarajan, *Appl. Phys. Comm.* 5(4), 263-283 (1985-86).
[4]   A. Bubenzer, B. Dishler, G. Brandt and P. Koidl, *J. Appl. Phys.* 54(8), 4590 (1983).
[5]   J. C. Angus, J. E. Shultz, P. J. Shiller, J. R. MacDonald, M. J. Mirtich and S. Domitz, *Thin Solid Films* 118, 311-320 (1984).
[6]   C. Wyon, R. Gillet and L. Lombard, *Thin Solid Films* 122, 203-216 (1986).
[7]   D. R. McKenzie and L. M. Briggs; *Solar Energy Materials* 6, 97-106 (1981).
[8]   A. Badzian, B.Simonton, T. Badzian, R. Messier, K. E. Spear and R. Roy, *Proc. SPIE* 682, (1986) in press.
[9]   S. Kaplan, F. Jansen and M. Machonkin, *Appl. Phys. Lett.* 47(7), 750-753 (1985).
[10]   M. P. Nadler, T. M. Donovan and A. K. Green, *Thin Solid Films* 116, 241-247 (1984).
[11]   L. Holland and S. M. Ojha, *Thin Solid Films* 38, L17 (1976).
[12]   L. Holland and S. M. Ojha, *Thin Solid Films* 58, 107-116 (1979).
[13]   B. Chapman, ed., *Glow Discharge Processes*, Wiley (1980).
[14]   B. Dischler, A. Bubenzer and P. Koidl, *Solid State Comm.* 48, N.2 105-108 (1983).
[15]   A. R. Nyaiesh and L. Holland, *Vacuum* 34(5), 519-522 (1984).
[16]   M. H. Brodsky, M. Carbona, and J. J. Cuomo, *Phys. Rev. B* 16(8), 3556-3571 (1977).
[17]   R. J. Gambino and J. A. Thompson, *Solid State Comm.* 34, 15-18 (1980).
[18]   N. F. Mott and E. A. Davies, eds., *Electronic Processes in Non-Crystalline Materials*, Clarendon Press, Oxford (1979).
[19]   B. Meyerson and F. Smith, *Solid State Communications* 41(1), 23-27 (1982).
[20]   D. C. Ingram, J. A. Woolam and G. Bu-Abbad *Thin Solid Films* 137, 225-230 (1986).
[21]   F. W. Smith, *J. Appl. Phys.* 55(3), 764 (1984).
[22]   J. Stöhr, R. Jaeger and S. Brennan, *Surf. Sci.* 117, 503-524 (1982).
[23]   P. A. Lee, P. M. Citrin, P. Eisenberger and B. M. Kincaid, *Rev. Mod. Phys.* 53, 769-806 (1981).

PART V

Frontiers of Materials Science in Infrared and Dilute Magnetic Semiconductor Materials

Variation in the Properties of Superlattices
with
Band Offsets

T. C. McGill, R. H. Miles and G. Y. Wu
T. J. Watson, Sr., Laboratory of Applied Physics
California Institute of Technology
Pasadena, California 91125

Abstract
The implications for recent reports of large valence band offsets at
the HgTe-CdTe heterojunction are examined. The variation of the
band gap and effective mass for transport normal to the layers in
the superlattice is examined in detail.

Introduction

Alloys of HgTe and CdTe are used extensively in the infrared.[1] These alloys have
the property that the band gap can be adjusted from zero up to the value of the CdTe
band gap (≈ 1.6 eV). In recent years it has been predicted that a number of advantages
result from layering the HgTe and CdTe into a superlattice structure like that shown in
Fig.1 instead of intermixing the Hg and Cd to make an alloy.[2,3] In the alloy the band gap
is controlled by the relative composition of Hg to Cd, while in the superlattice, the band
gap is controlled by the thickness of the layers making up the superlattice.[3] Studies of
the growth of alloys and superlattices by molecular beam epitaxy indicate that it will be
much easier to control the band gap of the superlattice than the band gap of the alloy.[4]
In alloys, the effective mass is strongly coupled with the band gap. The effective mass is
proportional to the gap. Hence, small band gap implies small effective mass. This has
lead to serious difficulties with leakage currents in p-n-junction-device structures.[3] In the
superlattices, the effective mass and band gap can be decoupled to some degree.[3]

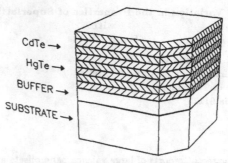

Fig. 1. A schematic diagram of a HgTe-CdTe superlattice. CdTe, CdZnTe and GaAs have all been used as substrates. The superlattices are characterized by the thicknesses of the HgTe and CdTe layers.

The properties of superlattices have been studied extensively both experimentally and theoretically in the last couple of years.[5, 6] The basic situation is that superlattices have been successfully fabricated by on the order of ten groups.[7] The optical properties of the superlattice including infrared absorption,[8, 9] photoluminescence[10] and photoconductivity[9] have been measured on a rather small number of superlattices. A number of other experimental studies have been carried out.[6]

BAND OFFSETS AT HETEROJUNCTIONS

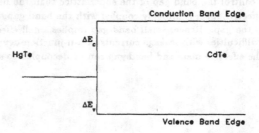

Fig. 2. Schematic diagram illustrating band offsets in the valence band ΔE_v and conduction band ΔE_c at a heterojunction between HgTe and CdTe.

One of the major unknowns comparing theory with experiment is the basic property of the electronic states at the heterojunction between the HgTe and CdTe, the band offset.[5] As shown in Fig. 2, the band offsets are the parameters which give the relative location of the valence and conduction band edges in the HgTe and CdTe as one goes across the heterojunction.

In this manuscript, we will concentrate on the role that band offsets have in determining the properties of superlattices. In Section II, we review the current situation on the values of the band offsets. In Section III, we present the results of some calculations of the variation of the band gap of the superlattices with band offset. In Section IV, we present the results of some recent calculations of the variation of the effective mass with band offset for transport by electrons normal to the layers. Finally in Section V, we will summarize our conclusions.

Band Offsets at the HgTe-CdTe Heterojunction

One of the basic concepts in heterojunction physics is that of a band offset. As illustrated in Fig. 2, the band offsets are the difference in position between the valence band edge or the conduction band edge as one moves across an interface between two semiconductors. Historically it has been difficult to obtain a precise value for this parameter. Theoretically it is difficult to predict the precise values of the band offsets. Experimentally, it has been difficult to come up with experiments that are sufficiently sensitive to the band offset and that do not contain a number of other parameters about which there is a great deal of uncertainity. There is a great deal of discussion about the appropriate value for HgTe-CdTe.

For some time the commonly accepted value for the band offset was based on the common-anion rule[11] and some theoretical considerations by Harrison.[12] These considerations, which at the time seemed to be consistent with the known values of the valence band offsets for heterojunctions involving III-V semiconductors, lead to the conclusion that $\Delta E_v \approx 0$. Early experiments in an attempt to measure the band offsets resulted in only an upper bound of the valence band offset of $\Delta E_v \leq 0.5$ eV.[13] More recent measurements on the magneto-optical properties of superlattices concluded that $\Delta E_v = .04$ eV.[14] Hence, most of the theoretical studies of superlattices have assumed that $\Delta E_v = 0$.

Recently the HgTe-CdTe heterojunction has been studied using x-ray photoemission spectroscopy (XPS).[15,16,17] These experimental groups conclude that $\Delta E_v \approx 0.3$ eV. New theoretical calculations[18] about the band offset have lead to the conclusion that the valence band offset is substantially larger than zero, $\Delta E_v \approx 0.5eV$. Hence, it has become important to consider the implications of larger valence band offsets and to make use the known properties of the superlattice as a check on the XPS values for the band offset.

Band Offsets and Band Gaps

In Fig. 3, we have illustrated the spatial variation of the band edges in the superlattice. The band gap occurs between the heavy hole band in the HgTe and the light electron like band in the HgTe. The confinement of the wavefunctions in the superlattice produces sub-bands like those illustrated in Fig. 3. The band gap is between the top of the valence like sub-band and the bottom of the conduction like sub-band. The major question is what is the variation of the relative position of these two sub-bands with the value of the valence band offset.

In some earlier theoretical calculations, it was found that the band gap was a

146

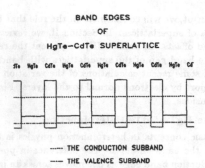

BAND EDGES

OF

HgTe-CdTe SUPERLATTICE

------- THE CONDUCTION SUBBAND

------ THE VALENCE SUBBAND

Fig. 3. A schematic diagram of the band edges in a superlattice. The effects of band bending due to doping and thermally excited charge carriers are not included.

maximum at $\Delta E_v = 0$ and decreased for $\Delta E_v < 0$ or $\Delta E_v > 0$.[19] In those calculations a single case was considered. Here, we report a more extensive study of the band gaps. The calculations were carried out using the model developed by Bastard.[20] The variation in the position of the heavy hole band due to confinement is neglected. The temperature dependence of the band gaps of the bulk materials[21] was included to model the temperature dependence of the superlattice properties.[22] In Fig. 4, we present contour plots of the band gap as a function of the number of atomic layers of CdTe and number of atomic layers of HgTe in each of the layers making up the superlattice. Three different values of the offset have been considered: $\Delta E_v = 0$ eV, $\Delta E_v = 0.3$ eV, and $\Delta E_v = 0.5$ eV. Each atomic layer is assumed to be 3.23 Å thick corresponding to growth on a (100) substrate. The results are actually only a function of the thickness of the layers in the superlattice. Hence, other orientations can be computed by computing the thickness of the layers along the (100) orientation and using those thicknesses for other orientations. The theory is based on an envelope function approximation which is suspect in the limit of thin layer thickness. Hence in the regions of the contour for thin CdTe and HgTe layers where the contours show a great deal of structure, the validity of the theory is somewhat in question and the structure may not be physical.

The results in Fig. 4 confirm overall results reported in Ref. 19. The band gaps of the superlattices tend to decrease by rather large amounts with increasing value of the band offsets. For the case of $\Delta E_v = 0.5$ eV, the band gaps are over a factor of two smaller than for the same layer thicknesses and a band offset of $\Delta E_v = 0$ eV. Comparison of the contours in the three figures indicates that while the region of small band gap always exists, an increase of the valence band offset from the zero value results in a substantial compression of the region of layer thicknesses producing band gaps of interest.

Band Offsets and Effective Masses

The larger effective masses in superlattices[3] are the result of carriers having to tunnel through the barriers provided by the CdTe layers (see Fig. 3). The tunneling amplitude is governed by the imaginary part of the complex wavevector of the CdTe at

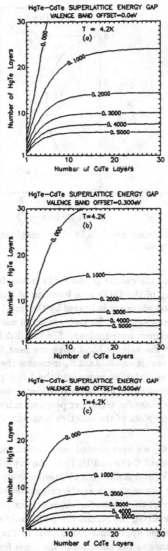

Fig. 4. Contour plot of the band gap of the superlattice in eV as a function of number of layers of HgTe and CdTe. The thickness of a single layer is 3.23 Å corresponding to a superlattice grown on the (100) plane of a substrate. The spacing between the contour lines is 0.1 eV. Three different values of the valence band offset are included: $\Delta E_v = 0$ eV in (a), $\Delta E_v = 0.3$ eV in (b), and $\Delta E_v = 0.5$ eV in (c).

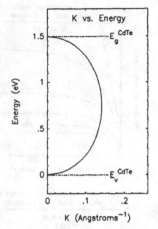

Fig. 5. The complex band structure of CdTe. The imaginary part of the wavevector as a function of energy is plotted.

the energy corresponding to the carrier in the CdTe layer. In Fig. 5 we have plotted the energy in the forbidden gap of the CdTe as a function of the imaginary wavevector. The conduction and valence band edges are indicated by the lines. The important point to note in this figure is that the decay constant for states is small near the valence and conduction band edge reaching a maximum near midgap. For small ΔE_v the states making up the conduction band edge and the valence band edge are near the valence band edge of the CdTe hence the decay constant is small. As ΔE_v increases the states at the conduction and valence band edges in the superlattice move into the band gap of CdTe hence increasing the decay constant for tunneling. This increase in decay constant would be reflected in an increase in the effective mass for transport perpendicular to the layers. This effect is shown clearly in the contour plots of the effective mass shown in Fig. 6.

In this figure (Fig. 6), we have plotted the contours of constant effective mass as a function of number of layers of CdTe and HgTe in the superlattice for the three values of $\Delta E_v = 0$ eV, $\Delta E_v = 0.3$ eV, and $\Delta E_v = 0.5$ eV, respectively. The theoretical calculations were carried out in exactly the same way as the calculations of the band gap presented in Fig. 4. Again one can convert the results for superlattices grown on (100) planes to those for other orientations by realizing the results are really only a function of the thickness of the layers. Since the envelope approximation is used in the theory, one must be careful in attributing any significance to the behavior of the mass for small CdTe and HgTe layer thicknesses.

These contour plots indicate the trend expected. With increasing band offset and fixed layer thickness the effective masses increase rather dramatically. The effective mass also becomes much more dependent on the thickness of the CdTe layer with increasing value

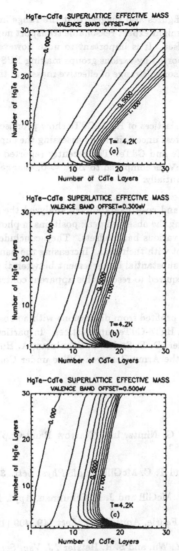

Fig. 6. Contour plots of the effective mass for electrons for transport normal to the layers as a function of the number of layers of HgTe and CdTe in the superlattice. The thickness of a single layer has been taken to be 3.23 Å corresponding to the growth of the superlattice on (100) planes. The spacing between the contour lines is 0.1 electron masses. Three different values of the valence band offset are included: $\Delta E_v = 0$ eV in (a), $\Delta E_v = 0.3$ eV in (b), and $\Delta E_v = 0.5$ eV in (c).

of the band offset. The figures also suggest that the range in which the effective mass is small enough so that we might expect reasonable transport normal to the layers is reduced with increasing band offset. It is important to note however that for band offsets like those currently being reported by various groups making XPS measurements,[15-17] $\Delta E_v \approx$ 0.3 eV, there is still a reasonable range of effective masses in the region of interesting band gaps.

Conclusions

In summary, superlattices of HgTe-CdTe show promise as a new infrared material. However, one of the greatest uncertainties in assessing the superlattices is the value of the band offset between HgTe and CdTe. The recently reported larger values of the valence band offset, $\Delta E_v \approx 0.3$ eV as compared to $\Delta E_v \approx 0$, change the values of the band gap and effective mass substantially.

Recently Baukus and co-workers[23] have analyzed their data on the band gaps of superlattices by comparing the observed peak positions in photoluminescence spectra with the values predicted for various band offsets. They conclude that a small valence band offset is in best agreement with their data. Increasing the band offset to values as high as $\Delta E_v = 0.3$ eV produces substantial disagreement between theory and experiment. More work will definitely be required to resolve this apparent contradiction.

Acknowledgment

The authors have profited from discussions with a substantial fraction of all of the investigators working on HgTe-CdTe superlattices. In particular, we have profited from discussions with S. Hetzler, D. L. Smith, J. Schulman, A. Hunter and J. P. Faurie. This work was supported by the Army Research Office under Contract No. DAAG-29-83-K-0104.

REFERENCES

1. R. Dornhaus and G. Nimtz, in: "Narrow Band-Gap Semiconductors" Springer-Verlag, Berlin (1983).

2. J. N. Schulman and T. C. McGill, *Appl. Phys. Lett.* **34**, 663 (1985).

3. D. L. Smith, T. C. McGill and J. N. Schulman, *Appl. Phys. Lett.* **43**, 180 (1983).

4. J. Reno and J. P. Faurie, *Appl. Phys. Lett.* **49**, 409 (1986).

5. T. C. McGill, G. Y. Wu, and S. R. Hetzler , *J. Vac. Sci. Technol.* **A4**, 2091 (1986).

6. See for example, the Proceedings of the HgCdTe Workshop October 1986 *J. Vac. Sci. Techn* July/August 1987.

7. J. P. Faurie, A. Million, and J. Piaguet, *Appl. Phys. Lett* **41**, 713 (1982);J. T. Cheung and T. Magee*J. Vac. Sci. Technol.*A1,1604(1983); P. P. Chow, D. K.

Greenlaw, and D. Johnson , *J. Vac. Sci. Technol.* **A1**, 562 (1983); K. A. Harris, S. Hwang, D. K. Blanks, J. W. Cook, Jr., J. F. Schetzina, N. Otsuka, J. P. Baukus, and A. T. Hunter , *Appl. Phys. Lett.* **48**, 396 (1986); J. P. Faurie, S. Sivananthan, and J. Reno, *J. Vac. Sci. Technol.*A4,2096(1986).

8. C. E. Jones, J. P. Faurie, S. Perkowitz, J. N. Schulman, and T. N. Casselman, *Appl. Phys. Lett.* **47**, 140 (1985).

9. J. P. Baukus, A. T. Hunter, O. J. Marsh, C. E. Jones, G. Y. Wu, S. R. Hetzler, T. C. McGill, and J. P. Faurie , *J. Vac. Sci. Technol.* **A4**, 2110 (1986).

10. S. R. Hetzler, J. P. Baukus, A. T. Hunter, J. P. Faurie, P. P. Chow, and T. C. McGill, , *Appl. Phys. Lett.* **47**, 260 (1985).

11. J. O. McCaldin, T. C. McGill and C. A. Mead, *Phys. Rev. Lett.* **36**, 56 (1976).

12. W. A. Harrison, *J. Vac. Sci. Technol.* **14**, 1016 (1977).

13. T. F. Kuech and J. O. McCaldin, *J. Appl. Phys.* **53**, 3121 (1981).

14. Y. Guldner, G. Bastard, J. P. Vieren, M. Voos, J. P. Faurie, and A. Millon, *Phys. Rev. Lett.* **51**, 907 (1983).

15. Steven P. Kowalczyk, J. T. Cheung, E. A. Kraut, and R. W. Grant, , *Phys. Rev. Lett.* **56**, 1605 (1986).

16. J. P. Faurie in Proceedings of the HgCdTe Workshop October 1986; *J. Vac. Sci. Technol.* July/August 1987 (to be published).

17. C. K. Shih, W. E. Spicer, J. K. Furdyna, and A. Sher, in Proceedings of the HgCdTe Workshop October 1986; *J. Vac. Sci. Technol.* July/August 1987(to be published).

18. J. Tersoff, *Phys. Rev. Lett.* **56**, 2755 (1986).

19. G. Y. Wu and T. C. McGill, *J. Appl. Phys.* **58**, 3914 (1985).

20. G. Bastard, *Phys. Rev.* **B25**, 7584 (1982).

21. J. Chu, S. Xu, and D. Tang, *Appl. Phys. Lett.* **43**, 1064 (1983).

22. Y. Guldner, G. Bastard, and M. Voos, *J. Appl. Phys.* **57**, 1403 (1985).

23. J.P. Baukus, A. T. Hunter, J. N. Schulman, and J. P. Faurie, *J. Vac. Sci. Technol.* July/August 1987 (to be published).

PHOTOCHEMICAL PROCESSES IN PHOTO-ASSISTED EPITAXY OF $Cd_xHg_{1-x}Te$

S J C IRVINE, J B MULLIN, G W BLACKMORE, O D DOSSER AND H HILL
Royal Signals and Radar Establishment, St Andrews Road, Malvern,
Worcestershire, WR14 3PS, UK.

ABSTRACT

Mechanisms for the low temperature photo-dissociation of alkyl precursors for the epitaxial growth of $Cd_xHg_{1-x}Te$ (CMT) are discussed. The roles of vapour and surface nucleation are considered in the light of the free radical model which also can provide methods for controlling the vapour phase photochemistry. Higher quality CMT has been grown at 250°C by using dimethyl mercury as a source of free methyl radicals. Problems encountered in reducing growth temperature for multilayer epitaxy are considered for CdTe and HgTe and conditions established for epitaxial growth at 200°C. The roles of alkyl concentration, substrate temperature, UV intensity and free radical concentration are explored. Results on the first reported growth of CdTe deposition using a cw UV laser source are compared with the arc lamp grown layers.

INTRODUCTION

The application of photo-dissociation to low temperature epitaxy is now widely recognised but the processes are not well understood and only a small proportion of the potential benefits has been realised so far [1,2]. The objectives for $Cd_xHg_{1-x}Te$ (CMT) low temperature photo-epitaxy are very specific, namely: (a) to achieve more abrupt heterostructure interfaces (b) control Hg vacancy concentration (c) achieve selective area deposition of detector structures (d) in-situ processing. Objectives (a) and (b) have been partially achieved with photo-epitaxial HgTe/CdTe interfaces of ~ 100Å at 230°C and Hg vacancy concentrations below 1×10^{16} cm^{-3} [1]. The photo-excitation of metal organics such as dimethylcadmium (Me_2Cd) and diethyltelluride (Et_2Te) can lead to either an enhancement of growth rate such as by a factor of 3 at 350°C for CdTe [3] or a purely photon initiated reaction such as HgTe epitaxy at 250°C.

The preferred photon sources have been arc lamps capable of illuminating large areas with high intensity UV radiation [2]. An excimer laser has been used for a purely vapour phase photo-dissociation by Morris [4] and in the present work some preliminary results on cw-laser irradiation will be reported. Although, all these approaches should be capable of fulfilling requirements (a) and (b), only the latter would be suitable for (c) and (d).

By non-thermal photo-dissociation of the alkyl precursors, the thermal stability is no longer a limitation to growth temperature. Furthermore, growth temperatures below 200°C have been demonstrated for both HgTe and CdTe with evidence of good crystallinity [4,5]. For growth temperatures above 200°C more detailed studies of epitaxy onto CdTe and InSb substrates have indicated a very high inherent epitaxial quality [1,3,6]. In this respect photo-excitation of the surface may actually aid the growth process by increased atom mobility or product desorption. Kisker and Feldman [3] showed that photo-enhanced CdTe growth at 350°C onto GaAs substrates using a low pressure Hg arc lamp gave improved near band edge photoluminescence. Photo-excitation may be very important for very low temperature growth, below the current MBE growth temperature of 180°C [7].

With very good interface abruptness of ~ 100Å already demonstrated for growth temperatures above 200°C the benefit of lower growth temperatures for CMT detectors is not clear. Indeed, current heterojunction detector requirements would need some grading of these interfaces in order to avoid conduction band spikes [8]. Temperatures below 200°C would be essential for superlattice or multi-quantum well devices. Predictions for the interdiffusion of photo-epitaxial CdTe/HgTe at 200°C and 150°C suggest that significant interdiffusion would occur even at 150°C [1]. This prediction may be pessimistic as the predicted interdiffusion coefficient of 4×10^{-17} cm^2 s^{-1} at 180°C is higher than that measured for a superlattice held at 180°C by Arch et al [9].

In this paper some new results on the free radical mechanism [10,11] whereby the vapour chemistry can be controlled by free alkyl radicals, will be discussed. The significance of this model in going to lower temperatures will also be considered. Failure to control the vapour chemistry can lead to vapour nucleation and a dust settling on the surface. The solutions to this problem can have a deleterious effect on growth rates and the significance of this for very low temperature growth will be explored. With adequate control over the vapour chemistry, the potential for very high quality epitaxy at low temperatures looks promising.

EXPERIMENTAL

All the photo-epitaxy experiments were performed in a horizontal, atmospheric presure reactor cell which has been described elsewhere [6,13]. The UV radiation sources were either a high pressure Hg arc lamp with 3kW total power dissipation, or a frequency-doubled argon ion laser green line yielding cw radiation at 257nm. The arc lamp was focussed using an elliptic reflector which gave an approximate power intensity of 1 W cm^{-2} at 254nm. Alternatively it could be used without the focus for lower intensity illumination of 12 mW cm^{-2}. The reactants were Me_2Cd, Et_2Te and Me_2Hg in He carrier gas. An additional source of Hg(V) was used for HgTe or CMT growth to promote a surface photosensitisation reaction for the decomposition of Et_2Te [6]. Substrates were either(100)CdTe, (100)$Cd_{0.96}$ $Zn_{0.04}$Te or (100)InSb using surface preparations described previously [6]. Matrix element depth profiles were determined using a Cameca 3F IMS for secondary ion mass spectrometry (SIMS) with an oxygen primary ion beam. Scanning electron microscope (SEM) micrographs were obtained on a Cambridge S250 SEM at 20kV.

FREE RADICAL MODEL

It has been shown in previous studies that the choice of carrier gas can dramatically effect the degree of vapour nucleation [12,13]. For example,using H_2, in combination with a high intensity UV source is likely to lead to dust formation in the vapour whereas the use of He carrier gas is less likely to lead to the conditions for vapour nucleation [12,13]. Even at growth temperatures, as low as 200°C, H_2 can play an important part in the decomposition processes, enhancing decomposition via a chain reaction caused by hydrogen radical formation [3,10,11]. This approach has been used successfully with low intensity UV lamps for CdTe growth but the high super-saturations when using high intensity sources give rise to homogeneous nucleation and dust formation [12,13]. This appears to place a fundamental limitation on growth rate as it would depend simply on the amount of supersaturation that can be sustained without homogeneous nucleation; defined as the critical chemical potential difference $\Delta\mu_{crit}$ [13].

$$\Delta\mu_{crit} = RT\ln(p_v/p_e) \tag{1}$$

where T is the vapour temperature, p_v the vapour pressure of the constituent atoms and p_e the corresponding equilibrium vapour pressure over the crystalline surface. Congruent evaporation conditions for the compounds are assumed which for

$$p_{Te_2} = 2\ p_{cd} \qquad (2)$$

However, better control of the growth process, while maintaining growth rate, can be achieved by suppressing vapour phase nucleation and encouraging surface reaction. The method adopted is the free radical mechanism which is the enhancement of vapour concentration of free Cd and Te atoms thus avoiding nucleation. The surface acts as a sink for alkyl radicals by catalysing the formation of stable alkanes; [11] by this means epitaxial growth can progress normally on the substrate. Homogeneous nucleation can still occur if $\Delta\mu > \Delta\mu_{crit}$ but previous reports have shown that using He carrier gas can avoid the alkyl free radical quenching mechanism with H_2. By this means both CdTe and CMT can be grown using a high intensity UV source [1,13]. Evidence for the free radical mechanism has been reported from (a) mass spectrometry of the stable alkane reaction products, (b) the confinement of CMT epitaxial growth within the UV focus and (c) the progressive increase in epitaxial area by adding Me_2Hg as an additional source of free Me radicals [11].

In this paper the consequence of the free radical mechanism for CdTe and CMT epitaxy will be explored further and used to explain some of the experimental observations.

PHOTO-EPITAXY OF CMT

CMT epitaxial layers have been grown using a Me_2Cd partial pressure of 0.05 torr, and a range of Me_2Hg partial pressures up to 10 torr at a substrate temperature of 250°C. The width of the epitaxial region, growth rate and composition profiles were determined for different Me_2Hg partial pressures. A plot of epitaxial width and growth rate is shown as a function of Me_2Hg partial pressure in Figure 1. The increase in the width

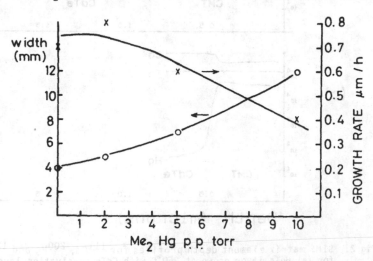

Fig 1. Growth rate and width of the epitaxial growth region for CMT on (100)2° → (110) CdTe substrates versus Me_2Hg partial pressure. The substrate temperature was 250°C and Me_2Cd partial pressure of 0.05 torr and Hg partial pressures of 30 torr.

156

of epitaxial growth can be explained by additional free Me radicals combining with Cd and Te atoms that diffuse out of the intensely illuminated region (within the lamp focus), thus avoiding the homogeneous nucleation of CdTe which would otherwise have occurred. A smaller concentration of Cd and Te atoms is generated within the diffusely illuminated region, sufficient to maintain the growth rate within a factor of 2 of that at the centre of the growth zone. It should be noted that the

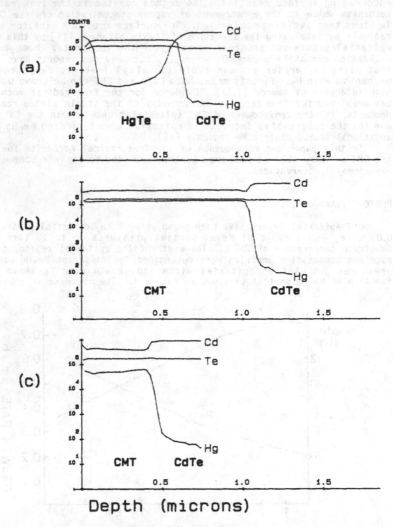

Fig 2. SIMS matrix element depth profiles for ^{111}Cd, ^{200}Hg and ^{125}Te, for (a) HgTe layer grown at 250°C with CdTe passivation layer. (b) CMT (x < 0.2) layer grown at 250°C with 0.05 torr Me$_2$Cd. (c) MCT (x = 0.61) layer growth at 250°C with 0.05 torr Me$_2$Cd and 10 torr of Me$_2$Hg.

enhancement in free radical lifetime brought about by using He rather than H_2 carrier gas did not significantly reduce the HgTe growth rate [13]. For the additional concentration of Me radicals, represented by Me_2Hg partial pressure in Figure 1, a significant decrease in growth rate is observed which suggests that some etching of the surface with Me radicals occurred. This may not be a disadvantage from the point of view of crystalline quality as the chance of nucleation on precipitated dust would be less and growth conditions would be closer to thermodynamic equilibrium. On the other hand, higher intensity UV irradiation could be used which might give a higher growth rate without dust nucleation.

Previous reports of photo-MOVPE CMT indicated the excellent control of composition with depth and good interface control [13]. In the present study a series of layers were examined with SIMS to determine depth compositional profile quality and to make a qualitative comparison of composition. In Fig 2 SIMS profiles for two CMT layers are compared with a photo-epitaxial HgTe layer. One CMT layer which was grown without Me_2Hg is shown in Fig 2(b). Infrared transmission spectra indicates that this layer has a composition x < 0.2 although Fig 2(b) clearly shows that uniform concentration of Cd was present within the layer. The HgTe layer in Fig 2(a) was grown with a 300Å passivating layer of CdTe which is shown clearly with the increase in Cd towards the surface, mirrored by a decrease in Hg. The composition of the CMT layer in Fig 2(b) can be compared with Fig 2(a) by looking at the ratio of the Hg counts to the Te counts. In Fig 2(b) this ratio is signficantly reduced and remains uniform with depth, consistent with good compositional control. The CMT layer in Fig 2(c) was grown with a Me_2Hg partial pressure of 7 torr and shows a much decreased Hg/Te ratio with an increase in Cd/Te ratio. The increase in Cd content was confirmed by infrared transmission spectra and Rutherford Back Scattering (RBS) [14]. The composition can be determined from RBS spectra by taking the ratio of the Hg backscattered counts to the total back scattered counts in a non-channeling direction. This gave a composition x of 0.61. The only change in growth conditions between the CMT layers in Figs 2(b) and 2(c) was the addition of Me_2Hg which therefore is responsible for the large increase in x.

PHOTO-EPITAXY OF HgTe

HgTe can be grown by photo-epitaxy using a surface photosensitisation process which has been described elsewhere [6]. The reaction does not yield HgTe in the vapour because the complex photo-dissociation of Et_2Te does not give a high vapour yield of Te, so the vapour super saturation is low [2]. At 250°C, high growth rates 2-4μm/h on (100) substrates have been reported but growth rates decreased at lower temperatures [6]. For abrupt structures with ~ 10Å interface widths growth at 200°C or below will be needed but for practical applications of photo-epitaxy, growth rates maintained at ~ 1μm/h would be desirable. Fig 3 shows results of HgTe growth rates measured at 200°C and 250°C versus $(Et_2Te$ p.p$)^{\frac{1}{2}}$. The square root relationship with diethyltelluride partial pressure has been shown previously by Ahlgren et al [5] and Irvine and Mullin [2] and was explained by surface decomposition of the alkyl.

For HgTe/CdTe multilayers it is desirable to keep growth conditions constant except for the Me_2Cd flow which can be switched on and off. Growing HgTe at Et_2Te partial pressures up to 7.5 torr does not appear to deteriorate the epitaxial quality but this is not the case for CdTe. For CdTe growth Kisker and Feldman [3] have shown that a Me_2Cd excess is required for good quality epitaxy. The additional constraint on alkyl partial pressure is the onset of premature reaction which depends on a number of factors, including Me_2Cd concentration, and will be discussed in more detail in the next section. The above constraints would in practice limit the HgTe growth rate at 200°C to around 0.1μm/h.

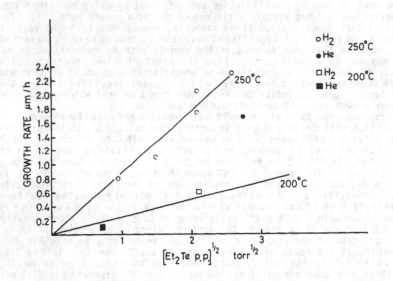

Fig 3. HgTe growth rates versus the square root of Et$_2$Te partial
pressure for H$_2$ and He carrier gases at 200°C and 250°C.

PHOTO-EPITAXY OF CdTe

The growth of CdTe using high intensity UV at 250°C has been shown to
be epitaxial and free from homogeneous nucleation provided that He carrier
gas is used [1]. Fig 4(a) shows a scanning electron micrograph (SEM) of a
0.3μm thick CdTe layer on (100)InSb with a Me$_2$Cd partial pressure of 0.18
torr. The surface shows few features and is indicative of surface
nucleated CdTe. However, reducing the growth temperature to 200°C
significantly degrades the surface as shown in Fig 4(b). Small
crystallites dominate the surface structure and are randomly aligned. It
can be seen from the facetted nature of these crystallites that they have
grown on the surface onto randomly orientated nuclei. This contrasts with
previous reports of homogeneously nucleated CdTe and CMT which resulted in
a porous structure with no clearly defined facets. The morphology in Fig
4(b) could be explained by a reduction in the amount of homogeneous
nucleation where most of the material nucleates on the surface but vapour
formed particles provide the random nuclei on the surface from which these
crystallites grow.

The chemical potential can be reduced below its critical value in
equation (1) by reducing the partial pressures of Cd and Te atoms released
in the vapour. This can be achieved in a variety of ways including
reducing alkyl concentrations, lamp intensity or by increasing free radical
concentrations. The result of reducing lamp intensity is shown in Fig
4(c). The lamp was defocussed which gave an estimated reduction in
intensity of ~ 90 to 12mW cm^{-2} measured with a photometer in a 10nm band
around 254nm. A specular surface, free from vapour nucleated dust is again
observed for a CdTe layer 0.14μm thick. The Cd and Te atomic
concentrations, according to the free radical model depend on the square
root of UV intensity. The above reduction in UV intensity would have had
the effect of reducing atomic vapour concentration by a factor of 9.5.
This decrease in vapour concentrations has been needed to balance the

decrease in p_e in equation (1) arising from the decrease in surface temperature.

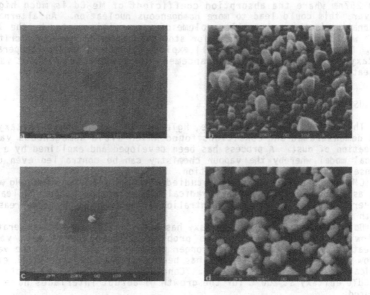

Fig 4.   Comparison of surface morphologies for CdTe photo-epitaxial layers grown with Me$_2$Cd partial pressure of 0.18 torr for(a) focussed lamp on InSb substrate at 250°C, layer thickness 0.3μm. (b) focussed lamp on InSb substrate at 200°C. (c) defocussed lamp on InSb at 200°C, layer thickness 0.14μm. (d) defocussed lamp on Cd$_{0.96}$Zn$_{0.04}$Te at 200°C with Hg partial pressure of 8 torr, layer thickness 0.10μm.

Figure 4(d) shows a CdTe surface after growth at 200°C with a defocussed beam and with a Hg partial pressure of 8 torr. These are conditions which could be used for the growth of multilayers of HgTe/CdTe with abrupt interfaces. Some vapour nucleation of CdTe was observed although not as much as in Fig 4(b). It has been observed previously with CMT photo-epitaxy that high Hg partial pressures can enhance not only the surface alkyl decomposition rate but also the vapour decomposition rate [11]. It is significant that in Fig 4(d) the growth rate had increased from 0.07μm/h in Fig 4(c) to 0.10μm/h. The correct balance of all the factors which influence vapour concentrations must be found to keep the super saturation below its critical value and avoid vapour nucleation.

A cw UV laser has been used to replace the UV arc lamp under conditions similar to Figs 4(b) and (c). CdTe was deposited with a thickness of 0.23μm over the beam area of approximately 2.6mm^2 at 200°C. Interference fringes were visible in white light over a large area which might be a thin film arising from some scattering of the beam in the silica reactor wall. No evidence of homogeneous nucleation was observed for an estimated UV intensity of $1W\ cm^{-2}$, although the growth rate was twice that using the Hg arc lamp. This intensity is comparable with that estimated for the focussed lamp. Comparisons become difficult when considering the broad spectral band of the Hg arc lamp which extends to shorter wavelengths than 257nm where the absorption coefficient of Me_2Cd is much higher, however, this could lead to more homogeneous nucleation. An alternative explanation is the much smaller volume of vapour illuminated by the laser which would make it less likely for atomic clusters to grow to a critical size. Future laser experiments will explore conditions for low temperature epitaxy and look for further enhancement of growth rates without vapour nucleated dust.

CONCLUSIONS

The epitaxial growth of CdTe, HgTe and CMT using photo-epitaxy has been demonstrated and conditions found which avoid homogeneous vapour nucleation of dust. A process has been developed and explained by a free radical model whereby the vapour chemistry can be controlled even under intense illumination with UV radiation.

CMT photo-epitaxy has been studied with the addition of Me_2Hg which acts as a source of free methyl radicals, suppressing vapour nucleation. However, increase in Me_2Hg concentration is accompanied by a decrease in growth rate and increase in x.

Both HgTe and CdTe photo-epitaxy has been investigated at temperatures as low as 200°C and although no problem is encountered with vapour nucleation of HgTe, CdTe has a stronger tendency to nucleate in the vapour at low temperatures. This effect has been explained by a critical excess chemical potential for nucleation. Conditions compatible with both HgTe and CdTe epitaxy at 200°C for the growth of abrupt interfaces have been explored.

Preliminary results on the first reported growth of CdTe using a cw UV laser source have been described. This method of illumination of the substrate does not induce strong vapour nucleation and would in the future be suitable for selective area epitaxy of CMT.

ACKNOWLEDGEMENTS

The authors wish to thank Mrs J Clements for her skilled assistance in the growth of epitaxial layers. Also, the authors gratefully acknowledge Dr A J Avery and Mr D J Diskett, RMCS, Shrivenham for the RBS measurements.

REFERENCES

1   S J C Irvine, J Giess, J S Gough, G W Blackmore, A Royle, J B Mullin, N G Chew and A G Cullis, J Crystal Growth 77 437 (1986).
2   S J C Irvine, J B Mullin, J Vac Sci Technol (to be published).
3   D W Kisker and R D Feldman, J Crystal Growth 72 102 (1985).
4   B J Morris, Appl Phys Lett 48 867 (1986).
5   W L Ahlgren, R H Himoto, S Sen and R P Ruth, Third International Conference on MOVPE, 13-17 April 1986, Universal City, California.
6   S J C Irvine, J B Mullin and J Tunnicliffe, J Crystal Growth 68 188 (1984).
7   J P Faurie, A Million and J Piaguet, J Crystal Growth 59 10 (1982).
8   P Migliorato and A M White, S S Electronics 26 65 (1983).

9   D K Arch, J P Faurie, J L Staudenmann, M Hibbs-Brenner, P Chow, J Vac
    Sci Technol A4 2101 (1986).
10 J B Mullin and S J C Irvine, J Vac Sci Technol A 4 700 (1986).
11 S J C Irvine and J B Mullin, J Crystal Growth (in press).
12 S J C Irvine, J Giess, J B Mullin, G W Blackmore and O D Dosser,
    Materials Letters 3 290 (1985).
13 S J C Irvine, J Giess, J B Mullin, G W Blackmore and O D Dosser, J Vac
    Sci Technol B 3 1450 (1985).
14 A J Avery and D J Diskett (private communication).

CONTROLLED SUBSTITUTIONAL DOPING OF CdTe THIN FILMS AND Cd$_{1-x}$Mn$_x$Te-CdTe SUPERLATTICES.

R.N. BICKNELL, N.C. GILES, AND J.F. SCHETZINA
Department of Physics, North Carolina State University
Raleigh, NC 27695-8202

ABSTRACT

We report the successful substitutional doping of CdTe epilayers grown by a new technique: photoassisted molecular beam epitaxy, in which the substrate is illuminated during the film deposition process. This new technique was found to produce dramatic changes in the electrical transport properties of the epilayers. In particular, highly conducting n-type and p-type CdTe films have been grown using In and Sb as n-type and p-type dopants, respectively. Photoassisted MBE has also recently been employed to produce for the first time highly conducting CdMnTe epilayers and Cd$_{1-x}$Mn$_x$Te-CdTe superlattices.

INTRODUCTION

CdTe is one of the most promising II-VI semiconductor compounds for applications in the areas of optoelectronics, gamma-ray detection and solar energy conversion [1-2]. There is at present also, considerable interest in the use of CdTe substrates and buffer layers for the growth of HgCdTe, an important infrared material [3-5].

The recent work of Kolodziejski et al. [6] and Bicknell et al. [7], who have grown multiple quantum well structures of Cd$_{1-x}$Mn$_x$Te-CdTe has further sitmulated interest in thin film CdTe. Stimulated emission has also been reported from optically pumped Cd$_{1-x}$Mn$_x$Te-CdTe, Cd$_{1-x}$Mn$_x$Te-Cd$_{1-y}$Mn$_y$Te, and Zn$_{1-x}$Mn$_x$Se-ZnSe multilayer structures [7-8]. Future applications of CdTe, HgCdTe, and Cd$_{1-x}$Mn$_x$Te-CdTe superlattices will require the precise control of their electronic transport properties through substitutional doping.

Incorporation of substitutional dopants in CdTe films grown by conventional molecular beam epitaxy (MBE) generally leads to poor activation of the dopant species [9]. Similar results have been obtained in bulk CdTe:In samples and has been attributed to self-compensation [10]. The development of II-VI semiconductor applications has been severely limited because of their strong tendencies towards self-compensation.

At North Carolina State University, we have developed a new doping technique, compatible with molecular beam epitaxy, which allows introduction of selected dopant species during thin film growth: photoassisted MBE. In the photoassisted MBE process, the substrate is illuminated during the thin film deposition. The incident photon beam provides high energy, low momentum particles (photons) at the film growth surface in order to influence surface chemical reactions during film growth. This may be accomplished in several ways: through enhancement of surface migration of constituent atoms, modification of surface bonds, conversion of surface molecules into atoms, etc. The net result is that the photoassisted process promotes the incorporation of the substitutional dopant species at the proper tetrahedrally coordinated lattice sites.

Mat. Res. Soc. Symp. Proc. Vol. 90. © 1987 Materials Research Society

A technique similar to photoassisted MBE has been recently reported by Yokoyama in which the substrate was illuminated with above band gap illumination during the deposition of evaporated ZnS [11]. As a result of substrate illumination an increase in growth rate, and an improvement in preferential crystalline orientation was observed. These changes were attributed to photo-induced surface adatom rearrangement and not to thermal effects.

The photoassisted MBE technique has been employed to grow highly conducting n-type and p-type CdTe films [12-16]. We have recently used the photoassisted MBE technique to prepare n-type CdMnTe films and also highly conducting $Cd_{1-x}Mn_xTe$-CdTe superlattices. The use of the photoassisted MBE technique not only results in dramatic changes in the electronic transport properties of the epilayers, but it also changes in a remarkable way the observed low temperature photoluminescence (PL) spectrum. These changes are a direct result of the changes in the point defect character of the growing epilayers resulting from the use of the photoassisted MBE technique. Results of detailed low temperature PL and excitation PL studies completed on doped epilayers grown by photoassisted MBE are reported in an accompanying paper [17].

EXPERIMENTAL DETAILS

The substitutionally doped CdTe thin film specimens were grown in a MBE system that has been described in earlier publications [7], using a new growth procedure which we call photoassisted molecular beam epitaxy. The system consists of a conventional MBE system that has been modified so that during the growth process the substrate can be illuminated. An argon ion laser was used as an illumination source in the present work, operating using broad-band yellow-green optics (488.0-528.7 nm). We believe that a conventional light source (noncoherent) may also be used in certain applications. The power density at the substrate during the deposition was about 150 mW/cm^2. In the present work four MBE effusion cells were employed. Two of these cells contained high purity polycrystalline CdTe, while the other two contained elemental In and Sb, respectively. For the deposition of the CdMnTe:In epilayers and the $Cd_{1-x}Mn_xTe$-CdTe superlattices, Mn rather than Sb was used in the fourth effusion cell. For the growth of $Cd_{1-x}Mn_xTe$-CdTe modulation doped multilayers, the Mn and In source shutters were opened and closed simultaneously in a cyclic fashion while maintaining a constant vapor flux at the substrate from the CdTe sources. The substrate was illuminated during the entire superlattice growth sequence.

Chemimechanically polished (100) CdTe wafers were used as substrates. Prior to insertion into the photoassisted MBE system the substrates were prepared as follows. First, they were degreased using standard solvents. Next, they were etched in a weak bromine-in-methanol solution and then rinsed in methanol. Finally, to remove any remaining oxide, they were dipped in concentrated hydrochloric acid and rinsed in deionized water. Immediately prior to growth, the substrates were annealed in the MBE system at a temperature of 300-400 $^\circ$C. Substrate temperatures ranging from 160 $^\circ$C to 320 $^\circ$C were employed.

Electrical characterization of the epilayers was carried by means of variable temperature van der Pauw Hall measurements. These measurements were performed using a computer-controlled Hall apparatus at temperatures between 300 K and 20 K. Contacting to the n-type samples was accomplished using an indium-based solder. P-type samples were contacted using gold.

RESULTS AND DISCUSSION

CdTe Substitutional Doping

Illumination of the substrate during the growth of substitutionally doped CdTe:In and CdTe:Sb epilayers was found to be of crucial importance since only those samples grown under illumination were conducting. All other samples were found to be semi-insulating. Substrate temperatures ranging from 180-250 ^{0}C have been investigated. A substrate temperature of 230 ^{0}C was employed to obtain the best results, although samples grown while under illumination at substrate temperatures as low as 200 ^{0}C exhibited carrier concentrations as high as 1×10^{17} cm^{-3}. This result is significant since this substrate temperature is in the range which is typically employed to grow HgCdTe and HgTe-CdTe superlattices [5].

All In doped samples were n-type as determined by Van der Pauw Hall measurements. CdTe:In films having 300 K carrier concentrations ranging from 6×10^{15} cm^{-3} to 6×10^{17} cm^{-3} were prepared by varying the indium flux density incident on the substrate during the epitaxial growth of the layers. Van der Pauw Hall measurements yield electron mobilities of 225 - 600 cm^{2}/V s at room temperature. At 77 K mobilities greater than 1200 cm^{2}/V s have been obtained. These mobilities are comparable to the best values reported in the literature for bulk CdTe:In samples [18-19]. Table 1 summarizes the results of Hall measurements obtained on a number of CdTe:In samples grown by photoassisted MBE.

Van der Pauw Hall effect measurements were also completed on several CdTe:Sb epilayers grown using the photoassisted MBE technique. All the Sb doped samples were found to be p-type and highly conducting. At room temperature, hole mobilities ranged from 40-45 cm^{2}/V s. These values are comparable to hole mobilities reported for bulk CdTe samples [20]. Hole concentrations in the CdTe:Sb epilayers were ~ 5×10^{16} cm^{-3} at 300 K. We have experienced considerable difficulty applying good ohmic contacts to the CdTe:Sb epilayers, thus, at this time we can report only accurate Hall measurements at room temperature.

Table 1

Film	T_{sub} (^{0} C)	T_{In} (^{0} C)	$\mu_{300 K}$ (cm^{2}/V s)	$n_{300 K}$ (cm^{-3})	μ_{peak} (cm^{2}/V s)	$n_{20 K}$ (cm^{-3})
CDTE1	230	375	600	2×10^{16}	1250	3×10^{15}
CDTE2	200	400	340	1×10^{17}	470	5×10^{16}
CDTE3	230	450	600	6×10^{17}	695	6×10^{17}
CDTE4	230	–	225	6×10^{15}	1290	9.8×10^{14}

CdMnTe:In Substitutional Doping

In order to demonstrate that the photoassisted MBE growth technique can be applied to other materials we have recently prepared conducting $Cd_{1-x}Mn_xTe$ (x ~ 0.15) thin films. The room temperature mobility in these samples are about 350 cm^2/V s and increase with decreasing temperature to a value near 900 cm^2/V s at 200 K. The room temperature carrier concentration in these samples are about 6 x 10^{15} cm^{-2} and decreases rapidly with decreasing temperature to 2.5 x 10^{14} at 170 K. The measured activation energy for carrier freeze out in these samples are ~ 130 meV. Assuming the presence of both donors and residual donors in this material, the activation energy determined from carrier freeze-out is equal to the ionization energy of the impurity level. This result indicates that In forms a deeper donor level in CdMnTe than in CdTe (14 meV), as one would expect in an alloy.

$Cd_{1-x}Mn_xTe$-CdTe Superlattice Doping

Encouraged by the results of the substitutional doping experiments carried out on the CdTe and $Cd_{1-x}Mn_xTe$ thin films, the growth and characterization of modulation doped $Cd_{1-x}Mn_xTe$-CdTe superlattices were undertaken. Again the photoassisted MBE technique was found to cause significant changes in the electronic transport properties of the $Cd_{1-x}Mn_xTe$:In superlattice structures. Only those structures which were grown using the photoassisted technique were found to be conducting, all others were semi-insulating.

Fig. 1 shows a double crystal rocking curve of one of these modulation doped superlattices. This particular sample consists of a fifty period superlattice with a period of 148 A, composed of equal well and barrier thicknesses. The main feature occurring at a relative angle of zero arc seconds is due to diffraction from the bulk CdTe substrate. To the right of

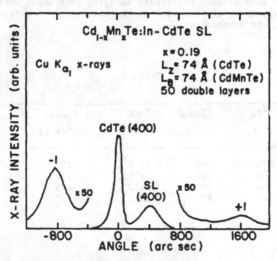

Fig. 1. Double crystal x-ray diffraction rocking curve for a $Cd_{0.81}Mn_{0.19}Te$ -CdTe superlattice with a period of 148 A.

this main feature, located at about 420 arc seconds is the diffraction peak due to the superlattice. It should be noted that superlattice diffraction peak is smooth and symmetric and shows no signs of multiple peaking which would be indicative of subgrain boundary formation. The problem of subgrain boundary formation has long plagued the growth of bulk II-VI materials. Located symmetrically about the superlattice diffraction peak are found satellites which are a direct indication of the quality of the superlattice. Analysis of the satellite peak spacings allowed a determination of the superlattice period.

Fig. 2 shows plots of carrier concentration and mobility for this $Cd_{1-x}Mn_xTe$-CdTe superlattice which was grown at a substrate temperature of 230 ^oC, using an In oven temperature of 400 ^oC. The photoassisted MBE technique has allowed the use of a significantly lower temperature than the 275-325 ^oC which has been previously employed to grow either $Cd_{1-x}Mn_xTe$ or $Cd_{1-x}Mn_xTe$-CdTe superlattices [6-7]. The mobility in this sample is greater than 250 cm^2/V s at 300 K, and increases with decreasing temperature to a maximum value of ~ 600 cm^2/Vs at 70 K. The carrier concentration is 3 x 10^{16} cm^{-3} at 300 K and decreases quickly to about 1 x 10^{15} cm^{-3} at 20 K.

Fig. 3 shows a double crystal rocking curve of a second $Cd_{0.81}Mn_{0.19}Te$ -CdTe modulation doped superlattice. This superlattice consists of 50 double layers of CdTe (thickness $L_z \simeq$ 155 A) alternating with CdMnTe (thickness $L_b \simeq$ 155 A). The main feature in the rocking curve occurs at an angle of zero arc seconds and is due to diffraction from the bulk CdTe substrate. The main superlattice diffraction peak, which has a FWHM = 2.9 arc minutes, occurs at about 390 arc seconds. In this sample two orders of satellites are observed to be centered about the zero order superlattice diffraction peak.

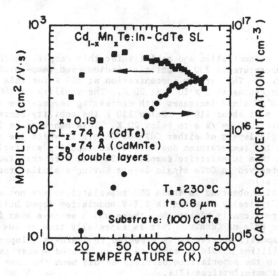

Fig. 2. Carrier concentration and mobility versus temperature for a n-type $Cd_{0.81}Mn_{0.19}Te$-CdTe superlattice

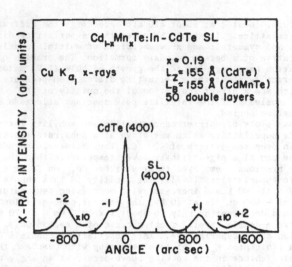

Fig. 3. Double crystal x-ray diffraction rocking curve for a $Cd_{0.81}Mn_{0.19}Te$ -CdTe superlattice with a period of 310 A.

   The carrier concentration and mobility for this sample, which was grown at a substrate temperature of 230 °C using an indium oven temperature of 400 °C is shown in Fig. 4. The carrier concentration at 300 K for this sample is 1 x 10^{17} cm^{-3} and decreases to 5 x 10^{16} at 20 K. The mobility at 300 K in this sample is 780 cm^2/V s and increases with decreasing temperature to greater than 1300 cm^2/V s at about 110 K. Below 110 K the mobility decreases slowly and remains above 1000 cm^2/V s to below 20 K. This behavior is in marked contrast to single layers of either CdTe or CdMnTe in which the mobility drops dramatically at low temperatures due to ionized impurity scattering. In fact, the mobility in this superlattice sample at 20 K is more than two times that which we have observed in CdTe single layers having a similar room temperature carrier concentration.
   The mobilities observed in these DMS superlattices are not as great as those which are typically observed in III-V modulation doped heterostructures, but they are significantly higher than those which we have seen in single layers of either CdTe or CdMnTe. This is especially true at low temperatures where ionized impurity scattering is expected to dominate. Improvement in low temperature mobilities should be achieved if an undoped spacer layer were incorporated into the superlattice barriers as has been the case with III-V modulation doped superlattices [21].

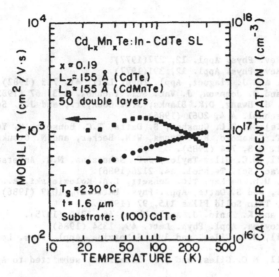

Fig. 4. Carrier concentration and mobility versus temperature for a n-type $Cd_{0.81}Mn_{0.19}Te$-CdTe superlattice with a period of 310 A.

SUMMARY

To summarize, we have used a new growth technique, photoassisted MBE, to grow highly conducting n-type and p-type CdTe. The n-type samples have room temperature carrier concentrations in the range 7×10^{15} cm^{-3} to 6×10^{17} cm^{-3}. P-type CdTe samples with room temperature carrier concentrations of ~5 $\times 10^{16}$ and mobitlities of 40-45 cm^2/V s have be grown using this new technique. The observed mobility in these samples show the classical temperature dependence. We have also used this new growth technique to grow conducting n-type CdMnTe. Highly conducting dilute magnetic semiconductor superlattices have been grown for the first time. These DMS superlattices have been grown with room temperature carrier concentrations in the range 3×10^{16} cm^{-3} to 1×10^{17} cm^{-3}, with low temperature mobilities greater than 1300 cm^2/V s.

ACKNOWLEDGMENTS

The work at NCSU was supported by Army Research Office contract DAAG29-84-K-0039, DARPA/ARO contract DAAL03-86-K-0146 and National Science Foundation grant DMR83-13036. We wish to acknowledge the assistance of R. Burns in substrate preparation and S. Hwang in performing some of the Hall measurements.

REFERENCES

1. F.V. Wald, Rev. Phys. Appl. **12**, 277(1977).
2. R.O. Bell, Rev. Phys. Appl. **12**, 391(1977).
3. J.P. Faurie, and J. Piaguet, Appl. Phys. Lett. **41**, 713 (1982).
4. P.P. Chow, and D. Johnson, J. Vac. Sci. Technol. A **3**, 67 (1985).
5. K.A. Harris, S. Hwang, D.K. Blanks, J.W. Cook,Jr., and J.F. Schetzina, J. Vac. Sci. Technol. A **4**, 2061 (1986).
6. L.A. Kolodziejski, R.L. Gunshor, S. Datta, T.C. Bonsett, M. Yamanishi, R.Frohone, T. Sakamoto, R.B. Bysma, W.M. Becker, and N. Otsuka, J. Vac. Sci. Technol. B **3**, 714 (1985).
7. R.N. Bicknell, N.C. Giles-Taylor, J.F. Schetzina, N.G. Anderson, and W.D. Laidig, J. Vac. Sci. Technol. A4, 2126(1986).
8. R.B. Bylsma, W.M. Becker, T.C. Bonsett, L.A. Koloziejski, R.L. Gunshor, M. Yamanishi, and S. Datta, Appl. Phys. Lett. **47**, 1039 (1986).
9. K. Sugiyama, Thin Solid Films 115, 97 (1984).
10. C.E. Barnes and K. Zanio, J. Appl. Phys. **46**, 3959 (1975).
11. Hiroyuki Yokoyama, Appl. Phys. Lett. **49**, 1354 (1986).
12. R.N. Bicknell, N.C. Giles, and J.F. Schetzina, Appl. Phys. Lett. **49**, 1095(1986).
13. R.N. Bicknell, N.C. Giles, and J.F. Schetzina, submitted to Appl. Phys. Lett.
14. R.N. Bicknell, N.C. Giles, and J.F. Schetzina, Proc. of the 1986 U.S. Molecular Beam Epitaxy Workshop, Boston, MA, J. Vac. Sci. Technol.
15. R.N. Bicknell, N.C. Giles, and J.F. Schetzina, Proc. of the 1986 U.S. Workshop on the Physics and Chemistry of MCT, Dallas, TX, to be published in the Nov. 1987 issue of J. Vac. Sci. and Technol. A.
16. N.C. Giles, R.N.Bicknell, and J.F. Schetzina, Proc. of the 1986 U.S. Workshop on the Physics and Chemistry of MCT, Dallas, TX, to be published in the Nov. 1987 issue of J. Vac. Sci. and Technol. A.
17. N.C. Giles, R.N. Bicknell, J.F. Schetzina, Proc. of the 1986 Fall MRS Symposia on Infrared Sources and Detectors, Boston, MA.
18. S. Yamada, J. of Phys. Soc. Jpn. 17, 645 (1962).
19. B. Segall, M.R. Lorenz, and R.E. Halsted, Phys. Rev. **129**, 247 (1963).
20. S. Yamada, J. Phys. Soc. Jpn. 15, 1940 (1960).
21. R. Dingle, H.L. Stormer, A.C. Gossard, and W. Weigmann, Appl. Phys. Lett. **33**, 665 (1978).

TAILORED MICROSTRUCTURES FOR INFRARED DETECTION

Quark Y. Chen and C. W. Bates,Jr.
Dept. of Materials Science & Engineering, Stanford University,
Stanford, CA 94305

Abstract

Using the effective medium approximation[1-4] and the theory of photoemission from small particles,[5] we look into the design-rules of infrared materials based upon the metal-semiconductor random heterostructures with Ag particles embedded in the semiconductor. It is found that semiconductor host matrices with higher dielectric constants show better optical absorption in the infrared and that optical properties are closely related to microstructural parameters such as volume fraction of metal particles, percentage of aggregation and particle size. Cu, Ag and Au particles all show similar characteristics. We synthesized materials with small silver particles embedded in Si. Their microstructures are analysized using X-ray diffraction, electron microscopy and sputter Auger profiling. Finally, we present the optical properties of these materials and make comparisons with our theoretical results.

Introduction

In modern technologies for military and optical communicational applications, optical detectors are especially demanded to be highly responsive at 1.06-1.55μm where optical fibers have low transmission losses. Particularly at 1.3 and 1.55μm, the glass fiber is free from dispersion and thus has large potential bandwidth. An efficient photodetector should exhibit high rate of photoexcitation, high mobility of electron transport, high speed of response, low noise and long lifetime of free carriers. Current efforts have been to seek new materials and devices suitable for various wavelength regions and meeting the above-mentioned requirements. Up to now, the fastest devices regarding the rise time are of the photoconductive type, however, values below 1ps have been reported.[6] Schottky junction diodes show fast rise times in the picosecond range, while it doesn't have the disadvantage of slow decay as most photoconductive detectors do.[7] Meanwhile, dark currents which may be substantial in extrinsic semiconductors can be alleviated by introducing Schottky junctions.

The major difficulties in the fabrication of the Schottky diodes, however, are the insufficient transparency of contacts and parasitic capacitances. These difficulties can be overcome by making these junction contacts in the form of small metal particles dispersed in the semiconductor which are supported by insulating substrates. The total Schottky junction area is greatly increased under such circumstances. The optical absorptions and the rate of internal photoemission are also tremendously enhanced due to the confinement of conduction electrons in the metal particles. On account of the high electron densities in metals, the photosensitivities of such structures will thus be much higher than those of extrinsic semiconductors while thermal noises can be significantly reduced at the presence of Schottky barriers.

The first material and device using small metal particles for optical detector applications is the S-1 photocathode which consists of small silver particles dispersed in the host matrix of cesium oxide or suboxides. For photocathodes so far available, it has the longest threshold wavelength and is sensitive to 1.47 and 2.94 μm lasers with picosecond durations.[8] The properties of S-1 photocathode, however, remained mysterious until recently because of its complicated microstructure. We have studied this system experimentally and theoretically to relate the optoelectronic properties to its microstructures.[4,9] Our theories involving the effective medium approximation and the diffuse surface scattering of photoelectrons and experimental results of X-ray and ultraviolet photoelectron spectroscopy on the S-1 are published elsewhere[4,9] which therefore won't be repeated here. Checked out successfully on the S-1 photocathode, these theories are now extended to calculate the optical properties of various types of random heterostructures. In this paper we shall first present the theoretical results on the properties of small particles of various types of microstructures in order to deduce some material design rules. We then report the experimental results on materials with Ag particles dispersed in Si.

Optical Properties of Heterostructures of Small Metal Particles

Fig.1 shows the optical absorption of a suspension of Ag particles in a medium of dielectric constant $\varepsilon=1$ and a continuous Ag film of the same mass thickness. The continuous and particulate films are, respectively, 50Å and 100Å thick, making equal mass thickness. Here, we haven't corrected the size effects of the 50Å continuous film. This is fine so long as the surface is smooth that the surface scatterings of electrons are specular.[10] The spherical particles are 50Å in diameters(D), 50% in volume fraction(p) and 2% aggregation(f2) which leads to some sort of void resonace.[3] We can see the enhanced photoeffects in the near IR region for the small particle system. Note that the peak at $\approx0.4\mu$m, sometimes called the dielectric anomaly, is due to isolated Ag particles (first microstructural unit)while that at $\approx0.8\mu$m due to aggregated ones (second unit). Reflectivity of the particulate film is greatly reduced as shown in Fig.2. Similar phenomena occur for Cu and Au particles with similar microstructures. Note that the second microstructural unit (aggregation) is essential to the enhanced absorption in the infrared. If there were no aggregated particles (f2=0) the peak at 0.8μm would have disappeared completely.

Our calculations predict a dielectric anomaly at ≈ 2.3 eV (0.53 μm) for isolated Au particles, close to the value of ≈2.2 eV (0.56 μm) measured by Yamaguchi et al.[11] As for Cu, the equivalent peak occurs at ≈2.2 eV (0.56 μm) and is obscured by the interband transition which takes place at about 2eV.[12] Meanwhile, the absorption peaks due to aggregated particles of both Au and Cu take place at ≈ 0.8 μm for particle volume fraction of 40%. At the same volume fraction, the second unit peak of Ag occurs at ≈ 0.6 μm as shown in the Fig.3 for an 100 Å thick layer of single size Ag-particle suspension with diameter 50Å and 2% of aggregation. In this figure, we see that, at constant fraction of the second unit, the increasing volume fraction of silver particles shifts the optical absorption into the infrared. The consistent line shape broadening at increasing volume fractions suggests the increasing electron-electron interactions between silver particles. Note that the first unit peak grows from p=10 to 40% and drops thereafter. This is related to both the percolation and lifetime effects.

The red shift also takes place when the dielectric constant of the host matrix semiconductor increases. This is given in Fig.4. The red shift in this case is related to the polarizability of semiconductors since a semiconductor with high ε is more polarizable. This result is consistent with previous work on alkali halides.[13] Our current goal has been to use semiconductors of high ε to embed small metal particles. As mentioned above, with controlled microstructural parameters, this will escalate photosensitivity in the infrared. Si with $\varepsilon \approx 12$ and many other semiconductors such as GaAs, InP are all potential candidates. In addition to enhancing infrared sensitivity, the semiconductor matrix either provides conducting paths for electrons to tunnel between metal particles via the impurities in it (hopping mechanism) or acts as the conduction channel of electrons internally photoemitted from metal particles at the excitation of light, as shown in Fig.5. The quantum yield of the Ag-Si system with a Schottky barrier height of 0.6 eV is given in Fig.6, calculated for a 100Å thick film with 50% Ag spherical particles embedded in Si. The diameter is 50Å and the percentage of aggregation is 2%. As photoelectrons have to travel to surfaces to emit, smaller particles incur more diffuse surface scatterings of photoelectrons at the boundaries. Within the mean free path of electrons, the surface scattering should be elastic and diffuse which upheave the escape probability.[5] With multiple surface scatterings, photoelectrons within the particles may reorient themselves and move in the directions more perpendicular to the surface. Hence the chance for them to overcome the energy barrier and emit is increased. However, particles too small in size may not be as effective radiation absorbers as larger ones. We found that 50Å is close to the optimal particle size for maximum photoabsorption and photoemission. Note that since the conduction electron density of Ag is 6.0×10^{22} cm-3, the density of photoelectrons at 1-3 μm, according to Fig.6, can be as high as 10^{20} cm-3, corresponding to the value for a heavily doped semiconductor.

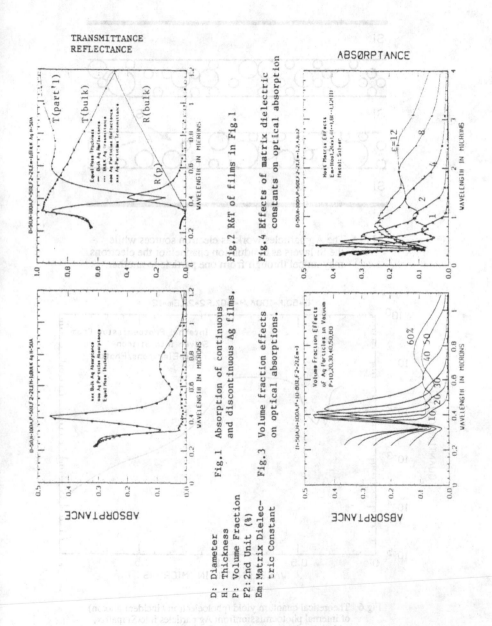

TRANSMITTANCE
REFLECTANCE

ABSØRPTANCE

Fig.2 R&T of films in Fig.1

Fig.4 Effects of matrix dielectric
constants on optical absorption

Fig.1 Absorption of continuous
and discontinuous Ag films.

Fig.3 Volume fraction effects
on optical absorptions.

D: Diameter
H: Thickness
P: Volume Fraction
F2: 2nd Unit (%)
Em: Matrix Dielec-
tric Constant

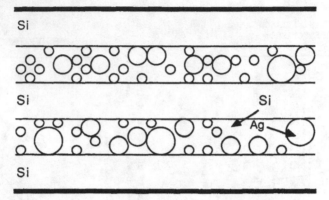

Fig.5 The Ag particles work as electron sources while
the Si layers as conduction channel; or the electrons
may tunnel through from one particle to another.

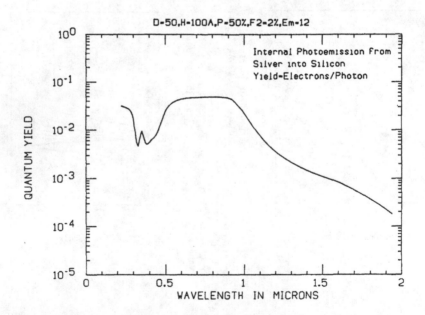

Fig.6 Theoretical quantum yield (photoelectron / incident photon)
of internal photoemissionfrom Ag particles into Si matrix,
with Schottky barrier height $q\emptyset_b = 0.6$ eV.

Experimentals on Ag Particles Dispersed in Silicon

The advantage of Ag-Si system is its simple eutectic nature with no reported stable compound formation and the negligible mutual solubilities at low temperature.[14] One can heat treat Ag-Si mixture to produce Ag particles as second-phase precipitates in the Si matrix, or deposit tandem layers of metals and semiconductors, as will be described later.

We first make thin film Ag-Si alloys by coevaporation of Ag and Si sources and then anneal them at relatively low temperatures (T<800°K) to force out second phase Ag-precipitates from the Si matrix. This approach has better control of spherical particle shape than the tandem deposition of Ag and Si. The tandem deposition of metal and semiconductor films are accessible since, at early stage of film deposition, metal first forms isolated islands which later on join up to form aggregates or a continuous film as aggregates further link to each other. The particle size and volume fraction can thus be controlled by adjusting substrate temperature and relative deposition rates of metal and semiconductor sources. This approach, however, has less control of spherical particle shape since small particles tend to agglomerate creating elongated islands or relatively large metal clumps of arbitrary shapes.

The Ag-Si alloy films are prepared by cosputtering dc-Ag and rf-Si sources facilitated with shutters. The compositions are determined by the respective deposition rates controlled by the relative positions and tilt angles of these sources as well as the sputtering voltage and current. The sputtering power is set at around 150W for Ag and 600W for Si sources. The sputtering rates range from 2-4 Å/sec for each source, depending on the composition desired. The base pressure of the deposition chamber is at 10-7 torr range while the working gas (Ar) pressure is 10-3 torr.

The Ag-Si films cosputtered on sapphire substrates at room temperature exist as separated Ag and Si layers. From TEM and sputter Auger analyses, we found that Ag atoms outdiffuse toward the free surface of the cosputtered film forming a semi-continuous Ag-film on top of the Ag-depleted Ag-Si film where only tiny amount of Ag particles are left behind in the amorphous silicon(a-Si) film. The surface segregation of Ag is partly attributed to the immicibility of Ag and Si. Meanwhile, the relatively large surface and strain energies of Ag precipitates in the Si matrix could further forbid the existence of Ag particles in Si. The competition amongst various free energy and kinetic activation energy terms determines the stable or meta-stable structure of this alloy film.

To overcome the surface segregation problem, we cover the surface of Ag-Si alloy film with either an additional Si- or carbon-layer in order to change the surface condition and favor the precipitations of Ag phase within Si matrix. Both types of overlayers seemed to work equally effectively. We found that the outdiffusion of Ag occurred concurrently with the deposition process. A delayed coverage of Si- or C-overlayer results in certain degree of clustering of Ag-particles. Fig.7 shows the plane view TEM micrograph of a 50:50 Ag-Si alloy film cosputtered at room temperature with a 60Å carbon overlayer covered right after the cosputtering process ended. The film was heated at 500 °C for about 30 minutes. After annealing, the electron diffraction pattern broke up from a continuous ring indicating the growth of Ag-particles. The matrix a-Si film on sapphire did not recrystallize under such annealing conditions judged from the persisting amorphous halos in the electron diffraction patterns. X-ray diffraction peak width analysis confirmed this result. What is important here is that the surface segregation has been prohibited by the timely introduction of an overlayer, carbon or silicon. The Ag-phase can now precipitate within the a-Si matrix as isolated small particles, since the free surface is now covered and no longer a thermodynamically favorable location to reside at. Fig.8 shows that a 10-minute deliberate delay in covering the similar Ag-Si film with silicon layer of 200Å results in a certain degree of Ag outdiffusion. The clustered particles should be formed by outdiffused Ag atoms before the Si-overlayer was put down. They thus sit at the interface of the silicon overlayer and the cosputtered Ag-Si alloy film. Auger profiling also confirmed this fact. In general, the precipitation method gives fairly spherical silver particles dispersed in silicon. Only when outdiffusion occurs do silver particles cluster and form the second microstructural unit. This segregation is a dynamic process which take place as the deposition goes on.

As for tandem depositions, we show in Fig.9 the SEM micrograph of a film with six alternating layers of 200Å Ag/200Å Si on sapphire deposited at 650°C. In this sample, the

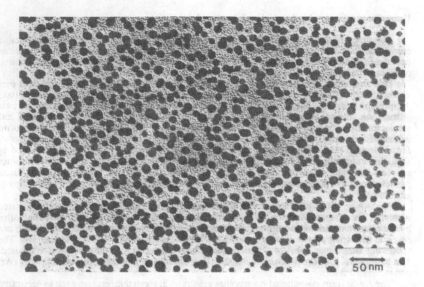

Fig.7 TEM micrograph of 50-50 Ag-Si alloy film cosputtered at room
     temperature with a C-overlayer, followed by 500°C anneal for
     30 min. Si matrix remains amorphous on sapphire substrate.

Fig.8 TEM micrograph of 50-50 Ag-Si alloy film cosputtered
     at room temperature with 200Å Si-overlayer deposited
     10 min after cosputtering ended. Surface segregation has
     already started and been observed.

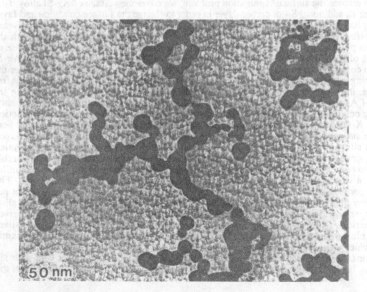

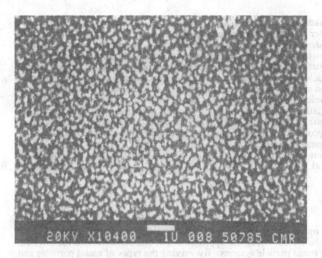

Fig.9
Tandem deposition
of Ag–Si at 650°C
with 200A layer for
each.
Si/Ag/Si/Ag/Si/Al₂O₃
SEM micrograph.
Note that the white
areas are Ag, contr-
ary to TEM graphs.
Second units are dark
areas surrounded by
white ones.

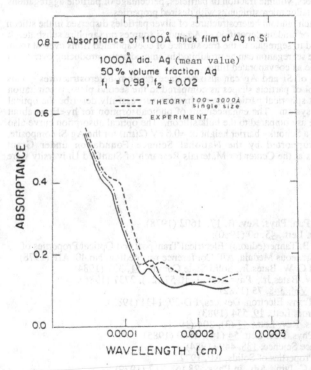

Fig.10
Optical absorption of
the sample in Fig.9.
Calculated with eddy
current term included.
Particle size distri-
bution obeys γ-distr-
ibution with mode ra-
dius r=D/2=500A.
The hump at 2μm is
associated with the
Ag–Si Schottky barrier
height.

178

matrix silicon is ploycrystalline according to X-ray diffraction studies. The average silver particle size is on the order of 1000Å and the particle shapes are not as regular as those made by the precipitation method. Note that in both Fig.8 & 9 we can identify the previously mentioned second microstructural units where the silicon matrix is surrounded by silver particles. Although the particle shapes are less regular for the tandem film, we may still model it with an equivalent particle size or particle size distribution to calculate its optical properties. Fig.10 gives the measured and calculated optical absorptance of such a film. The absorptance of the 1100Å film at wavelengths greater than 1.1 μm (hv<1.1eV) is enhanced by about three orders of magnitude relative to the bulk silicon. The hump at ≈2μm is believed to be associated with the Schottky barrier height. The effective medium theory with eddy current (magnetic dipole) term properly included nicely reproduces the absorption curve of such a system with microstructural parameters as indicated in the figure. Larger particles show higher optical absorption on account of magnetic effects, although the probability of photoemission from metal particles will be lower from the viewpoint of mean free path, as discussed previously.

Conclusion

We have shown that optical properties of small metal particles are closely related to their microstructures. Using the effective medium approximation, one can calculate the physical properties of these small metal particle systems. By varying the types of metal particles and host matrix semiconductors, volume fraction of particles, percentage of particle aggregations and particle size, we can tailor microstructures with desired properties.

We have prepared such random heterostructures of silver particles dispersed in the silicon matrix by coevaporation or tandem deposition of Ag and Si sources on sapphire substrate. It was found that silver tend to segregate to the free surface of coevaporated films even at room temperature. The surface segregation can be prohibited by immediatelyintroducing overlayers of silicon or carbon on the as-coevaporated films.

Tandem deposition of Si and Ag can also produce random heterostructures. This approach has less control of particle shapes as compared to the second phase precipitation method. The equivalent spherical particle approximation successfully describe the optical properties of the Ag-Si system. The enhancement of photoabsorption for hv<Eg is about three order of magnitude as compared to the bulk silicon. The optical absorption curve also suggests the existence of a Schottky barrier height of ≈0.6 eV (2μm) for the Ag-Si composite.

This research is supported by the National Science Foundation under Grant ECS-8520580. Facilities at the Center for Materials Research of Stanford University were used in this investigation.

References

1. D. Stroud and F. P. Pan, Phys. Rev. B, 17, 1602 (1978).
2. P. Sheng, Phys. Rev. Lett., 45, 60 (1980).
3. J. C. Garland and D. B. Tanner(editors), Electrical Transport and Optical Properties of Inhomogeneous Meduia, AIP Conference proceeding, No.40, AIP (1978).
4. N. V. Alexander and C. W. Bates,Jr., Solid State Comm., 51, 331 (1984)
5. Q. Y. Chen and C. W. Bates,Jr., Phys. Rev. Lett. 57 (21), 2737 (1986).
6. M. Y. Schelv, SPIE, Vol. 348, 75 (1982)
7. H. Beneking, IEEE Trans. Electron. Devices. ED-29, 1431 (1982).
8. W. Roth et al., Electron. Lett., 19, 554 (1983).
9. C. W. Bates,Jr., Phys. Rev. Lett., 47 (3), 204 (1981).
10. J. C. Hensel, et at., Phys. Rev. Lett. 54 (16), 1840 (1985).
11. S. Yamaguchi, Surface Science, 138, 449 (1984).
12. F. Wooten, Optical Properties of Solids, AP, 1972.
13. A. E. Hughes and S. C. Jain, Adv. in Phys., 28 (6), 717 (1979).
14. M. Arnold, et al., Crystal Res. Technol., 18 (8),1015 (1983).

LASER ANNEALING OF NARROW GAP HgTe-BASED ALLOYS*

R. E. KREMER, F. G. MOORE, M. R. TAMJIDI, AND Y. TANG
Oregon Graduate Center, 19600 NW Von Neumann Dr., Beaverton, OR   97006

ABSTRACT

We have used a cw CO_2 laser to study the effects of rapidly annealing HgTe-based alloys. Both as-grown and thermally annealed samples of HgCdTe, HgMnTe, HgZnTe, and HgMgTe have been examined for mercury loss and surface damage using energy-dispersive x-ray analysis and optical reflectivity measurements. Small, but systematic differences were found between the as-grown and the thermally annealed samples and among the various materials studied. No degradation of the material at all was observed when the samples were cooled to 77 K and exposed to the laser.

INTRODUCTION

HgCdTe has long been the material of choice for fabricating photodetectors for the 3 - 5 micron and 8 - 14 micron atmospheric windows. In order to make a photovoltaic detector, a p-n junction must be created in the material. One way to do this is to anneal p-type samples in a saturated mercury vapor atmosphere. The mercury diffuses into the material, annihilate mercury vacancy acceptors, and forming a thin n-type layer at the surface. By the carefully monitoring the temperature and time of the anneal, some control over the depth and abruptness of the is achieved [1]. More recently, efforts have been made to form the junction by implanting dopant ions into the material [2]. The major problem with this technique is that ion implantation always produces large amounts of damage. Unlike other semiconductors, such as silicon and GaAs, the volatility of mercury in HgCdTe prevents long-term thermal anneals to reduce or eliminate the damage. If mercury out-diffuses from the sample during annealing, the bandgap of the material will change.

We are exploring techniques by which the sample can be annealed rapidly enough to prevent significant mercury loss. This paper reports on the use of short exposures to relatively low power radiation from a cw CO_2 laser for this purpose. The use of a laser allows precise control of the annealing time, although at the expense of temperature control. A CO_2 laser was chosen because the energy of these photons (about 0.12 eV) is quite close to the bandgap of most of the materials that were to be examined. Thus, the absorption should be relatively small, and concentrated in the implant damage layer, rather than the bulk [3]. If absorption and subsequent heating can be minimized, mercury loss should also be greatly reduced compared to conventional thermal or laser annealing techniques.

The mercury loss problem stems from the weakness of the Hg - Te bond. When another element, e.g. cadmium, is added, the Hg - Te bond may change [4]. The presence of some elements tends to weaken the bond, while that of other elements can act to strengthen the bond. Laser annealing experiments can also be used to directly measure the stability of alloys made by substituting other elements into the mercury sublattice. The exposure time required to cause significant mercury loss can be easily determined. The samples used in this study contained substitutional cadmium, manganese, zinc, or magnesium. The presence of nearby Cd - Te bonds has been theoretically shown to destabilize the Hg - Te bond, while close Zn - Te bonds should stabilize the Hg - Te bond [4]. No theoretical studies have been made concerning either manganese or magnesium.

Mat. Res. Soc. Symp. Proc. Vol. 90. ᶜ 1987 Materials Research Society

EXPERIMENTAL

The crystals were grown using a vertical Bridgman process. Stoichiometric amounts of the three elements used were placed in a carbon-coated quartz ampoule, which was sealed off under vacuum. In each case, the amount of the third (not mercury or tellurium) element used was calculated to provide a material with a bandgap of about 0.1 eV. This fraction amounted to 0.20 cadmium, 0.11 manganese, 0.10 zinc, and 0.07 magnesium. The sealed ampoule was then placed in the upper zone of a two-zone furnace, the temperature of which was raised to above the melting point of the material. During the growth of all of the samples studied here, the furnace pulling speed was kept at 1.0 cm/hr, a rate at which previous work had shown produced high-quality crystals [5].

The as-grown material was sometimes n-type and sometimes p-type. In either case, it was highly compensated, yielding very low mobilities. Some of this compensated material was used in the laser annealing study. Slices of each sample were annealed in saturated mercury vapor at 300 K for 14 days, after which they were strongly n-type throughout the slice. These annealed wafers provided the uncompensated, n-type material for the laser annealing.

Samples are prepared for laser annealing by polishing the surface with successively smaller grit down to 0.05 micron. Each wafer is then diced into pieces roughly one mm square. These small pieces are then mounted on a microscope slide using orange dental wax. After mounting, each piece is given a final cleaning using a cotton swab soaked with high quality acetone to remove any excess wax or other surface particulates. The samples are then placed on a brass heat sink in a vented plexiglass chamber.

The laser used for this study is a Sylvania CO_2 laser currently operating at 5 W of cw power. Some of the early measurements reported here were taken with the laser operating at 7 W. The beam waist is about two mm, so the samples were completely exposed. Exposures are controlled using a timer with a resolution of about 0.5 seconds. Some of the samples were exposed for increments of two seconds, and then analyzed after each successive exposure. The majority, however, were continuously monitored, as discussed below.

Measurements of mercury loss and surface damage were taken using several techniques. The most quantitative mercury loss results were obtained using an energy dispersive x-ray analysis (EDX) attachment to a scanning electron microscope. It was observed that the onset of major mercury loss from the surface region of the crystals coincided with the amount of incident energy required to physically damage the surface. It then became much easier to simply monitor the surface roughness of the sample, and use this information as a qualitative measure of mercury loss. Surface roughness was monitored

Figure 1. SEM photograph of a sample after a long laser anneal showing surface damage. The marker is 1 mm.

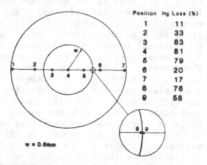

Position	Hg Loss (%)
1	11
2	33
3	83
4	81
5	79
6	20
7	17
8	76
9	58

w = 0.6mm

Figure 2. EDX measurements showing mercury loss taken from several positions on the crystal shown in figure 1.

using the reflectivity from a HeNe laser incident on the sample. A heavily damaged slice of crystal is shown in figure 1, with the corresponding mercury loss as measured by EDX shown in figure 2. As can be seen in the figures, the onset of major mercury loss and surface roughening is coincident and abrupt.

Inferring mercury loss from reflectivity data offers several advantages over conventional methods. One major benefit lies in the ease of application of this method. No items of large equipment are needed, and SEM time is not necessary. Probably the biggest advantage is that the reflectivity can be monitored continuously, even while the laser annealing is occurring. Thus the data obtained is continuous, as compared to our earlier EDX results which were usually measured after exposure increments of two or three seconds. Figure 3 shows a block diagram of the experimental arrangement used for the work reported in this article. An example of a reflectivity trace is exhibited in figure 4, showing the rapid drop in the reflected signal as both mercury loss and surface roughening occur.

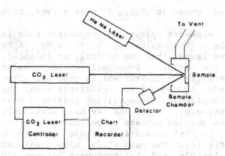

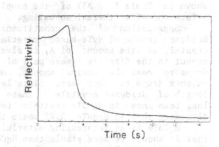

Figure 3. Block diagram showing the experimental arrangement used for the reflectivity measurements.

Figure 4. Example of a typical reflectivity trace.

RESULTS AND DISCUSSION

We have divided the discussion of our experimental results into two sections. The first will briefly describe the results of our crystal growth experiments. We have grown, or attempted to grow several materials which are relatively unusual. The second subsection will discuss the results of the laser annealing experiments, drawing comparisons between the different materials and different annealing conditions.

Crystal Growth Results

We have produced crystals of several HgTe-based alloys, all of which were designed to have energy bandgaps of approximately 0.1 eV. For the four materials reported on here, this requires compositions of $Hg_{0.8}Cd_{0.2}Te$, $Hg_{0.89}Mn_{0.11}Te$, $Hg_{0.9}Zn_{0.1}Te$, and $Hg_{0.925}Mg_{0.075}Te$. Bridgman-grown crystals of these materials can be either n-type or p-type depending on the details of the process. For most of the samples reported on here, the as-grown materials were n-type, but the electrons had very low mobilities, due to mercury vacancies which form during the Bridgman growth process. Long-term thermal anneals in saturated mercury vapor can be used to provide reasonably high-mobility, n-type material. To provide uniformity, all of the n-type materials reported on here consisted of 1 mm slices that had been annealed for 14 days at 300 C. Typical electrical properties for both the as-grown and the thermally annealed materials, as determined by Hall measurements at 77 K, are

Table I. Typical Electrical Properties at 77 K

| | As Grown | | After Thermal Anneal | |
	Concentration (cm^{-3})	Mobility (cm^2/Vs)	Concentration (cm^{-3})	Mobility (cm^2/Vs)
HgZnTe	3×10^{17}	1700	6×10^{16}	67,000
HgMgTe	7×10^{16}	4300	1×10^{17}	29,000
HgCdTe	5×10^{16}	6000	2×10^{16}	140,000
HgMnTe	8×10^{16}	2400	2×10^{15}	210,000

shown in Table I. All of the samples shown in Table I were n-type, both before and after thermal annealing.

Segregation of the constituent elements plays an important role in Bridgman growth of HgTe-based materials. The value of x in a $Hg_{1-x}A_x Te$ crystal, or the amount of A, is always larger than the nominal, or intended, amount in the first-to-freeze part of the ingot. Eventually, the x value drops to near the nominal amount, until the A constituent is used up, after which x drops to near zero. This leads to composition gradients along the length of Bridgman crystals. These longitudinal composition gradients have long been known in HgCdTe crystals (see, e.g., [6]), and we have previously discussed them in HgMnTe [5,7] where the problem appears to be less severe.

HgZnTe is an interesting material, in that theoretical studies have shown that it should be more stable than HgCdTe [4]. The same study also predicted that HgTe and ZnTe are not completely miscible, and that somewhere between 300 K and 600 K, spinodal decomposition occurs [4]. Our results show that even at the melt temperature, the segregation of ZnTe in HgTe is very large. Figure 5 shows the composition of a typical, Bridgman-grown HgZnTe ingot as a function of longitudinal position along the ingot, as obtained from density measurements. The nominal zinc content was 0.1 for this crystal. Only the first-to-freeze tip contained appreciable amounts of zinc, and then considerably less than the nominal amount. The remainder of zinc was segregated to the top of the ingot, which was composed almost entirely of zinc. Other growth techniques, such as the travelling heater method, have been used to successfully incorporate Zn into the Hg sublattice [8].

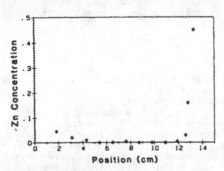

Figure 5. Zinc concentration as a function of position along the ingot.

Magnesium was chosen as a possible constituent due to its position in the periodic table. As a column IIA element, it should be isovalent with the mercury for which it substitutes in the lattice. As is the case for MnTe, naturally occurring crystals of MgTe form in the wurtzite structure. When the material is forced to assume a zincblende structure, it is somewhat uncertain what the value of the lattice parameter will be. Experimental work has determined the lattice parameter for the fictitious zincblende form of MnTe [9], but similar results do not exist for MgTe. Comparisons between the parameters of wurtzite MnTe and wurtzite MgTe are not helpful, as the

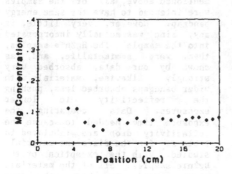

Figure 6. Magnesium concentration as a function of position along the ingot.

published parameters for MgTe are larger, but the ionic radius is smaller [10]. However, if we assume that the lattice parameter for HgMgTe does not change appreciably for small variations in magnesium content, density measurements can still be used to check for compositional gradients in Bridgman ingots. Figure 6 shows such results for a sample with a nominal magnesium concentration of 0.075. Somewhat less than this amount appears to be incorporated into the crystal, but the same characteristic Bridgman composition gradient is evident. Due to the lack of accurate information on the lattice parameter, the actual percentages of magnesium shown in figure 6 should be taken as qualitative, and only used for comparison purposes.

CO$_2$ Laser Annealing Results

In this section we will discuss the results of the laser annealing experiments. Some of the data reported here was obtained from EDX measurements. The bulk of the data, however, was obtained from reflectivity measurements such as that shown in figure 4. We annealed samples at both room temperature and at 77 K. The material annealed at the lower temperature showed no measurable effect, even after laser exposures of as long as 60 s. It is not certain whether enough absorption occurs to affect implant damage, but mercury loss and surface damage do not appear to be major problems. Further study of the low temperature annealing process is underway.

The reflectivity plots all have the general shape of that shown in figure 4. There is an initial rise in the reflectivity, which may, or may not be followed by an approximately level plateau, followed by a precipitous drop. All of the samples were polished prior to laser annealing. Although the surface of each sample was gently cleaned after the polishing and mounting procedure, some contaminants undoubtedly remained. In addition, a thin layer of native oxide may also be present after the final cleaning step. We interpret the initial rise of the reflectivity as being due to the surface being cleaned further by the laser beam. Any contaminants or particulates on the surface would be strongly absorbing, and thus quickly removed by the exposure. The fact that the time required for this rise to occur varies considerably from sample to sample of the same type of material lends credence to this belief, since it is quite possible for one sample to have more particles, etc. on its surface than the other following the final cleaning step. There is a danger in that the presence of these surface contaminants may shield the bulk of the crystal from the beam, thereby shifting the fall in reflectivity to somewhat longer exposure than would be the case for clean surfaces. Thus the most reliable data are those from reflectivity plots where the initial rise occurs soonest.

The final rapid drop in reflectivity is identified with the onset of surface damage. Our previous work using the EDX approach has shown that this surface damage threshold coincides with the threshold for major mercury loss from the surface region of the sample. The exposure time required to cause this drop, then, is a direct measure of the stability of the material if it is assumed that the absorption is the same for all the materials studied. As

Table II.   Average Exposure Time
             Required for Drop in
             Reflectivity

	As Grown	After Thermal Anneal
HgZnTe	1.5 s	1.8 s
HgMgTe	1.5	2.7
HgCdTe	3.0	3.6
HgMnTe	3.0	5.4

mentioned above, all of the samples were intended to have the same energy bandgap. However, very little, if any, zinc was actually incorporated into the sample. The HgZnTe samples, then, were semimetallic, and, as shown by our data, absorbed very strongly. Likewise, materials with wider bandgaps absorbed less, pushing the reflectivity to longer exposures. Data comparing the exposure time needed to cause the reflectivity drop are tabulated in Table II for the different materials studied. With the exception of the HgZnTe samples, all of the materials were assumed to have the same bandgap, and thus nearly identical absorption coefficients. It should be noted that samples of wider bandgap materials (such as 30% HgCdTe) were also studied. These samples showed much lower absorption, and the times required for the reflectivity drop were significantly longer. The times listed, in each case, represent the average from several trials, using different samples, but with identical values of x.

The as-grown materials, although usually n-type, contained a large number of mercury vacancies, which act as acceptors, in addition to the native donors. Slices of these materials almost always show strong absorption of all radiation due to intra-band transitions. It is, thus, not surprising, that the as-grown samples consistently showed surface damage/mercury loss after shorter exposures than did the thermally annealed samples. In all cases, the measured composition of the thermally annealed samples were very close to those of the as-grown samples.

Comparing as-grown or thermally annealed samples among different materials provides some interesting information. Both the HgZnTe and the HgMgTe samples showed reflectivity drops after fairly short exposures, even though the HgZnTe was predicted to be a more stable compound [4]. However, as noted above, and shown in figure 5, the large segregation between HgTe and ZnTe resulted in an ingot with very little zinc incorporated, and the samples examined had only about 1% zinc in them. The HgZnTe samples, then were semimetallic, and strongly absorbing, regardless of any post-growth annealing process. From these results it is not possible to speculate as to the stability of the material. The HgMgTe also gave rather disappointing results. This material, while perhaps a viable alternative to HgCdTe, does not appear to offer a significant advantage in stability over HgCdTe. Work is continuing on both of these compounds.

There appeared to be very little difference between HgCdTe and HgMnTe when the as-grown samples were being annealed. The major difference occurred in the materials after the thermal anneal. Here, the HgMnTe usually required a significantly longer exposure before surface damage and mercury loss became a problem. In this case HgMnTe may offer an advantage over the better known HgCdTe. From these results alone, it is not possible to state that HgMnTe is a more stable material. However, when combined with earlier results showing less segregation during growth for MnTe in HgTe than for CdTe in HgTe [5], it appears that HgMnTe has several superior qualities, and certainly warrants further study.

Finally, although this may seem like belaboring the obvious, the laser annealing reported on here was an extremely vigorous anneal. In all probability, the surface of the samples actually melted. At any rate, major surface damage and mercury loss occurred. Thus anneals of the power and duration reported on here are not suitable for ion implant activation. That task requires a much gentler anneal involving lower powers and shorter exposures.

REFERENCES

*This work was supported by the National Science Foundation through Grant No. ECS-8402080.

[1] C. L. Jones, M. J. T. Quelch, P. Capper, and J. J. Gosney, J. Appl. Phys. 53, 9080 (1982).
[2] L. O. Bubulac, W. E. Tennant, R. A. Riedal, and T. J. Magee, J. Vac. Sci. Technol. 21, 251 (1982).
[3] G. Bahir, R. Kalish, and Y. Nemirovsky, Appl. Phys. Lett. 41, 1057 (1982).
[4] A. Sher, A-B. Chen, and M. van Schilfgaarde, J. Vac. Sci. Technol. A 4, 1965 (1986).
[5] R. E. Kremer and M. R. Tamjidi, J. Crystal Growth 75, 415 (1986).
[6] B. E. Bartlett, P. Capper, J. E. Harris, and M. J. T. Quelch, J. Crystal Growth 46, 623 (1979).
[7] M. R. Tamjidi and R. E. Kremer, Mater. Lett. 4, 90 (1986).
[8] P. Becla, private communication.
[9] J. K. Furdyna, W. Giriat, D. F. Mitchell, and G. I. Sproule, J. Solid State Chem. 46, 349 (1983).
[10] B. R. Pamplin in CRC Handbook of Chemistry and Physics, 63rd Ed., ed. by R. E. Weast (CRC Press, Cleveland, 1982), p. E-98.

ACKNOWLEDGMENT

This work was supported by the National Science Foundation through Grant
DE-XOS-BA8XXXX.

[1] ...

[2] ...

[3] ...

[4] ...

[5] ...

[6] ...

[7] ...

[8] ...

[9] ...

[10] ...

PART VI

Characterization of Infrared Materials

CHARACTERIZATION OF INFRARED MATERIALS BY
X-RAY DIFFRACTION TECHNIQUES

W. J. TAKEI and N. J. DOYLE, Westinghouse R&D Center, 1310 Beulah Road,
Pittsburgh, PA 15235

ABSTRACT

X-ray diffraction techniques have proved invaluable in the
characterization of infrared materials, particularly those prepared by thin
film deposition techniques such as molecular beam epitaxy, MBE. The
techniques are sufficiently sensitive and rapid to provide the information
feedback required for efficient optimization of the growth process. They are
nondestructive and permit the correlation with results on the same sample
obtained by other characterization techniques such as those being described at
this Symposium. Depending on the development status of the growth technology,
the information to be acquired includes presence of twinning, quality, and
type of epitaxial orientation, strains, and compositional variations. A
critical issue in the application of these materials in detector arrays is the
question of uniformity control, both laterally and in depth. The techniques
to be described include not only modern x-ray topographic and multiple crystal
diffractometric techniques but particularly for the early stages of growth
process development, classical photographic ones such as the oscillation and
Weissenberg methods. Examples of these various aspects are presented with
emphasis placed on the characterization involved in MBE growth of HgCdTe
films.

INTRODUCTION

A viable IR materials and device technology depends critically on
characterization of the materials. Process optimization can be achieved only
with a knowledge of the effect of parameters on the structure and properties
of the materials which result. These factors are equally important in
developing the high performance IR detectors which are the ultimate goal of
this effort. This is particularly true in the case of HgCdTe with its
difficulty in control of composition, crystallographic quality, and electrical
properties. The promise of improved quality and practical requirements of
device configurations have led to a focus on epitaxial growth methods, LPE,
MOCVD, and MBE. This in turn has led to the need to study, not only active
layers and interfaces, but also the substrate which provides the foundation
for the entire process.

A wide variety of techniques, many described in other papers at this
Symposium, have been used to understand the material and process under study.
The integration of the techniques in the development of our MBE program has
been described previously [1-3]. In this paper one important class of
techniques will be reviewed -- those utilizing x-ray diffraction. The
approach adopted is a practical one, illustrating the type of information
achievable by example for various techniques. The intent is not to provide a
detailed understanding of various x-ray techniques but to give a general
survey of their utility in the study of IR materials. More details are given
in various surveys.[4-6]

GENERAL CONSIDERATIONS

The importance of x-ray diffraction characterization techniques is due to
several factors:
1)   Analysis of many materials parameters. Properties such as composition,
     presence of second phases including twins, epitaxial orientation,

crystallographic quality strain, and their distribution can be measured.
2) High accuracy in measurement of crystallographic parameters such as angles, strains, and approach to ideal perfection.
3) Nondestructive analysis of samples. This permits subsequent characterization of the same sample by other techniques.
4) Rapid information feedback, to the grower to evaluate the effect of growth parameter variations and occasionally, signal that a system malfunction has occurred.

As with any characterization technique, the limitations must be realized. In the case of x-ray diffraction these would include:
1) Limited sensitivity compositionally. Generally effects must be present well above doping levels.
2) Random fluctuations of many characteristics are averaged over the sample volume, e.g., different values of Hg-Te and Cd-Te bond lengths cannot be distinguished if atoms are distributed randomly.
3) Limited resolution-lateral dimensions sampled determined by beam size which usually is about 1×1 mm^2. This is increased on the sample surface (2X to 4X) since the beam generally strikes it at angle. Beam area usually can be reduced by one or two orders of magnitude at the expense of intensity.
4) The specific factors influencing the diffraction phenomena such as dislocations or precipitates causing line broadening can be difficult to distinguish.

Depth of Penetration

A common and important question is the sample depth from which information is being acquired. Penetration may be desirable to be small if interest is focussed on surface effects or to be large if data is required simultaneously from an epitaxial film and the substrate. It is a function of the wavelength, material, reflection plane, its orientation, and the crystal perfection. Two different mechanisms attenuate x-rays.
1) Mass or photoelectric absorption and scattering. This phenomena operates at all times during the passage of x-rays through matter, primarily by the photoelectric or fluorescence effect. The magnitude depends on the x-ray energy and the material, the higher the atomic number the greater the absorption as tabulated in the International Tables.[7] For HgCdTe and CuKα radiation, the linear absorption coefficient for $I/I_0 = \exp(-\mu t)$ is $\mu \sim 0.17$ for t in μm. Thus a pathlength of 14 μm would result in a 10X intensity reduction. For comparison, the corresponding lengths for GaAs and Si are 60 and 160 μm respectively. Use of Mo radiation would increase the HgCdTe value to 40 μm. When examining an epitaxial film of thickness d, d $\neq$ t: (a) a factor 2X is introduced because the x-rays must traverse the film twice, in and out; (b) the incident and diffracted beams traverse the film at some angle dependent on the reflection used. For the symmetric (400) the factor would be 2X. Thus a HgCdTe layer $14/2 \times 2 = 3.5$ μm has a 10X reduction in CuKα intensity available to study the underlying layer.
2) Extinction. If a crystal is set at a diffraction angle another attenuation mechanism is operative. As the beam transverses the crystal planes, diffraction occurs removing power from the beam. The energy loss through diffraction and interference phenomena is called extinction. Calculation is difficult because it is variable depending on the quality of the crystal and beam but for good crystals and strong reflections the 10X layer thickness is ~ 1 μm.

Generally extinction is the primary attenuation mechanism for penetration for diffraction within a layer. However, studying an underlying layer, since the upper layer is not at its exact Bragg angle, mass absorption is the relevant mechanism.

Film Techniques

Various advanced x-ray diffraction techniques have been developed using sensitive photon detectors, crystal monochrometers for high quality incident beams, precise mechanical structures, and computer control. However, the classical film techniques have proven to be very valuable, particularly in the early stages of growth optimization, when crystal quality is not high and information is desired very rapidly on various parameters, such as phases present and epitaxial orientation and quality. One reason is that the new techniques look at limited regions of reciprocal space, i.e., only limited combinations of crystal orientation and detector position. However, a film essentially surrounds the sample so that diffraction effects will be detected simultaneously at positions where a detector would not have been placed. Two valuable techniques are the rotation (or oscillation) and Weissenberg methods. These are discussed in most x-ray diffraction books and analyzed in detail by Buerger.[8] It should be emphasized that these discussions assume a classical sample, a small sphere or cylinder completely bathed in the beam with small absorption. This is far from the situation with IR samples. Half the expected pattern is not present because it is dependent on passage through the sample which, for epitaxial films, is opaque because of the thick substrate.

We take oscillation and Weissenberg photographs with a standard Nonius Weissenberg camera. Oscillation photographs are with a symmetry direction, e.g., [110], parallel to the rotation axis, a cylindrical film surrounding the sample and orthogonal to the incident beam, oscillation between 10 to 60° from the sample surface, and exposing for about 1 hour. Figure 1 shows two such photographs taken early in our MBE HgCdTe studies from HgCdTe/CdTe (001) films. The spot patterns from the CdTe crystal are arranged in layer lines. Since the rotation axis is [110], the central line contains reflections with h-k = 0, i.e., (004), (2$\bar{2}$0), (3$\bar{3}$1), etc. The central line is bracketed by layer lines for reflections with h-k = ± 1, ± 2. The length of the rotation axes can be derived from the layer line spacing. Figure 1a shows polycrystalline HgCdTe, with no indication of fiber orientation, i.e., no intensity variations along the arcs. Figure 1b shows epitaxial growth. The lattice parameter difference between HgCdTe and CdTe is difficult to resolve by this technique. Close examination shows weak reflections between the row lines. These are caused by twins in this film. Probably not evident on the reproduced figures are lines (1a), and spots (1b) from free Te, which in the latter case, subsequent cross sectional TEM showed was near the interface.

A surprising result was revealed by these techniques in the examination of an MBE CdTe film. It was grown on a GaAs substrate on which a GaAs buffer layer had been grown in another MBE system and protected by an As layer for the intersystem transfer. The As layer was removed thermally just before CdTe growth. The GaAs substrate had the common 2° off [001] toward [011] tilt. Figure 2a shows the [110] oscillation photograph. It is entirely as expected for completely parallel growth, CdTe (001) || GaAs (001). The layer lines for GaAs are outside those for CdTe since the [110] length is shorter for GaAs. A Weissenberg photograph was then taken. This only requires the addition of a cylindrical slit to allow passage of the zero level layer line and engagement of a motor drive to give camera translation synchronized to the crystal oscillation. The result is that the horizontal dimension (as the figures are oriented in this paper) corresponds to the crystal angular position while the vertical dimension corresponds to the Bragg angle. Therefore Weissenberg photographs are useful to give the crystal quality, the lateral extent of a reflection image giving the angular range over which reflection occurs, i.e., equivalent to a rocking curve. This is in addition to giving the various angular relationships involved. Before discussing the Weissenberg examination it may be clearer to illustrate a simpler case. Figure 3a shows a Weissenberg photo of CdTe/GaAs in which the GaAs is cut parallel to (001) and the MBE CdTe was grown directly without a buffer layer. The (002) reflections lie along a line, indicated by the arrow, with a slope equal to tan^{-1}2 or 63°. Starting

(a)   (b)

Figure 1. Oscillation patterns from early MBE $Hg_{0.8}Cd_{0.2}Te/CdTe(001)$. $CuK\alpha$ radiation. [110] rotation axes.

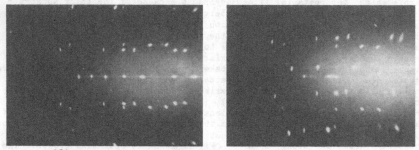

(a)   (b)

Figure 2. Oscillation patterns from CdTe/GaAs/GaAs $(001)_{2° off}$ $CuK\alpha$ radiation. Rotation axes (a) [110], (b) [1$\bar{1}$0].

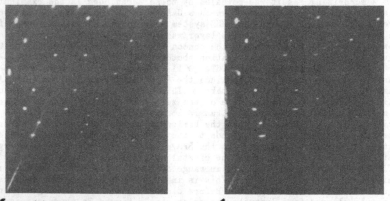

(a)   (b)

Figure 3. Weissenberg patterns $CuK\alpha$ radiation. Rotation [110] (a) CdTe/GaAs (001), (b) CdTe/GaAs $(001)_{2° off}$.

from the arrow left corner ($\omega = \theta = 0$), visible are a weak (002) pair from CdTe and GaAs, then the (004) pair, and finally (008) CdTe. All other reflections occur in CdTe + GaAs pairs (except near the high θ limit where GaAs reflections are not allowed since $\sin \theta > 1$) along the lines with 63° slopes. The two lattices are parallel. The Weissenberg photo, Figure 3b, based on the zero level of Figure 2a, shows a different pattern. Comparison shows that it can be derived from Figure 3a by shifting every CdTe spot to the right by a distance corresponding to 5°. Thus the oscillation and Weissenberg results show that for this sample, the CdTe has not grown exactly parallel but with [001] 5° toward [1$\bar{1}$0]. This was confirmed by the [1$\bar{1}$0] oscillation photo, Figure 2b, which shows the zero layer lines tilted at 5°. Figure 3b also shows that the crystal quality of the CdTe film is poorer than that of the substrate since lateral extent of the CdTe reflections is greater than those from GaAs.

Diffractometry

Diffractometry is the measurement of the diffracted intensity as a function of diffracted (powder 2θ) or incident (crystal ω-rocking curve) angle. The latter has proven important in development of epitaxial film growth technology. The reflection curve half-width (full angular width at half-maximum intensity) is a convenient and sensitive measure of crystal quality. In single crystal diffractometers one or more slits are used to define a collimated beam. We use a Siemans Omega Drive Diffractometer, Jarrell Ash Microfocus Generator and a 0.05 mm slit to give a divergence of 0.02°. Because intensity is high and sample alignment is easy, single crystal diffractometry is very useful and convenient to monitor the growth optimization process in early stages. Information feedback takes less than one hour. However, utility decreases as optimization proceeds. For the materials and reflections studied, the theoretical limit on half width is ~ 10 arc seconds.[9, 10] Thus at some stage, the instrumental broadening becomes of the same order or larger than the intrinsic value of the sample making interpretation difficult.

The majority of our rocking curve studies have been made using a double crystal diffractometer. In this instrument, made by Blake Industries, a highly perfect first crystal monochromatizes the x-rays from a standard sealed tube source. To avoid angular dispersion the so-called (+-) or parallel arrangement of the monochrometer and sample is used in which the two crystals are parallel to each other. For this it is not necessary that the two crystals be of the same material, only that the two interplanar spacings be the same. Therefore, as the monochrometer for studies in the HgCdTe system, we use an InSb crystal since it is lattice matched and is readily available in much higher quality. When (001) InSb is used as a sample, it is easy to achieve rocking curve widths < 13" for the symmetrical (400) reflection using CuKα radiation. The sample can be translated laterally and the beam vertically to permit uniformity studies.

A point for all potential users -- when reporting results it is important that the conditions used also be reported. $W_{1/2}$ is a strong function of the wavelength, reflection, and angles involved. This is particularly important for (111) orientations where, even if CuKα is assumed, it is ambiguous as to whether the (111), (333), or (444) reflections were used. Each is roughly equally convenient but give decreasing widths. We use the CuKα (333) reflection to give a good compromise between small half widths, 8" for InSb, and sufficient intensity. If results from various sources are to be compared, it is important that this be done on a common or known basis.

Figure 4 shows a CuKα (004) rocking curve from a high quality MBE CdTe/InSb (001) film showing both film and substrate peaks. The CdTe width (19") is very close to that of InSb (12.5"). In contrast the best width seen for equivalent films on GaAs is 100". Another indication of high perfection are oscillations in the region between the peaks, seen more clearly in the

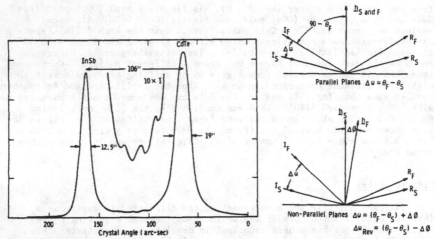

Figure 4. Double crystal rocking curve from MBE CdTe/InSb (001) CuKα (004) reflection.

Figure 5. Effect of tilt on diffraction angles.

expanded scale. These interference fringes are evident only in highly perfect structures and the spacing converts to a film thickness 1.1 μm which agrees with that measured directly by a Tallystep. The significance of the interpeak separation, Δw, 106", depends on two concepts: (1) the strain configuration of the film/substrate couple and (2) the effect of interplanar tilts.

1) The film and substrate generally have different lattice parameters in this case CdTe being larger than InSb (for HgCdTe/CdTe growth, the film is smaller). When the film couple is formed the state of strain can be described by two extreme cases:

a) The film and substrate are coherent, that is, the film is strained to match the substrate exactly. Analysis of this case has been discussed in detail.[11, 12] For CdTe/InSb the CdTe is compressed so $a_F^{\parallel} = a_S$. By a Poisson expansion the film expands so $a_F^{\perp} > a_F^{o}$. The film now has a tetragonal unit cell.

b) The film and substrate are incoherent. By formation of dislocations at the interface the film is uncoupled from the substrate and freed to assume its natural lattice parameter a_F^{o}.

c) It is possible that an intermediate state is achieved with incomplete strain relief.

2) The effect of interplanar tilts is shown in Figure 5. For parallel planes the crystal rotation angle Δw is a direct measure of the Bragg angle difference $\theta_F - \theta_S$. Reversing the x-ray beam direction, right to left, achieved experimentally by rotating the crystal 180° about h_S, has no effect. For non-parallel planes and the x-ray flow direction shown in the lower part of Figure 5, the tilt angle $\Delta\phi$ adds to $\Delta\theta_F - \Delta\theta_S$ while for the reverse direction it is subtracted. In either case the components can be extracted since

$$(\theta_F - \theta_S) = (\Delta w + \Delta w_{rev})/2$$

$$\Delta\phi = (\Delta w - \Delta w_{rev})/2$$

The Δw of 106" in Figure 4 is only one of the measurements needed. A 180° rotation also gave 106" showing there was no tilt. To study the state of

strain an oblique reflection plane (531) was measured. Table 1 shows the
results. The film has expanded in the surface normal direction to a 0.1%
mismatch but contracted within the plane to give zero mismatch, i.e., the film
has elastically strained into a coherent structure. Note that (531) planes
have a tilt. This tilt arises because oblique planes in the tetragonal film
and cubic substrate are not parallel to each other.

Table 1 — Coherency of Epitaxial CdTe/InSb

		(400)	(531)
	ω_1	106"	51"
	ω_2	106"	228"
	$\Delta\theta_B$	106"	140"
	$\Delta\phi_{oh}$	0	80"
$\Delta a/a_3^{\perp}$		0.000955	0.000959
	$\Delta a^{||}/a_s$	–	0.000002
	$(\Delta a/a)_o$	0.000404	0.000406

Numerous double crystal rocking curves have been measured for MBE
$Hg_{1-x}Cd_xTe$ films. Generally higher x values have sharper values. For an
(001) orientation on CdTe, x = 0.2, $CuK\alpha$ (004) widths range up from 130".
Coherency seems to be intermediate between the two extreme states but tends
toward the incoherent state. One interesting phenomena was noticed for growth
on slightly misoriented CdTe substrates $(001)_{2°off}$, similar to the GaAs
substrates discussed previously. Oscillation-Weissenberg studies showed no
anomalies but the much higher resolution double crystal studies showed a tilt.
Figure 6 shows the tilt angles from a series of 180° paired (004) curves at
different crystal azimuths. There is a tilt of 65" toward the [310]
direction. The direction is different and the magnitude 300X less than the
GaAs case.
An important application of diffractometry is characterization of
superlattice structures. This topic is discussed in other papers at this
Symposium and in the journal literature [13, 14] so only the principle of the
analysis will be described. X-ray diffraction is based on interference
effects caused by the periodic crystal lattice. The true unit cell of a
superlattice is not cubic but tetragonal with a large c value, ~ 50-100 Å.
Thus the true indices for the reflections usually analyzed are large, e.g.,
(0, 0, 200). Since the repeating layers are so similar, there is a pseudo
cubic unit cell equivalent to the original ones. Convolution of these two
periodicities results in a pattern which can be considered based on the simple
pseudo cubic one with simple indices, e.g., (004). Associated with each
widely space primary reflection is a set of closely spaced satellites which
can be indexed as ± 1, ± 2, etc. Since satellites depend on the differences
between the layers they are generally weak. Any degradation of the
superlattice such as variations in layer thickness and grading at interfaces
will weaken them further. Since superlattices are generally have a small
total thickness for practical reasons, substrate reflections are usually
superimposed. The separation of the substrate peak and the zero order
satellite contains the information on the lattice parameter of the pseudo
cubic unit cell from which the average superlattice composition can be derived
by application of Vegard's law. However, an analysis for the effect of
strains and a possible tilt must be performed as described earlier in this
section. The separation of the satellite peaks is directly related to the
true unit cell which is the period of the superlattice. Analysis of the
satellite intensity distribution gives the contents of the true unit cell,
i.e., the layer distribution. While x-ray analysis can characterize a
superlattice, it may be easier to combine it with other techniques such as
Rutherford Back Scattering and Electron Microscopy.

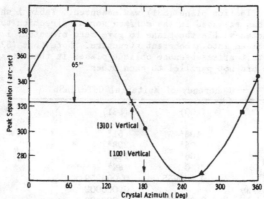

Figure 6. Peak separations as a function of azimuth for MBE $Hg_{0.8}Cd_{0.2}Te$ on CdTe (001)$_{2° off}$. CuKα (004) reflection.

Topography

By topography is meant the imaging of crystal defects and their distribution across the crystal.[10, 15] We generally use the Lang technique in the reflection mode since we are primarily interested in the near surface region. This avoids the tedious task of substrate thinning to permit transmission studies. The divergence is limited by use of a microfocus generator and a slit. Large area imaging is achieved by lateral sample translation to scan the x-ray line. Images are recorded on a nuclear emulsion plate. Exposure times depend on various factors but for the strong diffractors of interest here would be ~ 8 hours/inch of sample. This can be reduced considerably by use of a higher speed (lower resolution) plate. The technique has no inherent magnification which must be achieved by photographic enlargement during the subsequent printing process.

Crystal perfection imaging depends on two mechanisms, orientation contrast and extinction contrast. If a crystal is oriented to be in a reflection position, regions of the crystal which are sufficiently misoriented from the main crystal, such as twins, grains, or bends, will not diffract. Compositional variations with a large lattice parameter, and therefore Bragg angle change could also cause this effect. Such regions of the image will have low darkening due to misorientation contact. As discussed previously, a perfect crystal does not diffract strongly because of extinction. The strain field associated with a crystalline defect reduces extinction and results in much stronger diffraction intensity. Imaging by extinction contrast gives the variation of strain throughout the crystal. Uniform strain has no effect since it would look like a crystal with a different lattice parameter.

For simplicity, plates are printed with reversed contrast so that white regions correspond to high intensity. Figure 7 is a topograph of a (111) CdTe substrate with a CuKα (531) reflection. Better wafers are now available but this was selected to illustrate the capabilities of the technique. Misorientation contrast is shown across the sharp grain boundary running down the middle. The angular difference between the two regions is ~ 0.03° so both regions can be aligned to diffract. A grain at the upper left corner is so misaligned it does not diffract. Twins usually appear as black regions with linear boundaries. There is a gradual strain of the wafer which results in a gradual variation of intensity in the vertical direction. Extinction contrast images the dislocations, some of which bundle in a network of lineage structure slanting from the upper left to lower right. Close examination

Figure 7. Lang x-ray topograph from a (111) CdTe crystal. Cuα_1 (531) reflection. Area shown ~ 1.5 cm x 1 cm.

shows a polygonized pattern of dislocations at subgrain boundaries. These subgrains can also be evident on double crystal rocking curves.

While Lang topographs provide adequate sensitivity for the general status of the IR growth technology (except for MBE CdTe/InSb which show no features) future developments will hopefully require greater sensitivity. Such improvements can be provided by double crystal topography which has been pioneered by Kuriyama [16] for development of crystal growth technology and used by Qadri, et al.[17] for the study of ZnCdTe. Use of a crystal to monochromatize the analyzing x-ray beam results in greatly increased sensitivity to small crystalline strains. Fortunately, the increasing availability of synchrotron radiation facilities provides the high brilliance sources required to make this method a more useful and practical technique.

SUMMARY

Various methods of x-ray diffraction characterization techniques have been described. In conjunction with other methods, they have been invaluable in the development of IR materials growth technology and science. Particularly with regard to the HgCdTe system, by providing critical and timely information as to the factors influencing the properties of this difficult material, they have made an essential contribution to its present status. Undoubtedly they will continue to do so in this exciting and important area.

ACKNOWLEDGEMENT

The MBE films which were used as examples for various techniques were prepared by A. J. Noreika, R.F.C. Farrow, and F. A. Shirland. The technical assistance of C. M. Pedersen is gratefully acknowledged.

REFERENCES

1. S. Wood, J. Greggi, Jr., R.F.C. Farrow, W. J. Takei, F. A. Shirland and
   A. J. Noreika, J. Appl. Phys. 55, 4225 (1984).
2. R.F.C. Farrow, J. Vac. Sci. Tech. A3, 60 (1985).
3. A. J. Noreika, R.F.C. Farrow, F. A. Shirland, W. J. Takei,
   J. Greggi, Jr., S. Wood, and W. J. Choyke, J. Vac. Sci. Tech. A4, 2081
   (1986).
4. E. W. Nuffield, X Ray Diffraction Methods, (John Wiley and Sons, Inc.,
   New York, 1966).
5. B. E. Warren, X-Ray Diffraction, (Addison-Wesley Publishing Co., Inc.,
   Mass., 1969).
6. Brian K. Tanner and D. Keith Bowen, Editors, Characterization of Crystal
   Growth Defects By X Ray Methods, (Plenum Press, New York, 1980).
7. J.H. Hubbell, W. H. McMaster, N. Kerr Del Grande, and J. H. Mallet, in
   International Tables for X-Ray Crystallography, Vol. IV edited by J. A.
   Ibers and W. C. Hamilton (Kynoch Press, Birmingham, England, 1974)
   pp. 47-70.
8. M. J. Buerger, X Ray Crystallography, (John Wiley and Sons, Inc., London,
   1942).
9. W. J. Bartels, J. Vac. Sci. Tech. B1, 338 (1983).
10. B. K. Tanner, X Ray Diffraction Topography (Pergamon Press, Oxford,
    1976).
11. J. Hornstra and W. J. Bartels, J. Cryst. Gr. 44, 513 (1978).
12. W. J. Bartels and W. Nijman, J. Cryst. Gr. 44, 518 (1978).
13. V. S. Speriosu and T. Vreeland, Jr., J. Appl. Phys. 56, 1591 (1984).
14. J. Kervarec, M. Baudet, J. Caulet, P. Auvray, J. Y. Emergy and
    A. Regency, J. Appl. Cryst. 17, 196 (1984).
15. A. R. Lang, in Reference 6, p. 161.
16. W. J. Boettinger, H. E. Burdette, E. N. Farabaugh, and M. Kuriyama in
    Advances in X-Ray Analysis Vol. 20, Edited by H. F. McMurdic,
    C. S. Barrett, J. B. Newkirk and C. O. Rudd (Plenum Publishing Co., New
    York, 1977).
17. Syed B. Qadri, M. Fatemi and J. H. Dinan, Appl. Phys. Lett. 48, 239
    (1986).

ATOMIC STRUCTURE OF HgTe AND CdTe EPITAXIAL LAYERS GROWN BY MBE ON GaAs SUBSTRATES

F. A. Ponce and G. B. Anderson, Xerox Palo Alto Research Center, Palo Alto, CA 94304.

J. M. Ballingall, McDonnel Douglas Research Laboratories, St. Louis, Missouri 63166*

ABSTRACT

Using high resolution transmission electron microscopy (HRTEM) it is possible to directly image the projected structure of semiconductors with point resolutions at the atomic level. This technique has been applied to the study of interface and defect structures associated with molecular beam epitaxial (MBE) growth of HgCdTe layers on GaAs substrates. The structure of the CdTe/GaAs interfaces is described for (100) and (111) epitaxy. From the atomic structure, a model for the early stages of epitaxial growth is presented. The structure of HgTe-CdTe superlattices is discussed from the HRTEM point of view.

INTRODUCTION

The relatively successful growth of high quality, single-crystal infrared structures based on $Hg_xCd_{1-x}Te$ epitaxial layers on GaAs has created a large amount of interest in this system. In spite of the large lattice mismatch (14.6%) between CdTe and GaAs, single-crystal CdTe epitaxial layers have been grown in both (100) and (111) orientations on (100) GaAs substrates by molecular beam epitaxy [1-4]. The understanding of the growth mechanism and crystallographic defect generation requires an insight into the atomic arrangement of these structures.

The atomic arrangement in heteroepitaxial thin films and their interfaces can be studied by high resolution transmission electron microscopy [4,5]. With recent advances in electron optics, this technique can resolve the lattice structure with point resolution values below 0.20nm, which is of the same range as the interatomic distances in semiconductors [6].

In this paper, we present a study of the structural aspects of CdTe epitaxy on GaAs, using high resolution transmission electron microscopy. A description of the atomic structure of interfaces in the various epitaxial configurations is presented. Based on some of these observations, a model for the early stages of MBE growth is proposed.

GROWTH OF CdTe ON GaAs BY MBE

The growth of CdTe thin films on GaAs (100) substrates is well described elsewhere [1-3]. The materials described in this paper, were grown in a commercial molecular beam epitaxy system (Vacuum Generators V80H-MCT) using (100) GaAs substrates. The substrates were cleaned and etched using standard procedures [3]. The surface oxide was thermally desorbed *in situ* by heating, without an arsenic flux, up to 580°C. Desorption is verified by reflection high-energy electron diffraction (RHEED), which shows narrow, uniformly intense streaks. CdTe growth is accomplished at a substrate temperature of 300°C. Two types of epitaxial relationships, namely the (111) and (100) oriented CdTe layers, can be produced by varying the time at which the GaAs substrate is held at 580°C for oxide desorption, with longer times leading to (111) epitaxy. A single CdTe effusive source is used for the CdTe layers.

Mat. Res. Soc. Symp. Proc. Vol. 90. ©1987 Materials Research Society

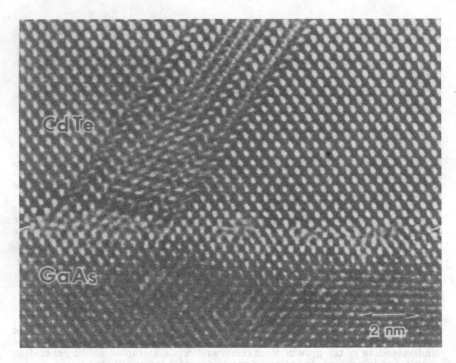

Figure 1. Cross-section lattice image of the <100>CdTe / <100>GaAs interface region. Notice the twinned region in the CdTe and the misfit dislocations present at the interface (noted by arrows).

PARALLEL EPITAXY: <100> CdTe / <100>GaAs

The most commonly achieved epitaxial configuration is the one in which both lattices are oriented in parallel along the [100] direction. Good quality, single-crystal films are obtained. Figure 1 is a cross section lattice image along a <110> projection, showing typical features found at the interface. The most common faults in the film are either stacking faults or microtwins. The interface between CdTe film and the GaAs substrate is locally coherent, with continuity of planes across the interface. Periodic arrays of misfit dislocations are observed at the interface, with a separation of about 3.1nm between dislocations, which corresponds exactly to the lattice mismatch between CdTe ($a_0 = 0.6481$nm) and GaAs ($a_0 = 0.5654$nm). The dislocations lie along <110> and <-110> directions, producing a two-dimensional periodic network along the interface. The dislocations are found to be at the exact CdTe/GaAs interface. This is probably due to the large difference in cohesive energy per bond between CdTe (1.03 eV) and GaAs (1.63eV) [7]; which happen to be the smallest and highest values, respectively, found for binary compounds with the zinc-blende structure. Figure 2 shows the two-dimensional dislocation network along the interface. It is a plane-view lattice image along a <100> projection. The structure of a single misfit dilocation is shown in Figure 3. The Burgers' vector is $\frac{1}{2}a_0$<110>, and corresponds to an extra {220} crystallographic plane in the GaAs side. The structure of these misfit dislocations is similar to the Lomer dislocations found in bulk GaAs [8].

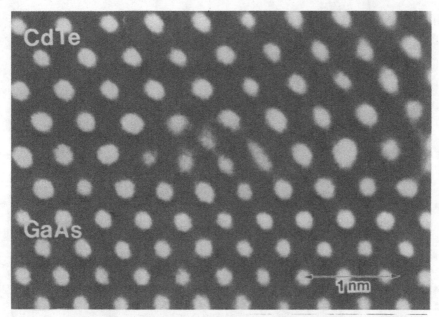

Figure 2. High resolution, phase contrast image of a misfit dislocation. Its Burgers' vector is $\frac{1}{2}a_0<110>$, and can be seen as two extra {111} planes in the GaAs.

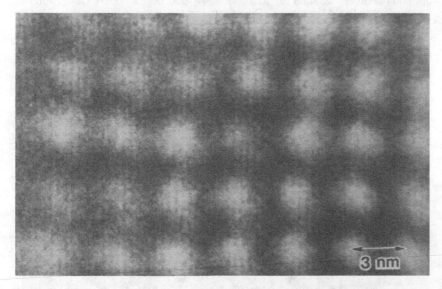

Figure 3. Plane-view lattice image of the interface region. The small fringes are the {220} planes with 0.40nm of spacing. The large dark-band network is the dislocation arrangement at the interface, with a period of 3.1nm.

Figure 4. V-shaped pits in the GaAs substrate. <100>CdTe / <100>GaAs epitaxy. Lattice viewed in <110> projection. Pit facets are {111}A GaAs planes.

The CdTe interface is typically flat following a {100} crystallographic plane, with very few steps. Using the 14% difference in interplanar spacing, the exact position of the interface can be determined by direct measurement of the {200} planes parallel to the interface.

Under certain conditions, the interface region shows pits in the GaAs surface. The pits shown in Figure 4 have a V-shape and are bound by {111} crystallographic planes. We have determined the lattice polarity by etching the substrate using a bromine-methanol solution [9], and concluded that the facets are Ga-terminated, i.e. correspond to {111}A planes. The origin and relevance of these pits is discussed later.

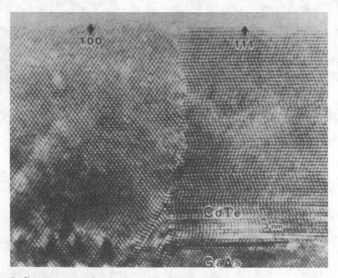

Figure 5. Coexistence of <111> and <100> epitaxy.

NON PARALLEL EPITAXY: <111>CdTe/<100>GaAs

Under certain conditions, CdTe can grow in both the <111> and <100> orientations. This is associated with the substrate surface preparation and growth conditions, as already mentioned above. Figure 5 shows a region where both of these epitaxial relationships are observed. <100> and <111> epitaxy is observed at the left and right hand sides, respectively. They are bound by an incoherent grain boundary. <100> epitaxial regions usually contain few defects, and these are usually single stacking faults. Regions with the <111> epitaxial orientation, on the other hand, are usually highly faulted close to the substrate interface. The majority of these defects are stacking faults and twins parallel to the growth direction. The density of twin boundaries diminish rapidly with distance down to values similar to those of the <100> case.

The <110>CdTe // <110>GaAs Projection

A complete <111> epitaxial layer is typically associated with the presence of small pits at the GaAs surface. Figure 6 shows electron diffraction patterns taken at the interface and in the immediate neighborhood in the GaAs and CdTe regions.

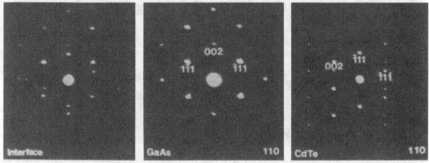

Figure 6. Electron diffraction pattern of the interface region for <111>CdTe / <100>GaAs epitaxy, along the <110>GaAs projection. The respective indices for GaAs and CdTe are shown.

Figure 7. Lattice image of the interface region under the same conditions as the diffraction pattern in Fig. 6. Notice flat-bottomed pit in GaAs and the grain on top with a <100> epitaxy. Outside this grain the CdTe has a <111> epitaxial orientation, and is being observed along a <110> projection.

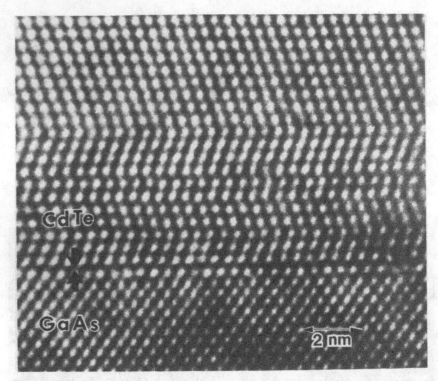

Figure 8. Lattice image of the interface region for the <111>CdTe / <100>GaAs epitaxial case. Notice the good lattice match (≦ 0.7% mismatch), and the twins parallel to the growth direction.

A corresponding image is shown in Figure 7. Here the lattices are viewed in the <110> projection, which is common to both. The key feature of this configuration is the shape of the pit and the CdTe grain that has grown on it. The pits, when viewed in this projection, have the shape of an inverted trapezoid with a flat bottom. From polarity determinations by etching (as described before), we have determined that in this configuration the {111}A faces point down. The CdTe grain on top of the pit has a <100> growth orientation; and the regions outside the grain have a <111> epitaxial configuration. In the <111> region, the defect structure is primarily stacking faults parallel to the growth direction. The density of these defects falls down sharply for thickness larger than the height of the <100> grains.

The atomic arrangement at the interface corresponding to this configuration is shown in Figure 8. It is interesting to note that the lattice mismatch is very small in this projection. The match is between the {220}GaAs and the {112} CdTe crystalline planes, with spacings 0.3998nm and 0.2646nm, respectively. The match occurs since $2\,d_{GaAs,220} \sim 3\,d_{CdTe,112}$; where the difference is ~ 0.7%. Misfit dislocations associated with this difference would occur every 109nm. The interface is planar with average separation of steps of about 20nm. Because of the presence of steps, misfit dislocations have not been observed in this projection. It is perhaps because of these characteristics of the interface, that the <111> epitaxy is favored under certain conditions.

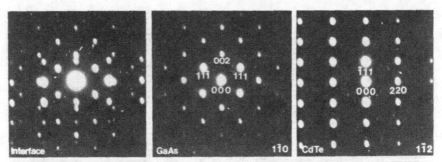

Figure 9. Electron diffraction pattern of the interface region for <111>CdTe / <100>GaAs epitaxy, along the <1-1 0>GaAs projection. The respective indices for GaAs and CdTe are shown.

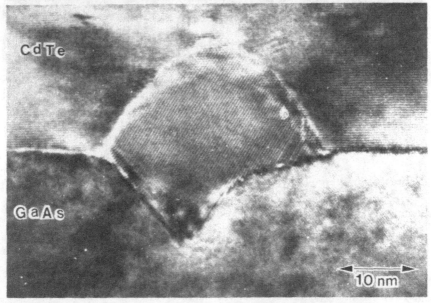

Figure 10. Lattice image of the interface region under same conditions as the diffraction pattern in Fig. 9. Notice V-shaped pit in GaAs and the grain on top with a <100> epitaxy. Outside this grain the CdTe has a <111> epitaxial orientation, and is being observed along a <112> projection.

The <112>CdTe // <1-1 0>GaAs Projection

The structure is quite different when viewed along a <1-1 0> projection (normal to the previously discussed configuration). This is due to the absence of 4-fold symmetry along the <100> GaAs axis. If in the <110> projection, {111}A planes face in the <100> direction (upwards); then in the <1-1 0> projection, normal to the previous one, it is the {111}B planes that face upwards. This leads to well known chemical etching anisotropy, which is associated with the pits formed during thermal annealing in the MBE chamber. The electron diffraction patterns corresponding to the <1-1 0>GaAs projection are shown in Figure 9.

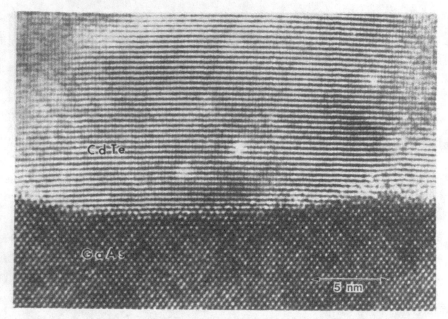

Figure 11. Lattice image of the interface region for the <111>CdTe / <100>GaAs epitaxial case. The interface is not planar. This view is equivalent to the one shown in figure 8, but rotated 90° along the axis perpendicular to the interface.

Notice that in this projection the CdTe is viewed along the <1-1 2> axis. The pits have a V-shape as shown in Figure 10, and the <100> grain is always present. The CdTe/GaAs interface in this projection is shown in Figure 11; and has an undulated appearance with a large density of steps.

A MODEL FOR EARLY STAGES OF GROWTH OF <111>CdTe

The existence of pits and their connection with <111> epitaxy is an interesting finding of this study. Although a direct cause-effect detemination has not been performed, the following considerations appear likely in view of our observations. (1) Pits form during the thermal-desorption step in the MBE chamber. A good control of the substrate surface temperature is usually very difficult, and overheating the surface beyond the nominal temperature reading should not be ruled out. In the absence of an As background pressure, it is expected that As preferentially desorbs, leaving a Ga-rich surface. In abundance, the excess Ga atoms cluster and form droplets which minimize their surface energy by the formation of pits into the GaAs substrate. (2) For temperatures above the congruent evaporation value ($T_c \sim 600°C$), the surface tends to be Ga-rich, whereas for T below T_c, the surface tends to be As-rich. Therefore, it is likely that the excess Ga will sublime during the cooling process prior to CdTe deposition, leaving an empty pit. (3) Under low growth rates, the empty pit acts as a nucleation site for <100>CdTe. After the pit is filled, the <100>CdTe grain acts as a nucleus for lateral growth of the {111} layer.

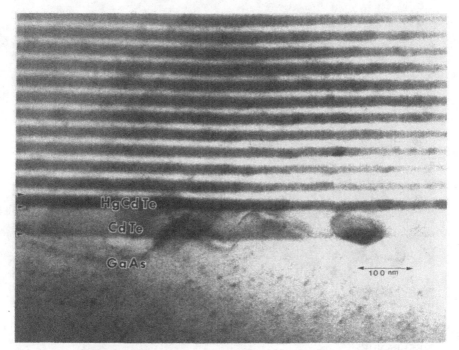

Figure 12.   Two-beam TEM image of a HgTe-CdTe superlattice, its CdTe buffer layer and the GaAs substrate.

GROWTH AND STRUCTURE OF CdTe/HgTe SUPERLATTICES

The HgTe-CdTe superlattice has been proposed as an alternative to $Hg_{1-x}Cd_xTe$ for infrared detectors [10]. The advantage of a superlattice structure is that the band gap and the effective mass can be nearly independently controlled by varying the individual HgTe and CdTe layer thicknesses, respectively. Also, higher absorption coefficients and lower interband tunneling are achieved by the larger value of the effective mass in the superlattice when compared with the alloy case.

The growth of superlattice structures are achieved by growing the CdTe layers from a single CdTe effusive source. Separate Hg and Te elemental sources are used for the HgTe layers. The substrate is held at a temperature of 170°C, and the growth rates are about 1micron/hour. The superlattices are capped with a CdTe layer of about 100nm in thickness [11]. Figure 12 shows a HgTe-CdTe superlattice grown on a CdTe/GaAs substrate. Other than for the typical defect structure of the CdTe/GaAs system, the superlattices are essential flat and abrupt within the resolution of the technique. The lattice image shown in Figure 13 seem to indicate some degree of diffussiveness at the interfaces. This is an artifact associated with specimen preparation. HRTEM requires samples which are about 5-10nm in thickness. This is achieved by Argon ion milling. It has been observed that ion milling tends to smear some thin film structures. Either low-magnification TEM, as in Fig. 11, or X-ray diffraction studies are much more reliable for this type of determination [12].

208

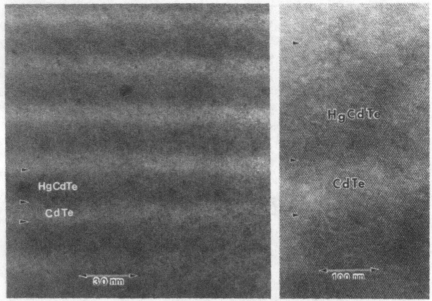

Figure 13.   High resolution TEM images of HgTe-CdTe superlattice.

REFERENCES

(*) Now at General Electric Company, Syracuse, New York.

1. H. A. Mar, K. T. Chee and N. Salansky, Appl. Phys. Lett. 44, 237 (1984)

2. R. N. Bicknell, R. W. Yanka, N. C. Giles, J. F. Schetzina, T. J. Magee, C. Leung, and H. Kawayoshi, Appl. Phys. Lett. 44, 313 (1984).

3. J. M. Ballingall, W. J. Takei, and B. J. Feldman, Appl. Phys. Lett. 47, 599 (1985).

4. F. A. Ponce, G.B. Anderson and J. M. Ballingall, Surface Science 168, 564 (1986).

5. N. Otsuka, L. A. Kolodziejski, R. L. Gunshor, S. Datta, R. N. Bicknell and J. F. Schetzina, Appl. Phys. Lett. 46, 860 (1985).

6. F. A. Ponce and M. A. O'Keefe, in Proceedings of the 44th Annual Meeting of the Electron Microscopy Society of America, edited by G. W. Bailey (San Francisco Press, San Francisco, 1986), p. 522.

7. W. A. Harrison, Electronic Structure and the Properties of Solids (Freeman, San Francisco,1980), p. 176.

8. F. A. Ponce, G. B. Anderson, P. Haasen and H. G. Brion, Materials Science Forum 10-12, 775 (1986).

9. Y. Tarui, Y. Komiya and Y. Harada, J. Electrochem. Soc. 118, 118 (1971).

10. J. N. Schulman and T. C. McGill, Phys. Rev. B23, 4149 (1981).

11. M. L. Wroge, D. J. Leopold, J. M. Ballingall, D. J. Peterman, B. J. Morris, J. G. Broerman, F. A. Ponce and G. B. Anderson, J. Vac. Sci. Technol. B4, 1306 (1986).

12. J. M. Ballingall, D. J. Leopold, M. L. Wroge, D. J. Peterman, B. J. Morris and J. G. Broerman, Appl. Phys. Lett. 49, 871 (1986).

SUBSURFACE MICRO-LATTICE STRAIN MAPPING

T.S. Ananthanarayanan, R.G. Rosemeier, W.E. Mayo* and P. Becla**

Brimrose Corporation of America
7720 Belair Road
Baltimore, MD 21236
(301) 668-5800

* Rutgers University
Piscataway, NJ 08854

** MIT, Francis Bitter National Magnet Lab
Cambridge, MA 02139

Synopsis

Defect morphology and distribution up to depths of 20um have been shown to be critical to device performance in micro-electronic applications. A unique and novel x-ray diffraction method called DARC (Digital Automated Rocking Curve) topography has been effectively utilized to map crystalline micro-lattice strains in various substrates and epitaxial films. The spatial resolution of this technique is in the the order of 100um and the analysis time for a 2cm^2 area is about 10 secs. DARC topography incorporates state-of-the-art 1-dimensional and 2-dimensional X-ray detectors to modify a conventional Double Crystal Diffractometer to obtain color x-ray rocking curve topographs.

This technique, being non-destructive and non-intrusive in nature, is an invaluable tool in materials' quality control for IR detector fabrication. The DARC topographs clearly delineate areas of micro-plastic strain inhomogeniety. Materials analyzed using this technique include HgMnTe, HgCdTe, BaF$_2$, PbSe, PbS both substrates and epitaxial films. By varying the incident x-ray beam wavelength the depth of penetration can be adjusted from a 1-2 micron up to 15-20um. This can easily be achieved in a synchrotron.

Background : II-VI Characterization and Use

II-VI compound semiconductors form a major family of compounds for advanced IR device applications. These include both active (laser generators) and passive (detector) devices. The compounds of interest may vary anywhere from a binary alloys to complicated multi-constituent epitaxies and heterostructures. The metallurgy involved in the crystallization of many of these multi-constituent, multi-layer materials is extremely complicated. Consequently, these materials tend to have higher defect structures than a single element material. Nevertheless, the consistency of any microscopic (electrical and mechanical) property is directly a function of the microlattice strain state. Hence there is increasing need for a quantitative, non-destructive microstructural characterization technique.

Mat. Res. Soc. Symp. Proc. Vol. 90. ⸱ 1987 Materials Research Society

Over the past three decades X-ray diffraction and topography techniques have emerged as sturdy tools for rapid quality control of crystalline materials(1). X-ray reflection topography has regained prominence in the recent past(2). This technique lacked quantification. Several attempts were made to digitize the photographic records of the diffraction events. Although these efforts yielded excellent results, they were unsuitable for rapid surface characterization required under production environments. This limitation affects the data collection abilities for research environments as well.

The use of electronic X-ray detectors is becoming increasingly popular in the field of X-ray diffraction. These detectors include: point counters, linear array detectors and 2-dimensional array detectors. The principal detector used in this study is a 2-dimensional X-ray detector. Several studies(3-4) have clearly established the use of 2-dimensional X-ray detectors for real time topographic inspection of various metallographic microstructures(5-11). The current study utilizes 2-dimensional X-ray detectors to quantitatively map micro-lattice strain state in II-VI materials.

EXPERIMENTAL METHODOLOGY

The technique used for surface characterization is known as X-ray rocking curve topography(6). The diffraction geometry used is shown in Figure 1.

The X-ray beam incident on the sample was not monochromated hence it contained $K\alpha_1$, $K\alpha_2$, $K\beta$ and some amount of Brehmstralung. The presence of other wavelengths cause multiple reflections ($K\alpha_1$, $K\alpha_2$, $K\beta$) and background noise. The signal to noise ratio and geometric divergence parameters diminish dramatically with increasing monochromation. Multiple crystal diffractometers can be used to achieve this(12).

The current study utilized no monochromators at all(13). Without monochromators the angular resolution of the micro-lattice misorientations achievable is limited. The specimen is then rocked (stepped) through its diffraction domain and every intersection of the Ewald sphere and the reciprocal lattice spot is recorded digitally. By tracking the X-ray intensity of each pixel, it is possible to record integrated pixel intensity, pixel rocking curve half width and pixel Bragg peak shift. Using these measurements the lattice parameters, dislocation density and epi-film thickness can be individually computed.

Figure 2 depicts the schematics of the rocking curve topography system and its configuration.

Figure 3 depicts the typical white beam rocking curve experiment performed on a VPE grown CdMnHgTe Epi/CdMnTe substrate. Angle of incidence was about 3-5o and the reflection topograph was obtained at about 54° two-theta. The reflection was found to be the 001 type. The angular increment between each frame was 0.1. Progressive frames begin left top corner and proceed right through the first row on to the second row (beginning left hand side again). The black and white images are translated to pseudo-color images and each pixel is tracked for intensity. Rocking curve half width for each pixel is obtained and displayed.

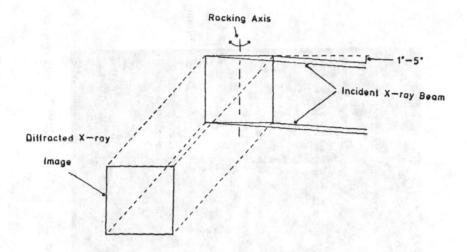

Fig. 1 DIFFRACTION GEOMETRY FOR WHITE BEAM ROCKING CURVE TOPOGRAPHY

DIGITAL AUTOMATED ROCKING CURVE TOPOGRAPHY

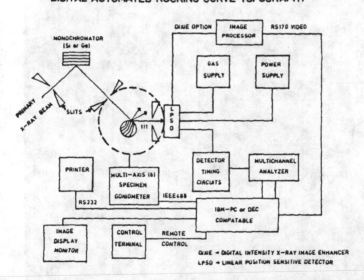

Fig. 2 Schematic of the Rocking Curve Topography System.

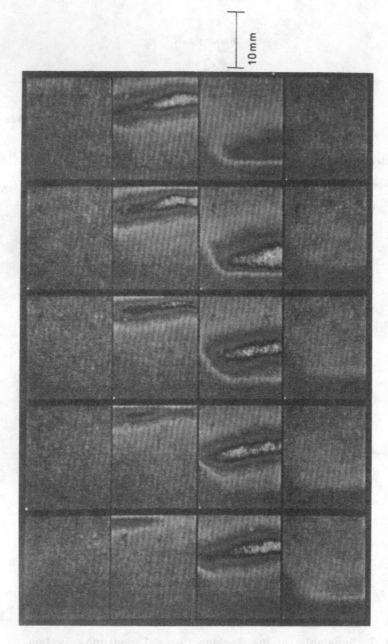

10mm

Fig. 3 Depicts Rocking Curve Experiment on a VPE Grown CdMnHgTe Epi/CdMnTe Substrate.

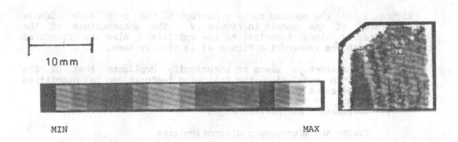

10mm

MIN   MAX

Fig. 4 Rocking Curve Topography of CdMnHgTe Epi/CdMnTe
Substrate, (100) Type Reflection.

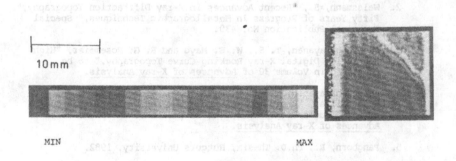

10mm

MIN   MAX

Fig. 5 Rocking Curve Topography of CdMnHgTe Epi/CdMnTe
Substrate, (100) Type Reflection, on the Back Side.

214

Figure 4 is the rocking curve topography of the CdMnHgTe Epi/CdMnTe substrate. This topograph clearly shows subgrain structure in the VPE film. This film was about 250 um thick. The damaged area towards the edge of the speciment is also evident.

Figure 5 is the rocking curve topograph of the back side (CdMnTe substrate) of the sample in Figure 4. The substructure of the substrate is almost identical to the epi film. Also the fractured piece (which was removed for Figure 4) is clearly seen.

The epi substructure seems to identically duplicate that of its substrate. In conclusion the following features can be quantified using rocking curve topographs.

1. Deformation at specimen edges.

2. Cracks and microscopic discontinuities.

3. Subgrain structure.

ACKNOWLEDGEMENTS

The authors wish to thank Alfred Wiltrout, Dawn Roche and Kathy Aversa for their assistance in compiling this manuscript. Special thanks to Dr. S. B. Trivedi for helpful discussions and Dr. P. J. Coyne for developing the software to implement digital rocking curve topography. This research effort was conducted under the sponsorship of DARPA, ONR and SPAWAR of the U.S. Dept. of Defense.

REFERENCES

1. Guinier, A., "X-ray Diffraction," W. H. Freeman & Co., 1963 ed., San Francisco.

2. Weissmann, S., "Recent Advances in X-ray Diffraction Topography," Fifty Years of Progress in Metallographic Techniques," Special Technical Publication No. 430.

3. Ananthanarayanan, T. S., W. E. Mayo and R. G. Rosemeier, "High Resolution Digital X-ray Rocking Curve Topography," to be published in Volume 30 of Advances of X-ray Analysis.

4. Ananthanarayanan, T. S., W. E. Mayo, R. G. Rosemeier and R. S. Miller, "Rapid Non-destructive X-ray Characterization of Solid Fuels/Propellants," to be published in Volume 30 of Advances of X-ray Analysis.

5. Pangborn, R., Ph.D. Thesis, Rutgers University, 1982.

6. Rosemeier, R. G., T. S. Ananthanarayanan and W. E. Mayo, "Feasibility Study on Real Time X-ray Topography - Phase I Final Report," DARPA, September 1984 - February 1985.

7. Mayo, W. E., R. Yazici, T. Takemoto and S. Weissmann, XII Congress of Int. Union of Crystallography, 1981, Ottawa, Canada.

8. Mayo, W. E., Ph.D. Thesis, Rutgers University, 1982.

9. Yazici, R., Ph.D. Thesis, Rutgers University, 1982.

10. Liu, H. Y., Ph.D. Thesis, Rutgers University, 1982.

11. Liu, H. Y., W. E. Mayo and S. Weissmann, Mat. Sci. and Eng., 63 (1984) 81.

12. Qadri, S. B., and J. H. Dinan, J. Appl. Phys. "X-ray Determination of Dislocation Density in Epitaxial ZnCdTe," p. 1066 (1985).

13. Ananthanarayanan, T. S., R. G. Rosemeier, W. E. Mayo and S. Sacks, "Novel Non-destructive X-ray Technique for Near Real Time Defect Mapping," submitted at the 2nd International Symposium on the Nondestructive Characterization of Materials, Montreal, Canada (1986).

STUDY OF HgTe-CdTe MULTILAYER STRUCTURES
BY TRANSMISSION ELECTRON MICROSCOPY

N. OTSUKA and Y. E. IHM
School of Materials Engineering
Purdue University, W. Lafayette, IN 47907

K. A. HARRIS, J. W. COOK, Jr., and J. F. SCHETZINA
Department of Physics, N. Carolina State University
Raleigh, NC 27695-8202

ABSTRACT

A transmission electron microscope study of HgTe-CdTe multilayer structures grown by molecular beam epitaxy (MBE) on (100) $Cd_{1-x}Zn_xTe$ is presented. Both cross-sectional and plain-view observations show highly regular structures of superlattices and tunnel structures. Dislocation densities estimated by plan-view observations are of the order of 10^4 cm^{-2} in these multilayer structures. A quantitative characterization of interface sharpness of superlattices has been carried out by intensity analysis of satellite spots in electron diffraction patterns. It is shown that interfaces in these superlattices are highly abrupt with a width of one or two monolayers. These observations suggest the effectiveness of the use of lattice-matched substrates to grow high quality HgTe-CdTe multilayer structures.

1. Introduction

Since the first growth of HgTe-CdTe superlattices [1], molecular beam epitaxy (MBE) has become a leading growth technique of Hg-based multilayer structures. A variety of novel multilayer structures of Hg-based materials have been grown by this technique, stimulating intensive experimental and theoretical studies for the development of optoelectronic devices of these structures. At present, two materials, GaAs and $Cd_{1-x}Zn_xTe$ (CdTe), are primarily used as substrates for the growth of Hg-based materials by MBE. GaAs is widely used because of the availability of high quality crystalline surfaces and the possibility of the monolithically integrating technology of Hg-based structures with that of GaAs. Recently, however, $Cd_{1-x}Zn_xTe$ is increasingly becoming an attractive substrate because of its lattice-matching with HgTe, and many high quality HgTe-CdTe multilayer structures have been successfully grown on this substrate [2,3]. In this paper, we present a transmission electron microscope (TEM) study of HgTe-CdTe multilayer structures grown on $Cd_{1-x}Zn_xTe$ substrates by MBE. Two aspects of microstructures, dislocation densities and interface sharpness have been primarily investigated in the study.

2. Experiment

Hg-based multilayer structures were grown in a MBE system specially designed for the growth of these materials. Film growth was carried out at 175°C on (100) CdTe or (100)

$Cd_{1-x}Zn_xTe$ substrates with a deposition rate of 1-2 A/sec.
Detailed growth procedures are described in earlier reports
[2,4]. Cross-sectional and plan-view specimens for TEM studies
were prepared by mechanical grinding followed by Ar ion
milling. A JEM 200CX electron microscope with a side-entry
goniometer was used for TEM observations.

3. Results and Discussion

Figure 1 (a) and (b) are bright field images of HgTe-
CdTe superlattices. The area shown in Fig.1(a) represents
a typical microstructure observed in these superlattices.
A highly regular superlattice layers with sharp interfaces
between HgTe and CdTe layers are seen in the image. No
threading dislocations nor misfit dislocations are found
in the observed areas of cross-sectional specimens. However,
dislocation loops are observed in a few areas of the superlattices. Figure 1(b) shows such an area. From the present
observations, it is difficult to identify the origin of
these dislocation loops; these defects may have been caused
by the Ar ion bombardment or may have formed during the
growth of these superlattices.

Figure 2(a) is a bright field image of a HgTe-CdTe
tunnel structure. In the image, CdTe layers appear as two
bright bands in the matrix of HgTe. Dark wavy lines are interference contours resulting from the bending of the thin area
which is caused by the epoxy glue used for the cross-sectional
specimen preparation. No defects such as dislocations and
precipitates are present in this area. Figure 2(b), which is
a bright field image of a tunnel structure grown under a non-
ideal condition, shows small Te precipitates along the tunnel
structure. These precipitates appear to have formed at the
interfaces between CdTe and HgTe layers and caused a distortion of the tunnel structure.

In order to estimate dislocation densities in these multilayer structures, plan-view specimens were examined. Figure 3
is a bright field image of a pla-view specimen of the tunnel
structure shown in Fig.2(a). For each plan-view specimen,
an area of 5,000~10,000 μm^2 was observed. These observations
show that dislocation densities in these samples are of the
order of $10^4 cm^{-2}$. This value is lower than those of typical
bulk crystals, indicating the effectiveness of the low-temperature growth of MBE to obtain epilayers with reduced dislocation
densities.

For the characterization of interface sharpness, selected
area diffraction patterns were taken from two HgTe-CdTe superlattices, BMCCT-1 and BMCCT-2. Two areas of each superlattice,
one near the CdTe buffer layer and the other near the free
surface, were examined. An interface width of each area was
estimated by analyzing intensities of satellite spots around
the 200 reflection. Figure 4(a) and (b) are intensity profiles
of the 200 reflection and satellite spots which were observed
from areas near the CdTe buffer layer and near the free surface of sample BMCCT-1, respectively. For the analysis of the
interface width, the composition profile at the interface is

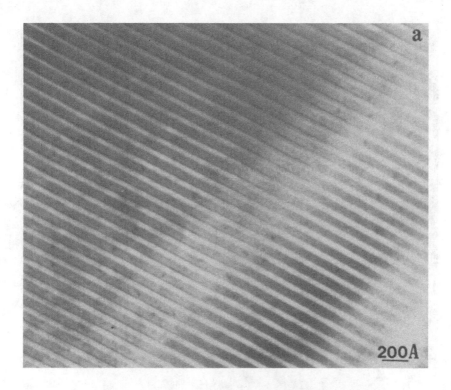

Fig. 1. Bright field images of HgTe-CdTe superlattices. The area shown in (b) has two small dislocation loops.

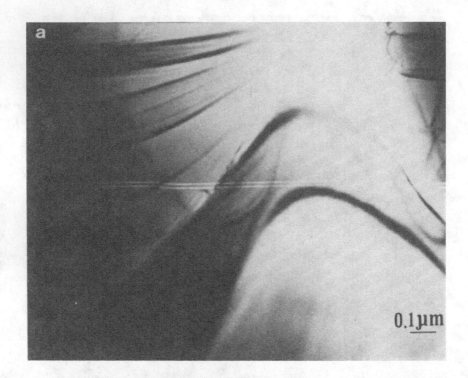

Fig. 2. Bright field images of HgTe-CdTe tunnel structures. The sample shown in (b) was grown under a non-ideal condition, resulting in the precipitation of metallic Te.

Fig. 3. Plan-view bright field image of a HgTe-CdTe tunnel structure. The image show a wide dislocation-free area.

Table I. Layer thicknesses and interface widths of HgTe-CdTe superlattices.

| | Layer thickness (Å) | | Interface |
	HgTe	CdTe	Width 2b(Å)
(1) BMCCT-1 Superlattice			
Near free surface	22	54	4.5 ± 2.0
Near buffer	77	29	5.0 ± 2.0
(2) BMCCT-2 Superlattice			
Near free surface	53	122	4.5 ± 3.0
Near buffer	53	122	4.0 ± 3.0

222

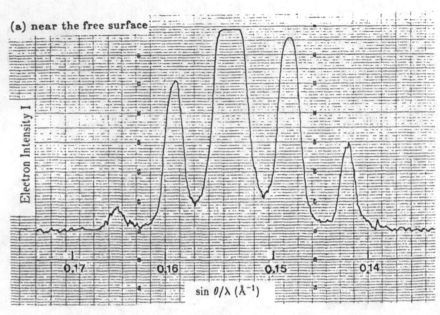

(a) near the free surface

Electron Intensity I

0.17   0.16   0.15   0.14

$\sin \theta / \lambda$ (Å^{-1})

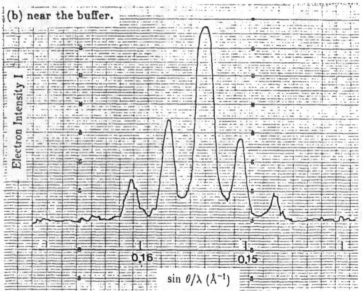

(b) near the buffer.

Electron Intensity I

0.16   0.15

$\sin \theta / \lambda$ (Å^{-1})

Fig. 4. Electron intensity profiles of the 200 reflection
and satellite spots observed from (a) an area near
the free surface and (b) an area near the CdTe buffer
layer of the sample BMCCT-1. The intensity is plotted
on a linear scale.

assumed to have the form of the error function. Details of the analysis as well as experimental procedures are reported elsewhere[5].

Results of the analysis are listed in Table I. As seen in the table, the interface width in all four areas studied is about 5A. This width corresponds to one or two monolayers of HgTe and CdTe crystals, the spacing of which is 3.24A. No appreciable difference in the interface width can be seen between areas near the buffer layer and those near the free surface, suggesting that the interface width in these samples is determined by the roughness of the growth front of an epilayer. With total thicknesses of superlattices and the deposition rate, it can be shown that the diffusion coefficient for inter-mixing is not greater than 5×10^{-20} cm^{2}sec^{-1}. This value is roughly two order of magnitudes smaller than that estimated by the X-ray diffraction study for this temperature range[6]. This discrepancy suggests that the rate of intermixing diffusion of HgTe-CdTe multilayer structures is highly sensitive to the structural quality of given samples.

In summary, this paper has presented a TEM study of HgTe-CdTe multilayer structures grown on (100) Cd$_{1-x}$Zn$_x$Te substrates by MBE. Both TEM images and electron diffraction patterns show low dislocation densities and highly abrupt interfaces in the multilayer structures. From these observations, it is suggested that the use of lattice-matched substrates may be one of key factors for the growth of Hg-based multilayer structures which can be employed for device applications.

Acknowledgements

Work at Purdue University was supported by ONR/DARPA contract N00014-86-K-0378. Work at North Carolina State University was supported by DARPA/ARO contract DAAG29-83-K-0201.

References

1. J. P. Faurie, A. Million, and J. Piaguet, Appl. Phys. Lett., 41, 713 (1982).

2. K. A. Harris, S. Hwang, D. K. Blanks, J. W. Cook, Jr., J. F. Schetzina, N. Otsuka, J. P. Baukus, and A. T. Hunter, Appl. Phys. Lett., 48, 396 (1986).

3. J. T. Cheung, G. Niizawa, J. Moyle, N. P. Ong, B. M. Paine, and T. Vreeland, Jr., J. Vac. Sci. Technol., A4, 2086 (1986).

4. K. A. Harris, S. Hwang, D. K. Blanks, J. W. Cook, Jr., J. F. Schetzina, and N. Otsuka, J. Vac. Sci. Technol., A4, 2061 (1986).

5. N. Otsuka, Y. E. Ihm, K. A. Harris, J. W. Cook, Jr., and J. F. Schetzina, (to be published).

6. D. K. Arch, J. L. Staudenmann, and J. P. Faurie, Appl. Phys. Lett., 48, 1588 (1986).

DYNAMICAL X-RAY ROCKING CURVE SIMULATIONS OF InGaAsP/InP DOUBLE HETEROSTRUCTURES USING ABELES' MATRIX METHOD

A. T. MACRANDER, B. M. GLASGOW, E. R. MINAMI, R. F. KARLICEK,
D. L. MITCHAM, V. G. RIGGS, D. W. BERREMAN, AND W. D. JOHNSTON, JR.
AT&T Bell Laboratories, Murray Hill, New Jersey 07974

ABSTRACT

Simulated rocking curves for a light-emitting diode structure are presented. Results for a structure containing uniform layers are compared to rocking curve data for a wafer grown by vapor phase epitaxy (VPE), and we conclude from the comparison that the VPE wafer closely approached the hypothetical ideal assumed in the simulations. Simulations illustrating difficulties in analyses and the effects of a graded active layer are also presented.

I. INTRODUCTION

Double heterostructures based on the quaternary alloy $In_{1-x}Ga_xAs_{1-y}P_y$ and on InP occur in device structures for the sources and detectors needed in optical fiber communications systems operating in the near infrared (1.3 and 1.5 μm). A high degree of structural perfection of the base material used to fabricate them is highly desirable since highly reliable devices are needed. We have been engaged in the evaluation of such base material via x-ray diffraction. In this paper both simulated and measured rocking curve data are presented for a light-emitting-diode (LED) base structure. LED's can provide an attractive alternative to laser based optical fiber communications systems[1].

Double crystal diffractometry (DCD) is a useful technique for characterizing the structural perfection of double heterostructure (DH) wafers. The structural perfection, mismatch, and thickness of the individual layers are obtainable from rocking curves. Because DCD is nondestructive and rapid it is an ideal tool for monitoring epitaxial crystal growth[2]. The technique is also well suited to measuring the uniformity of large area wafers.

A great deal can be learned by simulating double crystal rocking curves using dynamical diffraction theory. This can be done by assuming either an ideal structure or one with layers having a graded lattice parameter[3]. We present here our results for simulations of base structures for LED's not only for purely hypothetical cases, but also for comparison to rocking curve data obtained for a 2 inch diameter wafer grown by hydride-VPE[4]. Simulated rocking curves were obtained by convoluting the intrinsic reflectivities with those of the first crystal for both polarizations of the x-rays. For the reflecting power of the symmetric (400) reflection of the first crystal we used the solutions of Cole and Stemple[5] applied to InP[6]. Rocking curve data were obtained by using a <100> oriented InP first crystal[7] set to the (400) reflection, i. e., a wavelength dispersion free case.

II. HYPOTHETICAL STRUCTURES

The basic structure which we investigated is shown in Fig. 1. The quaternary active layer composition is determined both by the desired LED operating wavelength of $\lambda = 1.3\mu$m and the requirement that it be lattice matched to the substrate. For the quaternary capping layer the alloy composition was set for $\lambda = 1.1$ μm. The exact alloy compositions were calculated using Vegard's law[8] in a form amended to include tetragonal distortion.

In some instances the presence of a capping layer complicates the interpretation of a rocking curve, and it is useful to first examine rocking curves without such a layer. An example is shown in Fig. 2. The active layer Bragg peak is clearly discernible. The InP peak is due to both the substrate and the cladding layer. The InP peak has structure at its base which does not occur for single epitaxial layers[6]. If a 1.0 μm thick capping layer is added, then the simulated result shown in Fig. 3 is obtained. Aside from a reduction in intensity due to absorption in the capping layer, the result is almost a simple superposition of a Bragg peak for the capping layer in Fig. 2. This result holds generally provided the peaks are well separated.

If the peaks overlap, then interpretation is more complicated. This is shown in Fig. 4. The active layer and the capping layer peaks for the case of Fig. 4A are superimposed and the only clue to this are pronounced shoulders on either side of the combined peak.

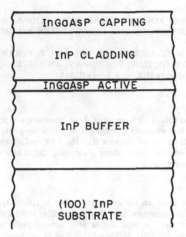

Fig. 1) Schematic LED base structure.

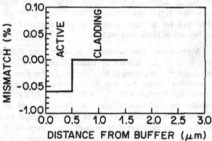

Fig. 2) Simulated rocking curve for a

DH without a capping layer.

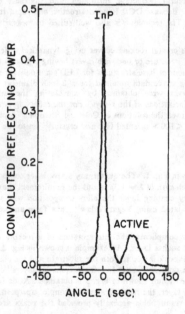

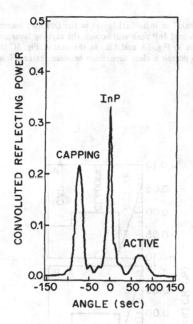

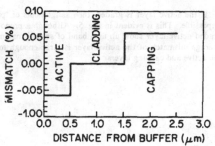

Fig. 3) Simulated rocking curve for the

full LED structure.

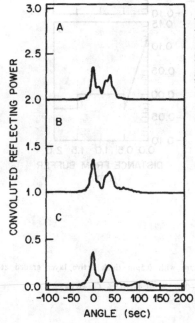

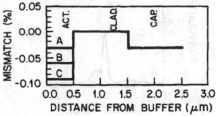

Fig. 4) Simulated rocking curves for the LED

structure with 0.5 μm thick active

layers and mismatches of A) −0.03%,

B) −0.06%, and C) −0.09%.

228

If the active layer is graded, then assignments of "peaks" to individual layers in the DH may become impossible. This is evident in Fig. 5. Although a pronounced InP peak still occurs, the capping layer and graded active layer show up as a band of several maxima in Fig. 5A and 5B. In the case of Fig. 5C the average mismatch of the active layer is large enough to permit a clear separation between features from the active and capping layers.

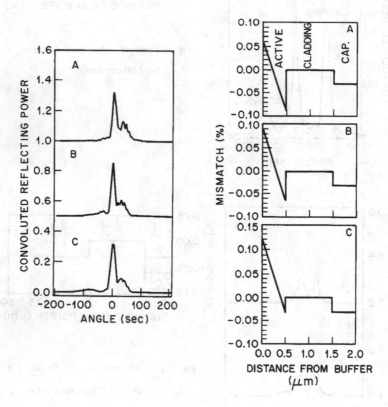

Fig. 5) Simulated rocking curves for the LED structure with 0.5μm thick active layer graded at 0.15%/0.5μm. The average mismatched are: A) −0.015%, B) +0.015%, C) +0.045%.

III. ROCKING CURVE DATA AND THEORETICAL FITS FOR A WAFER GROWN BY HYDRIDE-VPE

Double crystal rocking curves for an LED base material wafer are shown in Fig. 6 and 7. Dynamical diffraction theory fits to these data are also shown. We conclude from these comparisons that the LED structure is of high quality. Mismatches and layer thicknesses of all the layers were varied to obtain these fits. Compositions for the capping and active layers were determined by the mismatches and by the contraint that these have bandgaps corresponding to $\lambda = 1.1$ μm and $\lambda = 1.3$ μm emission wavelengths respectively. Ordinate values for the rocking curve data were obtained by dividing the observed x-ray count rate by an adjusted incident count rate. The measured incident count rate was decreased by 14% to obtain the best fit. The need for this adjustment is not fully understood at present.

The substrate plus cladding peak in Figs. 6 and 7 was found to be broadened and to have features which could not be fit with a pure binary composition for the cladding layer. A shoulder is visible in Fig. 6 and a pronounced additional peak is visible in Fig. 7. These features could be reproduced in the fits by invoking a very slight Ga contamination of the cladding layer. We obtained the fits by assuming the cladding layer to be $In_{1-x}Ga_xP$. The requisite values of x were 5.20×10^{-4} and 5.14×10^{-4} for Figs. 6 and 7, respectively. A clear additional peak resulted in Fig. 7 because in that case the cladding layer thickness was thinner (1.55 μm) than in the case of Fig. 6 (2.00 μm).

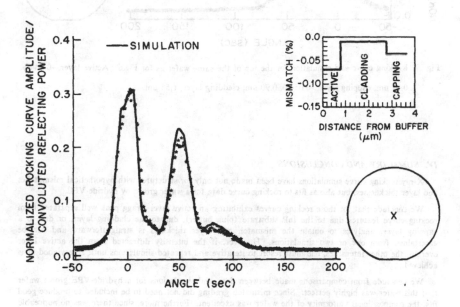

Fig. 6) Rocking curve data and simulations for the center of a 2 inch diameter hydride-VPE grown LED

   wafer. The following mismatch and thickness values resulted in the fit shown: active layer,

   −0.071%, 0.61μm; capping layer, −0.044%, 1.00 μm. The cladding layer was assumed to be

   2.00 μm thick.

230

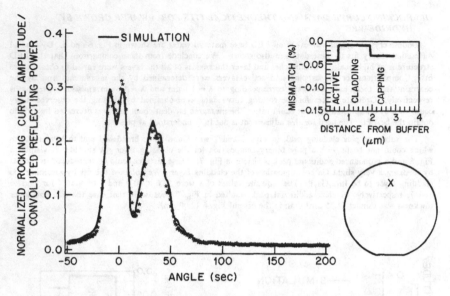

Fig. 7) Rocking curve and simulation for the top of the same wafer as for Fig.6. Active layer, -0.044%, 0.61 μm; capping layer, -0.034%, 0.90 μm; cladding layer, 1.55 μm.

IV. SUMMARY AND CONCLUSIONS

X-ray rocking curve simulations have been made not only for structures with hypothetical mismatches and layer thicknesses, but also as fits to rocking curve data for a wafer grown by hydride-VPE.

We conclude that for those rocking curves exhibiting an active layer Bragg peak well separated from rocking curve features due to the InP substrate (plus buffer), due to the cladding layer, or due to a capping layer, analyses to obtain the mismatch and layer thickness is straightforward and can be ascertained from one or two simulations. However, if the intensity diffracted from the active layer overlaps the other intensities, one must resort to iterative and repeated simulations until a very good fit is achieved.

We conclude from comparisons made between data and simulations for a hydride-VPE grown wafer that this wafer was highly perfect. Since mismatch grading did not need to be included to produce good fits, the compositional uniformity of the wafer was excellent. Furthermore, since there was no noticeable mosaic broadening, the LED structure was highly single crystalline.

ACKNOWLEDGEMENT

We would like to acknowledge useful discussions with K. Campbell of AT&T Technology Systems.

REFERENCES

[1] R. H. Saul, T. P. Lee, and C. A. Burrus, "Light-Emitting-Diode device design", Chap. 5 in Semiconductors and Semimetals, Vol. 22 ed. by R. K. Willardson and A. C. Beer (Academic Press, New York, NY).

[2] A. T. Macrander and K. Strege, "X-ray monitoring of InGaAs layers grown by VPE", MRS Symp. Proc. 56, 115 (1986).

[3] A. T. Macrander, E. R. Minami, and D. W. Berreman, "Dynamical x-ray rocking curve simulations of nonuniform InGaAs and InGaAsP using Abeles' matrix method", J. Appl. Phys. 60, 1364 (1986).

[4] Growth on 2 inch diameter rounds was performed in a reactor designed by R. F. Karlicek with contributions by W. D. Johnston, Jr. and K. Strege.

[5] H. Cole and N. P. Stemple, "Effect of crystal perfection and polarity on absorption edges seen in Bragg diffraction", J. Appl. Phys. 33, 2227 (1962).

[6] A. T. Macrander and K. Strege, "X-ray double crystal characterization of highly perfect InGaAs/InP grown by VPE", J. Appl. Phys. 59, 442 (1986).

[7] A. T. Macrander, W. A. Bonner, and E. M. Monberg, "An evaluation of (100) sulfur doped InP for used as a first crystal in a x-ray double crystal diffractometer", Materials Lett. 4, 181 (1986).

[8] R. E. Nahory, M. A. Pollack, W. D. Johnston, Jr., and R. L. Barns, "Band gap versus composition and demonstration of Vegard's law for $In_{1-x}Ga_xAs_yP_{1-y}$ lattice matched to InP", Appl. Phys. Lett. 33, 659 (1978).

Hg$_{1-x}$Cd$_x$Te NEAR SURFACE CHARACTERIZATION USING COMPUTER AIDED RUTHERFORD BACKSCATTERING SPECTROMETRY

T.-M. KAO AND T.W. SIGMON
Stanford Electronics Laboratories, Stanford, CA 94305

ABSTRACT

In this work, we report the use of Rutherford backscattering(RBS) measurements and computer simulations to provide accurate stoichiometry information and semi-quantitative defect densities for the near surface region of Hg$_{1-x}$Cd$_x$Te (MCT). The accuracy of the Hg$_{1-x}$Cd$_x$Te x-values determined by our method is found to be comparable to other commonly used methods, such as FTIR or the electron microprobe. The data obtained as structural defects from RBS channeling measurements are in basic agreement with other techniques, such as chemical etching. The sensitivity of the channeling measurement to uniformly distributed dislocations is found to be about 10^7 - 10^8 cm^{-2}, however, for dislocations forming subgrains, the detectable level of dislocation comes to 10^5 - 10^6 cm^{-2}. The depth profiles of lattice disorder resulting from ion implantation into MCT are also extracted from RBS channeling measurements using these simulation programs. These profiles are found to closely match the calculated profiles for the displaced atoms calculated using an implantation modeling program (TRIM). We also report on the use of channeling-in-grazing-angle-out technique for evaluating the stoichiometry of the first few monolayers of the MCT surface.

INTRODUCTION

Electrical properties of Hg$_{1-x}$Cd$_x$Te are controlled by defects, stoichiometry deviation and impurities, thus for improvement of device behavior improved the understanding and control of these factors is essential. Rutherford Backscattering spectrometry (RBS) is a physically non-destructive mass sensitive measurement technique, which can provide quantitative information on stoichiometry in the near surface region of materials [1]. Channeling of fast, light ions in crystals has been widely used as a tool to study crystal quality and to determine the amount, depth distribution, and annealing kinetics of disorder existing in crystalline materials[1,2]. Channeling studies on Hg$_{1-x}$Cd$_x$Te have also been reported by several authors[3 - 5].

The spectrum obtained using a beam without intentional alignment along certain low index crystal axis provides data from which the x-value of MCT in the near surface region can be extracted. This is a result of the spectrum height being proportional to the atomic fraction and Z^2 (Z is atomic number) of the species in the material. Usually, the surface energy approximation[1] is used to obtain quantitative information, such as the conversion between energy scale and depth scale. However, this method leads to errors if used to extract data from deeper regions or complex spectra. This is especially true for a ternary compound such as Hg$_{1-x}$Cd$_x$Te whose spectrum is constructed from the sum of the contributions of Hg, Cd and Te. Quantitative analysis of the channeling spectra is even more complex than this, since the dechanneling and backscattering processes is not simple and the spectrum height is not a direct indication of the amount of disorder. Hence to derive quantitative and extensive stoichiometry and defect information for Hg$_{1-x}$Cd$_x$Te from the raw RBS spectra, simple calculations do not suffice and a well-calibrated computer simulation program is required for rapid and accurate analysis.

In this work, we report using RBS measurements with the help of a well-calibrated simulation program to provide a precise way for studying the MCT near surface (~ 2 μm) stoichiometry and crystal quality. The simulation program can synthesize random and channeling RBS spectra and incorporates energy straggling. The comparison between the results extracted from RBS measurements and that obtained from chemical etching is reported.

EXPERIMENTAL

There are two major aspects of the RBS technique of interest in our work; rotational random backscattering and single aligned axial channeling. When using the "rotational random" technique, it is necessary to tilt the zone axis of the sample 6° from the beam and then rotate the sample through 360° to minimize the channeling effect of the low index planes. This effectively averages the measurement along all directions and provides a reproducible random spectrum. The "rotational random" measurements also provide the necessary information to locate the planar channels, from which the position of the axial channel can be determined. A detailed discussion of the alignment procedures can be found in Ref.[1]. The RBS sample goniometer used is similar to that described by Chu et al.[1]. A collimated $^4He^+$ beam generated from a van de Graaff accelerator with energy of 1.2 or 2.2 MeV is used in this work. All measurements are performed using a 1-2 nA beam current, 1x1 mm^2 beam spot and a total charge of 3.6 μC for $Hg_{1-x}Cd_xTe$. Two Ortec silicon Schottky barrier detectors are used, one at $\theta = 170°$ to retain the measurement sensitivity to the target mass, the other at $\theta = 94.5°$ for improved depth resolution in the near surface. The scattered particles are collected by these detectors which are interfaced to standard Ortec electronics and a Nuclear Data ND60 multichannel analyzer.

The measurements are performed on $Hg_{1-x}Cd_xTe$ LPE layers with > 4 μm of the top layer chemically removed and CdTe bulk crystals which have been chemi-mechanically polished and chemically etched. The samples are all etched with a HCl solution prior to the RBS measurements to remove the native oxide. Etch pit density (EPD) studies are performed on the CdTe crystals similar to those used as substrates for growing $Hg_{1-x}Cd_xTe$. Samples are firstly bromine-methanol etched to remove possible damage induced by the polish, then etched with the Nakagawa solution[6] for ~40 s, and immediately followed by a one-second dip in the E solution[7,8] to remove the dark stains that usually occurs when using Nakagawa solution.

RESULTS AND DISCUSSION

Stoichiometry evaluation

The computer code for analyzing the random spectra is first calibrated using Si and CdTe measurements prior to using it for characterizing the $Hg_{1-x}Cd_xTe$. The total integrated beam charge used for Si and CdTe are 18 and 3.6 μC, respectively. Since RBS has better mass resolution in low mass regime, we treat the Si target as a three elements (three isotopes) matrix. The CdTe is treated as MCT with x=1, the average atomic weight for each element is used since RBS cannot resolve small differences in this high mass region at the energy used. The simulation results (solid lines) are presented together with the experimental spectra (dotted lines) in Fig.1. The solid angle Ω of the detector and the system resolution used in this work are 1×10^{-3} steradian and 19.5 KeV, respectively. The experimental and simulated results for MCT (x=0.20,0.30,0.70) are presented in Fig.2 with the depth scale shown on the upper x-axis, which is refered to the Hg edge for x=0.20 material. To avoid overlapping the spectra, those for MCT with x=0.30 and x=0.70 are shifted 35.5 keV and 71.0 keV respectively. The resulting x-values determined by our measurements are compared with results obtained by other common methods, such as FTIR and electron microprobe. A comparison of these results is listed in Table I which shows a reasonable consistency between these various techniques. From our simulations for the various x-value samples, we estimate that the error in our x-value measurement is about ± 0.01. This can be improved if the scattering of the experimental data is narrowed by increasing the total dose of the probe ions.

Channeling

In order to check the validity of the channeling theory used in the our simulation program, we evaluate the experimental and theoretical values of the channeling parameters, $\psi_{1/2}$, χ_{min}, γ and Ω^2_s. Here $\psi_{1/2}$ is the angular half-width of the channeling dip at the yield value half way between the minimum yield and the yield of random incidence, χ_{min} is the normalized channeling yield obtained at the surface and is called the minimum yield[1,9], γ is the dechanneling coefficient[10,11], and Ω^2_s is the beam divergence resulting from the interaction between the channeled particles and surface atoms[12]. The measurement of $\psi_{1/2}$ is similar to that described by Chu et al.[1]. The evaluations of χ_{min}, γ and Ω^2_s are performed on spectra similar to those shown in Fig.3 which are obtained for "perfect crystals". It is very common for bulk-grown CdTe substrates or MCT epilayers grown on these substrates to have dislocations densities on

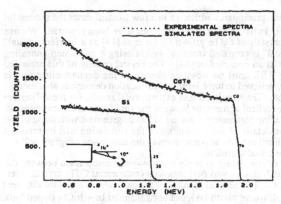

Fig.1 Simulated and experimental spectra obtained for Si and CdTe, total integrated beam charge used is 18 and 3.6 µC for Si and CdTe, respectively.

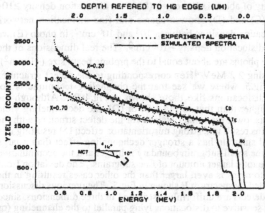

Fig.2 The simulated and experimental spectra for MCT (x= 0.20, 0.30, 0.70). The spectra from x= 0.30 and x= 0.70 samples have been shifted 35.5 KeV and 71.0 KeV respectively for clarity. The depth scale is refered to the Hg edge of the x= 0.20 sample.

Table I Comparison between x-values measured by different methods.

SAMPLES	X-value	
	RBS(± 0.01)	FTIR OR MICROPROBE
VPE213TP	0.15	0.157
TI205	0.20	0.205
VPE196	0.28	0.29
VPE195A	0.36	0.347
VPE428E	0.33	0.326
TIBULK	0.28	0.28
VPE321AP	0.30	0.297
JIMVPE1	0.27	[a] 0.265
JIMVPE2	0.70	[a] 0.70
JIMVPE3	0.40	[a] 0.40

[a] x-value measured by electron microprobe

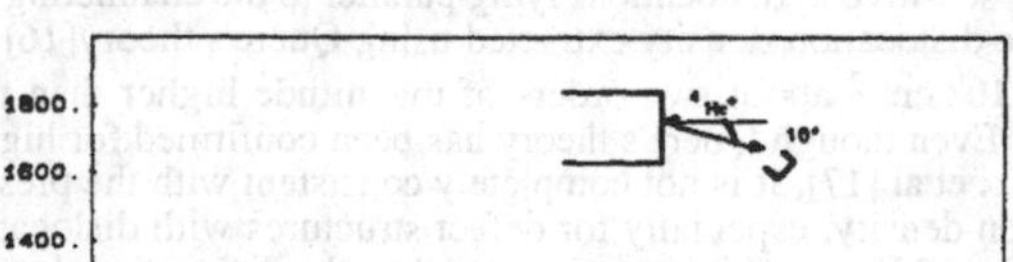

Fig.3 The experimental and simulated results obtained for "perfect crystals" of Si, CdTe and $Hg_{0.7}Cd_{0.3}Te$. The total beam charge used for Si is ten times of magnitude larger than that used for CdTe or $Hg_{0.7}Cd_{0.3}Te$.

the order of 10^5 - 10^6 cm^{-2} and some precipitates, although in a few isolated cases the successful growth of low-dislocation density (~1x10^4 cm^{-2}) CdTe substrates have been reported. We use the low-dislocation density bulk crystals of CdTe grown by Lu et al.[13] as a "perfect crystal" standard. The MCT "perfect crystal" is prepared from epi-layers using a multi-step annealing process with a MCT native oxide as an encapsulant[14]. The crystal quality of this annealed MCT epi-layer is checked using RBS until no obvious change in the dechanneling rate is observed. A commercial Si wafer, believed to have very low dislocation density, is also used as a reference to check our simulation. The simulated and experimental results for these "perfect crystals" plotted in Fig.3 show an excellent agreement between the simulation and measurement. The theoretical values of the channeling parameters used are in basic agreement with those of the experimental results, however, the details of these results and the simulation will be reported elsewhere. Consequently, the application of these parameters to the simulation program will not cause significant deviation from the actual situation.

To investigate the sensitivity of channeling to study dislocations, a comparison between the channeling and etch pit density (EPD) studies was performed on commercial CdTe crystals used as the substrates for growing $Hg_{1-x}Cd_xTe$. Several features of the etch pits obtained for different CdTe wafers are presented in Fig.4, where photo (A), obtained from an In + InCl$_3$ doped bulk CdTe[13] shows a dislocation density of about 1x10^4 cm^{-2}, (B) has a dislocation density $\geq$10^7 cm^{-2}, and the etched surfaces in (C) and (D) show a coarse and fine dislocation network respectively with both having dislocation densities between 10^5 and 10^6 cm^{-2}, in photo (E) we see several twin boundaries with a dislocation density ~10^6 cm^{-2}. The real dimensions of the individual sample areas shown in the photos are about equal to the probing beam size (~1 mm^2). The channeling spectra obtained using 2.2 MeV ^{4}He$^+$ corresponding to the defect structures shown in Fig.4 are presented in Fig.5, where we see that the channeling technique is not sensitive to uniformly distributed dislocations like those shown in Fig.4(B). Although the dislocation density in Fig.4(B) is higher than that in Fig.4(C) or (D), the dechanneling yield obtained corresponding to Fig.4(B) is lower than that obtained for the defect structure with the subgrains. This is due to the subgrains result in a strong misorientation effect[15] resulting in a larger dechanneling effect. A small subgrain has a stronger dechanneling effect than a large subgrain because the total length of the small subgrain boundary is longer and the accumulative misorientation effect is bigger for the total larger amount of fine subgrains. The deviation from the initial channeling direction due to twins is even larger than the other cases resulting in the highest dechanneling effect as indicated by spectrum E shown in Fig.5. The previous discussion is based on the assumption that the defects are uniformly distributed in three dimensions, since the channeling technique is not very sensitive to dislocations lying parallel to the channeling (or incident beam) direction[2,16]. The dislocation density extracted using Quere's theory[16] for spectrum B shown in Fig.5 is ~1x10^9 cm^{-2} about two orders of magnitude higher than that obtained from EPD measurements. Even though Quere's theory has been confirmed for highly damaged single crystal Al by Picraux et al.[17], it is not completely consistent with the present case where there is a low dislocation density, especially for defect structures with dislocation networks. There are several reasons could cause this deviation: a) when the dislocation density is high, the elastic strain field arround a given dislocation tends to decrease due to the interaction with other dislocations[14] and the resulting dechanneling effect of individual dislocation decreases, therefore, Quere's theory give a good estimation for the high dislocation case, but probably underestimates the dechanneling width (or dechanneling factor) in the low dislocation density situation; b) the existence of less obviously polygonized dislocations (cannot resolve in Fig.4(B)) could enchance the dechanneling effects, and c) the dechanneling effect resulting from a high dislocation density dominates over that of the subgrains.

Further channeling results obtained from these samples using a 1.2 MeV primary beam energy are presented in Fig.6. Here, we find that the difference between the spectra obtained for the defect structures shown in Fig.4(A)-(C) is within the statistical error of the measurements (represented by spectra A in Fig.6), and we are just able to identify the channeling result (represented by spectrum B) obtained for the sample with the largest subgrain density, however. Obviously a low primary energy probing beam is less sensitive in discriminating between the different defect structures. That is caused by the acceptable angle for channeling increasing with $E^{-1/2}$ while the dechanneling factor of the dislocation decreases as $E^{1/2}$, where E is the incident

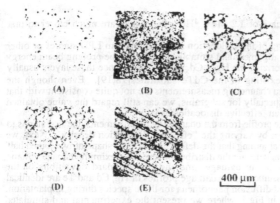

400 μm

Fig.4 Etch pit features on the (111) surface of CdTe, (A) In and InCl$_3$ doped CdTe with ~1x10^4 cm^{-2} dislocations, (B) uniformly distributed dislocations, >10^7 cm^{-2}, (C) dislocations, 10^5~10^6 cm^{-2}, forming coarse subgrains, (D) fine subgrain structure with 10^5~10^6 cm^{-2} dislocations, (E) twins and uniformly distributed dislocations.

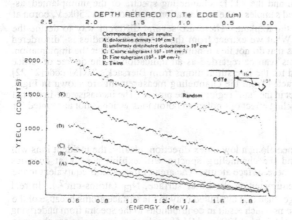

Fig.5 The channeling spectra corresponding to the defect structures shown in Fig.4. Using an aligned 2.2 MeV ^{4}He$^+$ beam the channeling measurements can discriminate between the twin, subgrain, etc., the channeling results are in basic agreement with the results for etch pit studies.

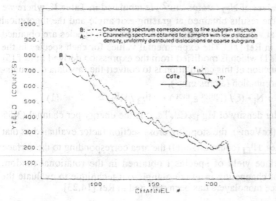

Fig.6 Channeling measurements performing on samples with the defect structures shown in Fig.4 using an aligned 1.2 MeV ^{4}He$^+$ beam do not have the same sensitivity as that (Fig.5) of using 2.2 MeV ion beam.

beam energy. Therefore for the case of 1.2 MeV ^{4}He$^+$, the results are much closer to that predicted by Quere's theory.

Hence, with an understanding of the dislocation distribution from EPD studies or other methods, channeling with precise alignment and suitable selection of the probing beam energy can be a powerful tool for characterizing the $Hg_{1-x}Cd_xTe$ epi-layers since these layers usually reproduce the defect structures existing in the CdTe substrates[18,19]. Even though the dislocation densities extracted from channeling measurements are not quite consistent with that obtained from the EPD studies, especially for subgrains, we can still regard the value obtained using the channeling simulation as an "effective dislocation density".

To extract the implant damage profile from the channeling measurements, the first step is to simulate the unimplanted spectrum by varying the "effective dislocation density". Next, we simulate the implanted spectum by assuming that the defect profile has a shape of two joined half Gaussians and the fitting parameters defining the distribution are, the maximum damage volume fraction, the position of the maximum damage, and the two standard deviations of the asymmetric Gaussians. We also assume that the damage profiles of Hg, Cd and Te are identical and neglect the different enchanced diffusion phenomena for these species during implantation. The simulation results are shown in Fig.7, where we present the experimental and simulated results for the rotational random and the <111> channeling spectra of the unimplanted, as-implanted, and implanted-annealed samples. The implantation conditions are 250keV, boron at 10^{15} cm^{-2} held at 77°K during the implant. There are several different way to define the implantation induced damage. What we extract from the channeling studies is "disordered atoms"[20], this means the atoms that do not lie on normal lattice sites after the implantation. The number of "disordered atoms" can be regarded as an indication as the degree of lattice disorder. The profiles calculated for the interstitial atoms from Biersack's TRIM code[21,22] and the "disordered atoms" extracted from the channeling measurement are shown in Fig.8. Excellent agreement is seen expect in the "tail" region of the profile. The inconsistence in the tail region is mainly due to the channeling effect of implanted boron ions which is not accounted for in the TRIM program.

Channeling-in-grazing-angle-out

If the probing beam is aligned along a low index direction, ideally, the incident ions will only see the outmost atoms and the channeling spectra will exhibit a peak at energies corresponding to scattering from these surface atoms. The area of this peak is equivalent to the number of atoms per square centimeter on the outer surface, $N_\square$ (atoms-cm^{-2}). In real situations, the surface peak is larger than that attributed only to the surface atoms because of the thermal vibrations of the lattice atoms which result in contributions to the spectra from underlying atoms. The experimental spectra obtained at 170° and 94.5° backscattering angles are shown in Fig.9. The grazing exit geometry has better resolution of the surface peak intensity because the longer outgoing path of this geometry results in a better depth resolution. A comparison between the experimental and theoretical values of $N_\square$ (atoms-cm^{-2}) is tabulated in Table II, where we see a better consistency between the results obtained at grazing exit angle and the theoretical values as expected. The theoretical values of $N_\square$ (atoms-cm^{-2}) for each species are obtained following the approach described in Ref.[1]. The experimental values for each species in the target can be evaluated using Eqn. (1) which is modified from the expression derived in Ref.[5] for calculating the Hg loss. The function of this calculation is to convert the raw data of counts-channel into the surface atomic concentration (atoms-cm^{-2}).

$$N_{\square,i} \text{ (atoms-cm}^{-2}) = N_i \cdot (\xi / (N^{mct} \varepsilon_i^{mct}) \cdot (Hi_\square / H^R_i) \qquad (1)$$

where N^{mct} is the volume molecular density of $Hg_{1-x}Cd_xTe$, ξ is the energy per channel (3.55 keV/channel for our setup), ε_i^{mct} (keVcm^2), the stopping cross section factor evaluated as that described in Ref.[1] for each species, $Hi_\square$ (count-channel) the area corresponding to the surface peak i, and H^R_i (counts) the surface yield of species i obtained in the rotational random measurements. The application of channeling-in-grazing-angle-out technique to evaluate the stoichiometry of the first few surface monolayers has been reported in Ref.[14,23].

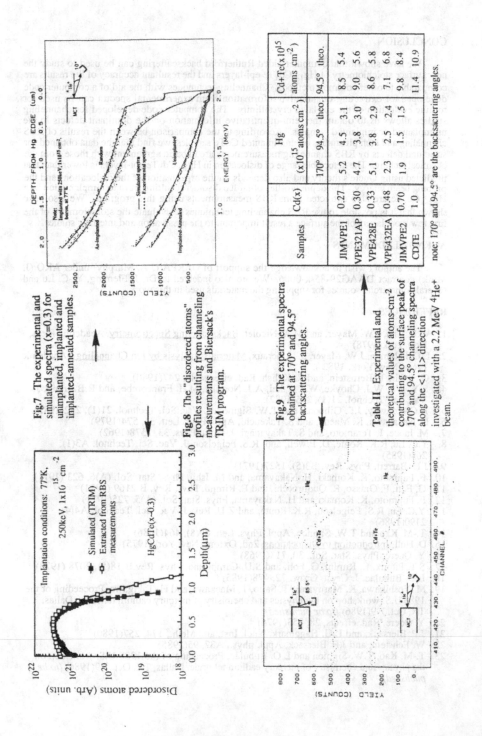

Fig.7 The experimental and simulated spectra (x=0.3) for unimplanted, implanted and implanted-annealed samples.

Fig.8 The "disordered atoms" profiles resulting from channeling measurements and Biersack's TRIM program.

Fig.9 The experimental spectra obtained at 170° and 94.5° backscattering angles.

Table II Experimental and theoretical values of atoms-cm^{-2} contributing to the surface peak of 170° and 94.5° channeling spectra along the <111> direction investigated with a 2.2 MeV ^{4}He$^+$ beam.

Samples	Cd(x)	Hg (x10^{15} atoms cm^{-2})			Cd+Te(x10^{15} atoms cm^{-2})	
		170°	94.5°	theo.	94.5°	theo.
JIMVPE1	0.27	5.2	4.5	3.1	8.5	5.4
VPE321AP	0.30	4.7	3.8	3.0	9.0	5.6
VPE428E	0.33	5.1	3.8	2.9	8.6	5.8
VPE432EA	0.48	2.3	2.5	2.4	7.1	6.8
JIMVPE2	0.70	1.9	1.5	1.5	9.8	8.4
CDTE	1.0				11.4	10.9

note: 170° and 94.5° are the backscattering angles.

CONCLUSION

We have shown that computer aided Rutherford backscattering can be used to study the near surface stoichiometry of $Hg_{1-x}Cd_xTe$ epi-layers and the resultant accuracy of the results are comparable to other common methods. Channeling techniques with the aid of a computer code developed for extracting quantitative information from experimental spectra can be used for routine evaluation of the epi-layer crystallinity. The computer code developed for channeling studies allows us to obtain the semi-quantitative information on the dominant defects in the unimplanted or implanted samples. According to the comparison between the results of RBS channeling and etch pit studies on unimplanted CdTe samples, we find that the data obtained for structural defects by RBS channeling measurements are in basic agreement with those from the chemical etching. The detectable range of dislocation in $Hg_{1-x}Cd_xTe$ epi-layers is lower than that predicted using Quere's theory and also depends on the organization of the dislocations and the probing beam energy. The depth profiles of lattice disorder resulting from ion implantation into $Hg_{1-x}Cd_xTe$ are also extracted from RBS measurements using this program. We also have shown that it is possible to use RBS channeling techniques to evaluate the stoichiometry of the first few monolayers of the surface a result important to the passivation and interface studies.

ACKNOWLEDGMENTS

The authors wish to acknowledge the support of DARPA (J.A. Murphy) under ARO (J. Mink) contract DAAG29-85-k-0006. We are also indebted to Dr. J. Fleming, Y.-C. Lu and several commercial sources for supplying the materials used in this work.

REFERENCES

1. W.K. Chu, J.W. Mayer, and M.A. Nicolet, Backscattering Spectrometry, Academic Press, New York(1978)
2. L.C. Feldman, J.W. Mayer, S.T. Picraux, Materials Analysis by Ion Channeling (Academic Press, New York, 1982).
3. G. Bahir, T. Bernstein, and R. Kalish, Rad. effects, 48, 247(1980).
4. S.Y. Wu, W.J. Choyke, W.J. Takei, A.J. Noreika, M.H. Francombe, and R.B. Irwin, J. Vac. Sci. Technol. 21(1), 255(1982).
5. K.C. Conway, J.F. Gibbons, and T.W. Sigmon, J. Vac. Sci. Technol. 21(1), 212(1982).
6. K. Nakagawa, K. Maeda, and S. Takeuchi, Appl. Phys. Lett. 34, 574(1979)
7. M. Inoue, I. Teramoto, and S. Takayanagi, J. Appl. Phys. 33, 2578(1962)
8. Y.C. Lu, R.K. Route, D. Elwell, and R.S. Feigelson, J. Vac. Sci. Technol. A3(1), 264(1985)
9. J.H. Barrett, Phys. Rev. B3(5), 1527(1971)
10. F. Fujimoto, K. Komaki, H. Nakayama, and M. Ishii, Phys. Stat. Sol. (A)6, 623 (1971)
11. G. Foti, F. Grasso, R. Quqttrocchi, and E. Rimini, Phys. Rev. B 2169 (1971)
12. F. Fujimoto, K. Komaki and H. Nakayama, Phys. Stat. Sol. (A)5, 725(1971)
13. Y.C. Lu, R.S. Feigelson, R.K. Route, and Z.U. Rek, J. Vac. Sci. Technol. A4(4), 2190(1986)
14. T.-M. Kao and T.W. Sigmon, Appl. Phys. Lett. 49(8), 464(1986)
15. D. Hull, Introduction to Dislocations, 2nd, Oxford, New York(1975)
16. Y. Quere, Phys. Stat. Sol., 30, 713(1968)
17. S.T. Piraux, E. Rimini, G. Foti, and S.U. Campisano, Phys. Rev.B 18(5), 2078 (1973)
18. L.O. Bubulac, J. Cryst. Grow. 72, 478(1985)
19. M. Yoshikawa, K. Maruyama, T. Saito, T. Maekawa, and H. Takigawa, Proceeding of the 1986 U.S. workshop on the physics and chemistry of mercury cadmium telluride, Dallas, TX, Oct. 7-9(1986) (to be published)
20. Y. Quere, Rad. effects, 28, 253(1976)
21. J.P. Biersack, and L.G. Haggmark, Nucl. Inst. and Meth., 174, 257(1980)
22. W. Echstein, and J.P. Biersack, Appl. phys., A37, 95(1985)
23. T.-M. Kao, T.W. Sigmon and L.O. Bubulac, Proceeding of the 1986 U.S. workshop on the physics and chemistry of mercury cadmium telluride, Dallas, TX, Oct. 7-9(1986) (to be published)

DEEP LEVEL DEFECTS IN CdTe

W. B. LEIGH AND R. E. KREMER
Oregon Graduate Center, 19600 NW Von Neumann Dr., Beaverton, OR   97006

ABSTRACT

We have used two complementary techniques to study deep trapping levels in CdTe single crystals. Samples of undoped, semi-insulating material and p-type material doped with phosphorus or cesium have been examined using transient spectroscopic techniques. Both capacitance transients (DLTS) and photocurrent transients (PITS) have been measured. The DLTS measurements showed several trapping levels in all of the specimens, while the PITS data usually revealed only a single level.

INTRODUCTION

CdTe is a material that has shown promise for a wide variety of applications. Among these are use in gamma ray detection, solar cells, and, more recently, as a substrate for the epitaxial growth of HgCdTe. For any of these applications, material that is as defect-free as possible is required. This is especially important if the CdTe is to be used as a substrate for epitaxial growth, as defects in the substrate tend to propagate up into the epitaxial layer during the growth. Knowledge of what defects are present in the material, and to what they are due, then, is necessary before they can be controlled and eliminated.

In CdTe, as in most wide gap II-VI compounds, the defect structure is strongly dependent on the chemistry of native defects, such as Cd or Te vacancies. The concentration of these native defects can be affected by the activity of Cd or Te during growth or some postgrowth anneal process [1,2]. The position of the energy levels corresponding to these defects and their relative concentrations determine basic material properties such as the conductivity type, and properties such as absorption spectra, or the sample resistivity.

For wide bandgap semiconductors such as CdTe, the concentration of shallow and deep levels are often found to be related. This complicates the identification of an electron or hole trap as being either impurity or native defect related, since either impurities or native defects may act as shallow levels in CdTe. For example, a change in the concentration of certain deep levels may be observed as CdTe is annealed in various Te or Cd overpressures. But this change can be accomplished in two ways: the anneal can change the native defect concentration in the material, or the shallow impurity can be changed due to dopant out-diffusion during the anneal. The actual case may even be a combination of the effects.

We have used the complementary techniques of deep level transient spectroscopy (DLTS) and photo-induced current transient spectroscopy (PITS) to study trapping levels due to defects and impurities that lie deep in the band of the material. The samples examined have come from undoped, semi-insulating ingots, and from p-type ingots doped with phosphorus and cesium. In all cases, the ingots were grown in quartz ampoules, using the vertical Bridgman process. The samples were not annealed, to ensure that the same Cd and Te activity was present for all samples. Both electron and hole traps were observed using the PITS method, while the DLTS spectra showed hole traps only.

While many studies of deep levels in CdTe have been made on n-type material, relatively little information exists on p-type material. A wide variety of experimental techniques have been used to collect this information. DLTS measurements have been made on copper-doped samples [3], and thermally stimulated current experiments have been performed on undoped p-type materials annealed in various atmospheres [4]. Temperature-dependent Hall measurements have been used to study phosphorus doped samples [5], and PITS experiments were performed on antimony-doped material [6]. Comparison between the levels observed by this earlier work and the results presented here will be made below. For ease of presentation, the results obtained from our PITS experiments will be discussed separately from the DLTS measurements.

PITS MEASUREMENTS

Photo-induced transient spectroscopy (PITS) was developed as a technique to observe deep levels in highly resistive materials when other methods, such as DLTS, become insensitive [7]. Ohmic contacts are applied to the sample, and a bias is applied, providing a small "dark current." Deep trapping levels are populated via optical excitation, and the carriers are emitted by thermal activation. By creating the carriers optically, rather than attempting to inject carriers, the method retains its sensitivity even for materials with relatively high resistivity. The resistivity is limited only by the ability to make ohmic contacts with the material.

For the samples discussed here, contacts were made by evaporating a gold-germanium eutectic onto portions of the materials. The samples were transferred to a hydrogen atmosphere with as little exposure to air as possible. They were then heated to 250 C and annealed in flowing hydrogen for 30 min. Following the anneal, small pieces of indium were pressed onto the surface. The samples were then heated in an argon atmosphere to a temperature high enough to melt the indium (about 160 C). Tinned copper wires were then soldered to the indium. While this procedure did not always produce ohmic contacts, it did work the majority of the time.

A block diagram of the PITS spectrometer used for this study is shown in figure 1 [6]. The system consists of three parts: temperature control, optics, and electronics/data recording. The cryostat is a small Joule-Thomson refrigerator, capable of keeping a sample within 0.1 degree of any setpoint between 78 and 350 K. The temperature could be set either manually or by computer. The optical source was a tungsten lamp, which was chopped by an electromechanical shutter. A constant voltage is applied across the sample, and the current is monitored by measuring the voltage

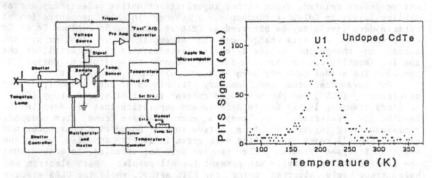

1. Block diagram of the experimental arrangement used for the PITS measurements.

2. PITS spectrum for an undoped CdTe sample. The time constant was 34.3 ms.

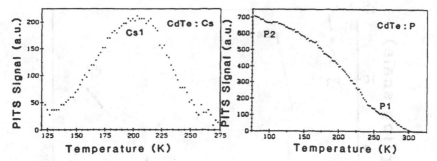

3. PITS spectrum for a sample of CdTe doped with Cs. The time constant was 0.635 s.

4. PITS spectrum for a P-doped sample of CdTe. The time constant was 1.65 s.

across a resistor in series with the sample. This signal is fed through a variable gain, high impedance operational amplifier, and then digitized and stored on disk.

Figure 2 shows the PITS spectrum for an undoped sample of CdTe. Thermal probe measurements showed that this sample as high-resistivity p-type. This undoped material gave relatively little photoresponse, i.e., the application of a light pulse generated only a few carriers, and the transient decay back to the dark current occurred very rapidly, typically in a few tens of milliseconds. As seen in figure 2, only one peak could clearly be resolved.

The spectra measured for Cs-doped and P-doped materials are shown in figures 3 and 4 respectively. Only one, rather broad peak is seen in the Cs-doped material (figure 3). Although this peak appears at roughly the same temperature as the single peak of the undoped material shown in figure 2, the time constants involved indicate that different defects are present. Two peaks appeared in the P-doped material, one corresponding to a fairly shallow trap, and the other to a very deep level. In both of these doped materials, the amount of electrical activity, and the length of time it persisted, were quite long. The decay of the photoconductivity for both materials required several seconds.

The data consist of current transients recorded at a series of fixed temperature points. The stored transients are sampled at different times, providing different rate windows, and the resulting information analyzed in a manner analogous to that used for DLTS measurements [9]. Although this type of analysis is problematic for several reasons [8,10], the results obtained provide emission signatures for the traps present in the material. The emission rate of carriers from deep traps observed by the PITS method can be expressed in the usual Arrhenius-type form to determine the emission rate parameters:

$$e_p/T^2 = A\sigma_p \exp(-E_A/kT) \qquad (1)$$

for holes, where e_p is the hole emission rate, σ_p is the temperature independent capture cross section, E_A is the activation energy, and $A = N_p v_p/T^2$, is a constant for the material which contains effective mass and density of state parameters. A similar equation holds for electrons. PITS measurements are sensitive to both electron and hole traps, so it is not possible to distinguish between donor and acceptor levels. Emission rate data taken with several different time constants for the peaks shown in figures 2 - 4 has been fitted to equation (1) and is shown in figure 5. It

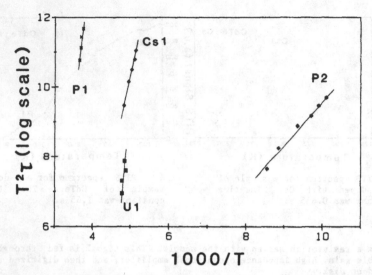

5. Arrhenius-type plot for the peaks observed in the PITS spectra shown in figures 2 - 4.

should be noted that the range of time constants used in figure 5 is quite small. As has been discussed elsewhere [8], this limitation is inherent to the technique, and places some restrictions on the accuracy of the trapping parameters obtained. Table I lists the traps observed using the PITS methods, and the activation energy measured from figure 5.

Several comments can be made about the traps listed in Table I. The peak seen in the undoped material, U1, is seen in most undoped samples. As discussed below, it is also observed using DLTS. We believe that it is related to a native defect in the material. The traps measured in the doped materials are assumed to be related to the particular dopant involved. It should be pointed out, however, that a level similar to the deep level seen in the P-doped material has also been observed elsewhere, and attributed to the presence of nickel [11]. The Ni (presumably present as an impurity in the Cd) sits substitutionally on the Cd sublattice, and acts as an acceptor. We did not, however, observe this peak in any of our other material.

Table I. Activation Energy of the Traps Observed by PITS for Doped and Undoped CdTe

Trap	Activation Energy
U1	0.78 eV
Cs1	0.46 eV
P1	1.00 eV
P2	0.11 eV

DLTS MEASUREMENTS

The DLTS measurements were performed on evaporated Au-Ge Schottky barriers to CdTe, and capacitance transients were recorded using a transient capacitance spectrometer constructed at OGC. The sample diode sits within a liquid nitrogen cooled cryostat which has the capability of scanning temperature from 77 to 500 K. For the investigation reported here, however, sample temperatures were kept below 340 K. The capacitance signal was measured using the standard 1 MHz signal from a modified Boonton 72BD capacitance meter. The output signal from the meter is digitized using a storage scope (Tektronix 7D20) and then sent to a computer for analysis and storage. The system is capable of analyzing capacitance transients in one temperature scan, using a variety of rate window methods. For this study, all rate windows were set using the two-channel boxcar method as originally proposed by Lang [9].

Figure 6 shows a DLTS spectrum for an unintentionally doped sample of CdTe. Thermal probe measurements on this material revealed it to be p-type, with a net acceptor concentration of 1×10^{15}, as measured from C-V analysis. Since minority carrier injection is expected to be insignificant in a Schottky diode fabricated on CdTe, only hole traps can be observed in this p-type material. These hole traps are represented by negative peaks in the DLTS spectra. As seen in figure 6, in the undoped material, several peaks were observed around 300 K, which correspond to traps quite deep in the bandgap. These deep defects were observed in all samples, whether doped or undoped, and in most cases, only the peak labeled U1 in figure 6 could be resolved. The

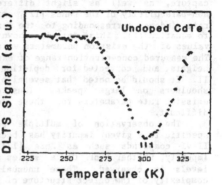

6. DLTS spectrum for undoped CdTe. The rate window was 5.86 x 10^{-2} s^{-1}.

traps in this figure are not observed to be related to interface states in the diode, as evidenced from pulse-height experiments conducted on the sample. No other shallow hole traps were observed for undoped material.

The DLTS spectra observed for Cs- and P-doped CdTe are shown in figures 7 and 8, respectively. These figures show the shallow levels which are observed in DLTS measurements of doped material, as opposed to the undoped. In addition to these hole traps, the deep levels observed around

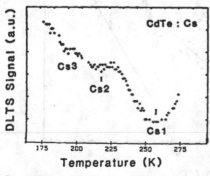

7. DLTS spectrum for CdTe : Cs, measured with a rate window of 1.95 x 10^{-2} s^{-1}.

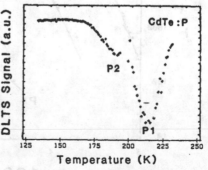

8. DLTS spectrum of CdTe : P, measured with a rate window of 3.246 x 10^{-3} s^{-1}.

room temperature in figure 6 are also found in the doped materials. The peaks identified in figures 7 and 8 are defects which are specific to the Cs or P impurities, and are not observed in any other samples. This suggests that in addition to the shallow hydrogenic acceptor levels that these impurities add to the material, there are also deep defects introduced simultaneously during crystal growth.

The emission rate of holes from these traps is expressed in the usual Arrhenius form of equation (1) above to allow the determination of the trap parameters. Recognizing the errors involved in the measurement of the thermal emission parameters without correction for temperature-dependent capture, as well as slight differences in A due to anisotropy and non-parabolicity of the bands [12], emission rate data was taken at various rate windows corresponding to the peaks shown in figures 6 - 8 and fitted to equation (1). This data is shown in figure 9. From this data, the values of the emission parameters were determined, and listed in Table II. The measured concentration range of each trap, as determined from DLTS peak heights, and corrected for depletion edge effects, is also given in Table II. It should be noted that several of the observed peaks appeared only as shoulders on larger peaks. Thus obtaining accurate measurements of emission rate parameters for these low-concentration traps often proves difficult.

The observation of multiple peaks in the DLTs spectrum which are specific to a given impurity has been made before in other wide bandgap II-VI compounds such as ZnSe [13]. In each case, for a certain added impurity, whether P or Cs, a series of two or more closely spaced deep levels is observed. These unusual defects are illustrative of the complexity of the defect structure of wide gap II-VI materials. Although not investigated in this study, it is expected that, in addition to the appearance of extra hole traps during the introduction of acceptor impurities, there will also be electron trap introduced. This so-called "self compensation" mechanism directly involves native defects [14]. Although the original mechanism for self compensation involved single

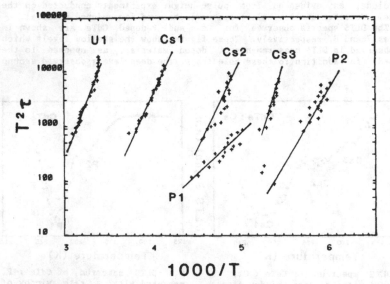

9. Arrhenius plots of the traps observed in figures 6 - 8.

Table II. Results of C-V and DLTS Measurements of Doped and Undoped p-type CdTe.

Dopant	$N_A - N_D$ Range (cm^{-3}) x 10^{15}	Trap Label	Activation Energy (eV)	σ_{P2} (cm^2) x 10^{-18}	N_t Range (cm^{-3}) x 10^{14}
none	1 - 5	U1	0.78 + 0.02	32.4	5 - 7
Cs	5 - 10	Cs1	0.64 + 0.02	1.6	5 - 6
		Cs2	0.65 + 0.07	0.013	2 - 3
		Cs3	0.6 + 0.2	-	1 - 2
P	5 - 8	P1	0.32 + 0.05	1.3	5 - 8
		P2	0.4 + 0.2	-	2 - 3

unpaired vacancies, we observe that there are other defects also present in the material which may create hole traps when paired with the impurity atoms. In this case, a simplistic argument of defect chemistry and even defect stability would be difficult to defend.

Finally, we comment on the agreement between the results of the PITS and DLTS experiments. Both techniques found a deep level at about 0.78 eV above the valance band. This peak was also present in the DLTS spectra for the doped samples. However, the levels observed in the doped materials that were attributed to the presence of the dopant atoms were different for the two techniques. As mentioned above, the PITS method does not allow one to distinguish between electron and hole traps. It is thus highly probable that the levels labeled Cs1 and P2 in figures 3 and 4 correspond to electron traps also introduced by the addition of the dopants. The PITS level P1 was also not seen in the DLTS spectra. For the DLTS rate windows used, however, a trap this deep would have required a temperature higher than those used in the experiment. Thus the lack of this peak in the DLTS measurements is not disturbing.

ACKNOWLEDGEMENTS

This work has been supported by Tektronix, Inc. and NERCO, Inc., as member of the OGC Solid State Consortium, and by Research Corporation. Thanks are due also to R. H. Bube for supplying some of the samples, and to J. L. Steiger for assisting in the PITS measurements.

REFERENCES

[1] T. Takebe, J. Saraie, and H. Matsunami, J. Appl. Phys. 53, 457 (1982).
[2] C. B. Norris and K. Zanio, J. Appl. Phys. 53, 6347 (1982).
[3] R. T. Collins and T. C. McGill, J. Vac. Sci. Technol. A1, 1633 (1983).
[4] K. Yokota, S. Katayama, and T. Yoshikawa, Jpn. J. Appl. Phys. 21, 465 (1982).
[5] F. A. Selim and F. A. Kroger, J. Electrochem. Soc. 124, 401 (1977).
[6] Y. Ivamura, S. Yamamori, H. Negishi, and M. Moriyama, Jpn. J. Appl. Phys. 24, 361 (1985).
[7] C. Hurtes, M. Boulou, A. Mitonneau, and D. Bois, Appl. Phys. Lett. 32, 821 (1978).
[8] J. C. Abele, R. E. Kremer, and J. S. Blakemore, submitted to J. Appl. Phys.

248

[9] D. V. Lang, J. Appl. Phys. 45, 3014 (1974).
[10] R. E. Kremer, J. C. Abele, and M. C. Arikan, in Semi-Insulating III-V
     Materials: Kah-nee-ta 1984, edited by D. C. Look and J. S. Blakemore
     (Shiva Publishing, Nantwich UK, 1984), p. 480.
[11] W. Jantsch and G. Brunthaler, Appl. Phys. Lett. 46, 666 (1985).
[12] W. B. Leigh, J. S. Blakemore, and R. Y. Koyama, IEEE Trans. Elec. Dev.
     ED-32, 1835 (1985).
[13] W. B. Leigh and B. W. Wessels, J. Appl. Phys. 55, 1614 (1984).
[14] G. Mandal, F. F. Morehead, and P. R. Wagner, Phys. Rev. 136, 826
     (1964).

IDENTIFICATION AND SOURCES OF IMPURITIES IN
InGaAs GROWN BY LIQUID PHASE EPITAXY

D.G. KNIGHT, C.J. MINER AND A. MAJEED
Bell-Northern Research Ltd., P.O. Box 3511, Station C, Ottawa, Ontario,
Canada, K1Y 4H7

ABSTRACT

High purity In$_{.53}$Ga$_{.47}$As and InP with carrier concentrations $[N_D-N_A] < 5x10^{15} cm^{-3}$ has been grown by the LPE technique on both n-type and semi-insulating substrates to detect and identify trace donor and acceptor impurities. Acceptor impurities have been detected in low temperature photoluminescence spectra where LPE melt baking and growth programs indicate a melt origin for two of these species, one of which is zinc. Data from semiconductor profiles provides evidence for sulfur and tin donor impurities, which comes from the rinse melt used to etch back substrates doped with the respective contaminants. Silicon and sulfur contaminants have been detected by SIMS measurements; and may arise not only from the indium and III-V materials, but also the graphite boat used to grow the epilayers. Volatile sulfur-containing compounds have been detected during high temperature bake-out of high purity graphite boats.

INTRODUCTION

For the production of avalanche photodiodes (APD's) and PIN photodiodes for $\lambda = 1.0-1.6\mu$m fiber optic communication systems, it is necessary that In$_{.53}$Ga$_{.47}$As with $[N_D - N_A] < 5x10^{15} cm^{-3}$ be grown on InP as a light absorbing material. However, the stringent purity requirements makes it necessary to identify the contaminants and their probable sources, so that production of wafers for PIN diodes can be achieved repeatably. The wafers grown by liquid phase epitaxy (LPE) in this work consist of an n$^-$In$_{.53}$Ga$_{.47}$As layer lattice matched to an n$^-$InP buffer layer, which is in turn grown on an n$^+$InP substrate. Unintentionally doped layers grown without precautions are always n-type, and are typically 1-5x10^{16} cm^{-3} for InGaAs and $\sim 10^{17}$ cm^{-3} for InP. The two donor impurities most often implicated for this unintentional doping are sulfur and silicon, which have distribution coefficients of .4-10[1] and 30[2] respectively. There is a difference of opinion in the literature as to which impurity is more dominant and exactly what precautions should be taken during crystal growth to produce high purity epilayers[3-11]. However, it is reasonable to assume that the high distribution coefficient for these elements makes them both suspect as contaminants if it can be shown that they are present in trace amounts in either the crystal growth materials or the LPE system. The most commonly reported acceptor in LPE grown InGaAs is zinc [12, 13]. Its most probable sources are the InAs and GaAs used in melt preparation. When present, acceptors degrade the mobility of the InGaAs layers and thereby reduce device speed.

All authors agree that baking of the indium and III-V materials is necessary before crystal growth. Oliver and Eastman[3] were the first to propose the neutralization of silicon by H$_2$O as the purification mechanism. This is reasonable since silicon is present at $\sim$.1ppm$_a$ in the indium used for growth which would yield a carrier concentration of 10^{17} cm^{-3} in epilayers if it is not neutralized. The reaction predicts that the equilibrium concentration of silicon should be lower at lower melt baking temperatures, which was found to be true by Penna et al.[4] and Amano et al.[5]. Penna et al. also report that higher purity layers can be grown if the indium is baked by itself before the III-V compounds are added. As a result, a reasonable baking scheme for this work would be to initially bake indium at 700°C for ~ 48 hours which would quickly convert most Si to SiO_2, add the ternary melt III-V compounds, and then bake for 72 hrs. at ~ 675°C to reach a low equilibrium value of silicon. Bhattacharya et al. also found dramatic reduction in donor and acceptor impurities due to indium baking[14], but also showed that the concentration of zinc acceptors could be reduced by baking the III-V compounds. They demonstrate that the limit to this procedure is the eventual incorporation of carbon acceptors from the boat.

Several authors report the use of an RF furnace operated at 1400-1500°C to purify the graphite boat used to produce the high purity epilayers[6-8], while others[4] find this procedure unnecessary. The need for this procedure may be dependent on the purity of the graphite boat, since Groves and Plonko[9] note that 150ppm sulfur was detected for a boat which could not produce high purity layers while ~ 1ppm was detected for a high quality boat. The depth of the graphite layer that is purified by this technique may be shallow, since Cook et al.[6] note that a temporary decrease in epilayer purity is observed if the surface layer of graphite is peeled away from the boat. Leaching out of impurities by subsequent melts restores initial purity levels.

For high purity $10^{14} cm^{-3}$ range InP material the major contaminant was found to be sulfur, both by high resolution photoluminescence spectroscopy[10] and secondary ion mass spectroscopy (SIMS)[11]. Lower levels of Si were seen by Skolnik et al.[10], which is reasonable since long baking times were employed for the InP growth melts. For the $In_{.57}Ga_{.47}As$ epilayers, signals of approximately equal magnitude were seen from the two donor species by far infrared photothermal ionization measurements[6].

In this work, the sources and types of donor impurities existing in n⁻InP and n⁻InGaAs layers will be determined for LPE growth on conducting InP substrates. Melt baking as outlined above will be employed for growth of all epilayers, and special attention will be devoted to the substrates themselves since substrates containing dopants must be used to make PIN diodes and APD's. The latter issue is of particular concern since sulfur or tin from the conducting substrates will be incorporated into the rinse melt and may be carried over into the melts from which the high purity epilayers are grown. Also, the graphite boat used to grow the epilayers will be baked in a specially constructed vacuum audio frequency induction furnace, capable of operating at a temperature of 2200°C. The presence of volatile sulfur or silicon containing species as monitored by a mass spectrometer would lend strength to the argument that the LPE boat is a source of contaminants, and the temperature at which these species are evolved would suggest the required temperature for a baking program for boat purification. Finally, grown epilayers will be analyzed using SIMS and low temperature photoluminescence to determine which impurities are actually present. The above measures should help to identify some of the impurity sources which are still in doubt in the literature, and to lead to methods of growing purer InP and InGaAs epilayers.

EXPERIMENTAL

The two layer structures described in the introduction were grown on sulfur and tin doped (100) oriented InP substrates using a combined ramp and step cooling LPE technique, where the n⁻InP buffer layer was grown in the 634-629.5°C range and the n⁻$In_{.53}Ga_{.47}As$ layer at 629.5°C. A conventional, horizontal multi-well graphite boat of POCO DFP-3-2 carbon was used in a Pd purified H_2 ambient to produce these wafers, where the bakeout schedule previously described, was carried out before crystal growth. The

InP for the buffer layer melt was added immediately before crystal growth to avoid loss of phosphorus. Carrier concentrations for the two layer structures were then determined using a Polaron semiconductor profile plotter. Occasionally, single epilayers of either InP or InGaAs were grown on InP:Fe substrates so that Hall mobility measurements at 77 and 300°K could be taken. 7K photoluminescence from each sample was excited by a He-Ne laser (632.8nm) focussed down to a 250μm spot size, and analysed by a 0.75m Spex 1702 monochromator equipped with a thermo-electrically cooled PbS detector using standard lock-in techniques. The excitation power density was varied up to 50 W/cm^2 by the use of neutral density filters. The spectra are uncorrected for system response. In this experiment the system spectral resolution was approximately 2.5 meV.

The high temperature graphite boat baking furnace with attached quadrupole mass spectrometer is shown schematically and described in Figure 1. The furnace is a vacuum baking chamber manufactured by Tek Specialties Inc., Winchester Mass., attached to a 10kHz 20kW induction power supply from PPS Manufacturing Inc. The graphite felt and plugs act as thermal insulators for the susceptor, since radiation loss from the susceptor and boat would otherwise limit the operating temperature of the furnace. With the design described, operation at 2400°C is possible although sublimation of graphite limits the practical operating temperature to ≤2200°C. The temperature of the graphite boat in the susceptor was monitored using an optical pyrometer. Using a cryopump, a base pressure of 3x10^{-6} torr is possible despite the large surface area of the graphite felt. Effluent gases from the induction furnace were analyzed using a quadrupole mass spectrometer. Gases were admitted through a valve which connected the two systems, and the signal at each mass number was recorded after subtracting the background signal obtained when the valve was closed.

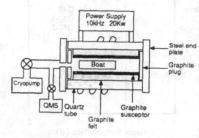

Figure 1: Schematic diagram of the high vacuum audio frequency induction furnace used to bake graphite boats. The boat is placed inside a cylindrical graphite susceptor, which is held within the vacuum chamber by purified graphite felt. The audio frequency magnetic field from the coil couples to the susceptor through the quartz tube of the vacuum chamber. The system is kept under vacuum by a cryopump, where part of the exhaust gases from the furnace enter a quadrupole mass spectrometer (QMS) via a valved vacuum line.

RESULTS AND DISCUSSION

The net carrier concentrations as determined by the profile plotter was 2±1x10^{15} cm^{-3} for InGaAs, and 5±3x10^{15} cm^{-3} for the n⁻InP layers. These averages are the result of profiles for 23 wafers, with two profiles taken per wafer. A part of the contaminants that reach the ternary melt via carryover likely comes from the substrate itself. Semiconductor profiles typical of two layer structures grown on sulfur and tin doped substrates are shown in Figure 2. It can be seen that a grading defined as the number of decades decrease in carrier concentration per unit thickness of epilayer was noted near the substrate, where typical values were 1.4 ±0.2 and 0.56 ±0.2 decades/μm for growth on sulfur and tin doped substrates respectively. If carryover from the rinse melt to the InP melt occurs, this difference in grading can be explained in terms of the differing distribution coefficients of the substrate dopants, which are 0.4-10 for sulfur and 3x10^{-3} [15] for tin. The tin will linger in the melt, and Figure 2 shows that the InP epilayers grown on tin doped substrates have high impurity levels as well as the shallowest gradings. The gradings are not caused by diffusion of substrate dopant, since MOCVD grown n⁻InP:n⁺InP(S) interfaces have gradings of 3.4 decades/μm despite an InP growth temperature of 625°C for >5 hours. The presence of dopant in the InP epilayer is confirmed by SIMS analysis, as shown in Figure 3. A definite lack of abruptness is noted for two of the wafers analyzed, which is similar to the deterioration in abruptness observed for SIMS profiles of III-V materials when carryover takes place[16]. Occasionally a semiconductor profile of a wafer on an InP:S substrate has a grading as high as 5 decades/μm, which is consistent with the observation of some abrupt n⁺InP:n⁻InP interfaces. There will be a residual grading due to unevenness in the etching crater of the profiler, and an intrinsic limit to the depth resolution governed by the Debye length of the semiconductor[17]. It is therefore reasonable to conclude that substrate dopant which is etched off in the rinse melt is carried over into the InP melt, which is then incorporated into the InP epilayer during crystal growth. Any residual dopant can then be carried forward into the ternary melt, since carryover between these melts is observed as well. The effect of InP melt material on the purity of InGaAs epilayers will be described in a forthcoming report.

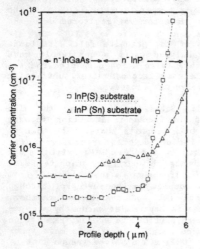
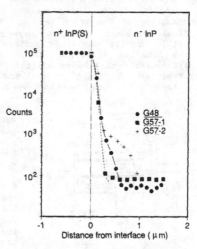

Figure 2: Polaron profiles for two layer structures grown on S-doped InP (dashed line) and Sn-doped InP (solid line).

Figure 3: SIMS depth profiles of ^{34}S for the n⁺InP(S) substrate : n⁻InP epilayer interface for three different PIN detector wafers. The high initial reading corresponds to the substrate doping of ~ 2×10^{19} cm^{-3}. Depth measurements were obtained by normalizing sputtering times to SEM n⁻InP layer thicknesses.

The results of the mass spectrometric analysis for the high temperature baking of the graphite boat is shown in Figure 4. Sulfur containing compounds ^{34}H$_2$S and ^{76}CS$_2$ were detected in the effluent gas for the bakeout of a boat loaded into the furnace, where the boat and furnace had been previously baked at high temperature to remove N$_2$ adsorbed on the graphite felt insulation. The level of H$_2$S is higher when the boat is in the furnace when compared to an empty furnace baking. This type of behaviour is noted for a freshly loaded furnace but more uncertainty in the data exists due to N_2 outgassing from the graphite felt. The largest component of the furnace gas was $^{28}N_2$ or ^{28}CO, where the mass 14:28 ratio drops from ~ 7% during the initial low temperature start-up to ~ 3% for constant baking at ~ 1800°C. This indicates that a significant amount of ^{28}CO is evolved at the baking temperature, since the mass 14 signal from N_2 has decreased relative to the mass 28 signal. The typical operating pressure during high temperature baking was $2-6\times10^{-5}$ torr, where ~ $10^{-7}-10^{-6}$ torr of H_2S and CS_2 are present since the signal ratios of mass 34 and 76 to 28 are ~ 4% and 1% respectively for baking at 1700°C.

From Figure 4 it is evident that a threshold baking temperature for removing sulfur from graphite exists, which is ~ 1500°C for a furnace that has been kept under vacuum after a previous bake. This threshold temperature may explain why some authors find boat baking ineffective in reducing contamination of high purity epilayers; since sulfur has barely begun to be evolved at 1400-1500°C, which is the operating temperature of most RF bakeout furnaces[6-8]. The rise in CO content at high temperature also suggests a mechanism for graphite purification if water is adsorbed onto the graphite felt, which is the case for most vacuum systems that are not hard baked. That is the reaction:

$$H_2O_{(g)} + C(S)_{graphite} \longrightarrow H_2S_{(g)} + CO_{(g)} + C_{(graphite)}$$

The production of CS_2 need not involve any compound other than contaminated graphite, and its volatility (boiling point = 46.3°C) ensures that it will be driven away from the graphite once formed. No marked improvement in the epilayer purities are observed after boat baking but these purity levels are already very high. The POCO graphite was probably of high enough quality not to contribute significant amounts of sulfur previous to the boat baking. Penna et al.[4] had similarly good results from POCO boats that were not baked at high temperature.

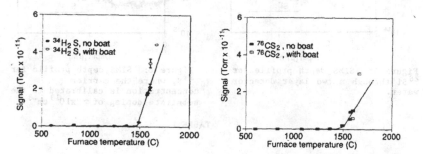

Figure 4: Plot of mass spectrometer signal at mass 34(H_2S) and 76(CS_2) as a function of the induction furnace temperature for a boat that had been previously baked, and kept under high vacuum. Also shown is data for an empty furnace, where the error bars for baking at 1700°C for H_2S represents fluctuations in the signal over a three hour period at this temperature.

The results of high sensitivity SIMS analysis for selected wafers is shown in Figures 5 and 6, which show that low levels of ^{28}Si and ^{34}S are detected in both epilayers as expected. Unfortunately, calibration standards are not yet available for silicon in InGaAs, and sulfur in InP and InGaAs, so quantitative analysis can only be applied to silicon in InP at present. Gauneau et al.[11] note the presence of localized silicon rich features in LEC grown InP. These may occur in LPE epilayers as well, which would explain high levels of silicon observed for the ternary layer in some SIMS results (not shown). Semiconductor profiles have occasionally shown anomalously high carrier concentrations in n^-InP or n^-InGaAs, which suggests that pockets of silicon contaminants can be randomly placed in either epilayer. It should be noted that the silicon pocket has caused the silicon signal to increase in the substrate material with respect to a SIMS profile with no pocket, suggesting a "memory effect" in the apparatus or a continued low level contribution from the silicon pocket. Consequently, the silicon signal from the substrate will be subtracted from the epilayer signal when evaluating carrier concentration for the wafers studied. Results for the analyses of several wafers are shown in Table 1, where count rates were converted to carrier concentrations using a silicon implant standard. All results are in the 10^{15} cm^{-3} range as expected, and the correspondingly high or low carrier concentration as measured by the profile plotter strongly suggests that silicon is a major contaminant in n^-InP. Part of this silicon probably originates from the InP used in the melt, since it was not baked prior to crystal growth.

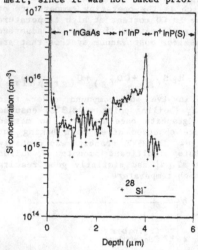

Figure 5: SIMS depth profile for ^{28}Si through a two layer detector wafer.

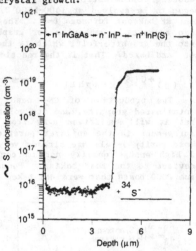

Figure 6: SIMS depth profile for ^{34}S, where the carrier concentration is calibrated to the substrate doping of $\sim 2 \times 10^{19}$ cm^{-3}.

TABLE 1

wafer	InP epilayer silicon (cm^{-3})	substrate silicon reading (cm^{-3})	net silicon in InP epilayer (cm^{-3})	average silicon in InP epilayer (cm^{-3})	$[N_D - N_A]$ (cm^{-3})
G42	5.5x10^{15}	1.0x10^{15}	4.5x10^{15}	3.9x10^{15}	2.5x10^{15}
	5.5x10^{15}	2.2x10^{15}	3.3x10^{15}		
G54	8.0x10^{15}	0.9x10^{15}	7.1x10^{15}	7.1x10^{15}	7x10^{15}
G57	9.0x10^{15}	1.0x10^{15}	8.0x10^{15}	8.0x10^{15}	6x10^{15}
	1.3x10^{16}	5.0x10^{15}	8.0x10^{15}		

Concentration of silicon in the n^-InP epilayer as determined by SIMS depth profile measurements. The substrate silicon reading was subtracted from that of the epilayer to compensate for background (see text). For comparison, the profile plotter carrier concentration for the InP epilayer is also shown.

The low temperature photoluminescence spectra of these layers (Figure 7) are typically dominated by a strong transition, near the band gap energy predicted by x-ray diffraction analysis, and a weaker line located 18 meV to lower energy. Excitation intensity analysis has verified that the high energy peak is due to exciton recombination [18]. From the width of the line (3-5 meV), we believe it to be a superposition of donor and acceptor bound excitons, which are widened somewhat by alloy fluctuations in agreement with other workers [12,13]. The lower energy peak is attributed to band-to-acceptor transitions involving zinc [12-14, 19, 20]. Under extended baking conditions, the intensity of the zinc transition decreases until a weaker line located 28±1 meV away from the bound excitons, which is normally hidden under the zinc peak, is the only acceptor left (Figure 8). The identity of this residual line is not known at present. The reduction of zinc contamination by extended baking of the III-V material was also found by Bhattacharya et al.[14]. The intensity of the transition located 28 meV from the bound excitons was found to decrease monitonically for layers grown sequentially from the same melt. The electrical properties of these layers indicates gradually increasing carrier concentrations and correspondingly reduced compensation as the melt is reused. The initial value of $[N_D-N_A]$ of 1.3×10^{15} cm^{-3} increases to 3.1×10^{15} after the fourth epilayer is grown, where the decrease in the 28 meV acceptor for the same wafers is clearly shown in Figure 8. Carbon acceptors were identified as a line 9 meV away from the main line [12,14,21] but this contaminant was only found at the periphery of the wafers. The origin of the carbon may be the graphite boat. Finally, in experiments comparing growths on sulphur and iron doped substrates, and most clearly in an experiment where the normal substrate was replaced by two half-sized substrates of each type, it was obvious that a transition 22 meV away from the exciton line is associated with the use of iron doped InP substrates. The identity of this acceptor may be silicon[14] although the transition energy usually quoted is 25 meV from the bound exciton line. It is more probable that there is another shallow acceptor originating from the iron doped substrates.

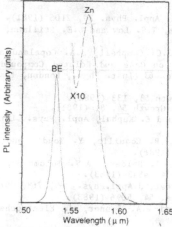

Figure 7: High purity LPE samples typically exhibit a sharp peak with a FWHM of 3.8meV due to bound excitons (BE) and a weaker peak 18 meV into the band gap due to transitions to zinc acceptors.

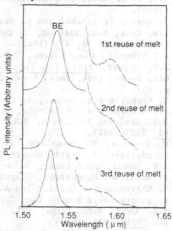

Figure 8: Low temperature PL spectra showing the residual 28 meV transition after bakeout of Zn. Repeated use of the same ternary LPE growth melt results in a decrease in signal for this acceptor, indicating that it is depleted from the melt.

CONCLUSIONS

Both sulfur and silicon have been detected in n^-InP and $n^-In_{.53}Ga_{.47}As$ epitaxial layers in wafers used to fabricate PIN photodiodes. The average carrier concentration $[N_D - N_A]$ for the $In_{.53}Ga_{.47}As$ layer is $2 \pm 1x10^{15} cm^{-3}$, and $5 \pm 3x10^{15}$ for InP as measured by the semiconductor profile plotter. The chief source of silicon is likely the indium, which remained unreacted from the melt baking. Sulfur from the substrate is a contaminant in n^-InP layers which may then be carried forward into the ternary layer melt during crystal growth. Tin may contaminate epitaxial layers in a similar manner if tin doped substrates are used. The graphite boat is also a source of sulfur, but the graphite used is of high enough purity to make the contamination of epilayers from this source negligible. For sulfur contaminated graphite boats, purification is possible at $T > 1500°C$ in vacuum by evolution of CS_2 and H_2S. The major acceptor found by low temperature photoluminescence was zinc, originating from the melt. Other acceptors, as yet unidentified, were found. One was in association with the use of iron doped substrates, and another is depleted from the melt.

ACKNOWLEDGEMENT

The assistance of F.R. Shepherd for SIMS analysis, is gratefully acknowledged.

REFERENCES

1. K.E. Brown, Solid State Electron. 17, 505 (1974).
2. G.G. Baumann, K.W. Benz and M.H. Pilkuhn, J. Electrochem. Soc. 123, 1232 (1976).
3. J.D. Oliver and L.F. Eastman, J. Electron. Mater. 9, 693 (1980).
4. T.C. Penna, M.C. Tamargo and W.L. Swartzwelder, J. Cryst. Growth 67, 27 (1984).
5. T. Amano, K. Takahei and H. Nagai, J.J. Appl. Phys. 20, 2105 (1981).
6. L.W. Cook, M.M. Tashima, N. Tabatabaie, T.S. Low and G.E. Stillman, J. Cryst. Growth 56, 475 (1982).
7. A.G. Dentai, C.A. Burrus, T.P. Lee, J.C. Campbell, J.A. Copeland and J.D. Oliver, Proc. 9th Intern. Symp. on GaAs and Related Compounds, Oiso Japan, 1981, Inst. Phys. Conf. Ser. 63 (Inst. Phys., London, 1982) p. 457.
8. E. Kuphal and A. Pocker, J. Cryst. Growth 58, 133 (1982).
9. S.H. Groves and M.C. Plonko, J. Cryst. Growth 54, 81 (1981).
10. M.S. Skolnik, P.J. Dean, S.H. Groves and E. Kuphal, Appl. Phys. Lett. 45, 962 (1984).
11. M. Gauneau, R. Champlain, A. Rupert, R. Coquille, Y. Toudic and G. Grandpierre, J. Cryst. Growth 76, 128 (1986).
12. K.H. Goetz, D. Bimberg, H. Jorgensen, J. Selders, A.V. Solomonov, G. Glinskii and M. Razeghi, J.Appl.Phys. 54, 4543 (1983).
13. C. Charreaux, G. Guillot and A. Nouailhat, J.Appl.Phys. 60, 768(1986).
14. P. Bhattacharya and M. Rao, J.Appl.Phys. 54, 5096 (1983).
15. B.H. Chin, R.E. Frahm, T.T. Sheng and W.A. Bonner, J. Electrochem. Soc. 131, 1373 (1984).
16. P. Besomi, R.B. Wilson and R.J. Nelson, J. Electrochem. Soc. 132, 176 (1985).
17. P. Blood, Semicond. Sci. Technol. 1, 7 (1986).
18. D. Bimberg M. Sondergeld, W. Schairer and T.O. Yep, J.Lumin. 3, 175 (1970).
19. P.W. Yu and E. Kuphal, Solid State Comm. 49, 907 (1984).
20. C.P. Kuo, J.Electron.Mater. 14, 231 (1985)
21. J.P. André, E.P. Menu, M. Erman, M.H. Meynadier and T. Ngo, J.Electron. Mater. 15, 71 (1986).

INFRARED REFLECTANCE CHARACTERIZATION OF A GaAs-AlAs SUPERLATTICE

J. M. ZAVADA*, G. K. HUBLER**, H. A. JENKINSON***, W. D. LAIDIG****
*US Army Research Office, Research Triangle Park, NC 27709
**Naval Research Laboratory, Washington, DC 20375
***US Army Armament R&D Center, Dover, NJ 07801
****North Carolina State University, Raleigh, NC 27695

ABSTRACT

The optical properties of a GaAs-AlAs superlattice have been examined using the non-destructive technique of infrared reflectance spectroscopy. Through this technique, the absorption edge, the effective superlattice refractive index, the thickness, and the optical grading of the superlattice-substrate were determined. Location of the absorption edge was made from the reflectance spectrum and showed general agreement with photoluminescence measurements. A more detailed analysis of the infrared spectra indicated the presence of a transition region between the substrate and the superlattice. Based on a non-linear least squares method for fitting the experimental data, a dispersion relation for the dielectric function was obtained. This dielectric function yielded a value for the superlattice refractive index that was lower than that of the corresponding, homogeneous, AlGaAs alloy film for wavelengths between 1.0 and 2.5 micrometers.

INTRODUCTION

The synthesis of high quality GaAs-AlAs superlattices has initiated a new class of infrared materials with important consequences in the area of optoelectronics. These materials have found a wide variety of applications in optical components such as sources, photodiodes, modulators, and switches. For many of these applications it is critical to accurately characterize the optical properties of these materials in frequency regions of interest.

In the present investigation the non-contact technique of infrared reflectance spectroscopy was used to determine the optical properties of a GaAs-AlAs superlattice in the near infrared region (4,000 to 13,000 cm-1). While reflectance spectroscopy has been used with considerable success to examine optical properties of bulk and ion implanted GaAs specimens [1,2], little prior work has centered on applying this method to the characterization of superlattice films. Here it is shown that reflectance spectroscopy can be used to determine the absorption edge, the effective refractive index, the thickness, and the optical grading of the GaAs-AlAs multilayer film.

EXPERIMENTAL PROCEDURES AND RESULTS

The superlattice for this study consisted of alternating layers of GaAs and AlAs grown on an undoped, semi-insulating GaAs substrate. The total thickness of the superlattice was estimated to be 6 micrometers with the individual layers being 73 A for GaAs and 80 A for AlAs. From these thicknesses the average aluminum concentration x of the superlattice was 0.523. Growth conditions relating to the superlattice as well as characterization of the layer parameters have been previously reported [3].

258

The infrared reflectance measurements were performed on a Perkin-Elmer Lambda-9, double-beam spectrophotometer and extended from the visible to beyond 2.5 micrometers. A polished wafer of semi-insulating GaAs was used to calibrate the instrument over this spectral region. Figure 1 contains the reflectivity spectrum that was obtained for this superlattice. The curve shown in this figure is normalized to the reflectance of GaAs over this region. Several features in this spectrum were noteworthy. First, since the film was quite thick, there were many fringes in the spectral region. Second, the amplitude of the fringes decreased steadily with increasing wavenumbers. There was a sudden drop in the reflectance at 11,640 cm-1 due to a detector change in the instrument. Above this wavenumber, one additional fringe minima was observable at 11,680 cm-1.

In most situations the disappearance of the interference fringes gives a reliable estimate of the absorption edge of the epilayer. For the reflection pattern in Figure 1, the cutoff wavenumber for the superlattice was identified as occurring at about 11,700 cm-1 (0.855 micrometers). Room temperature photoluminescence measurements using an Argon laser showed an absorption peak with an edge at approximately 11,830 cm-1 (0.845 micrometers). Both of these techniques provided evidence that the cutoff wavelength of the superlattice is lower than that of bulk GaAs (0.870 micrometers). These results are consistent with the prediction of quantum size effects in such multilayer films [4].

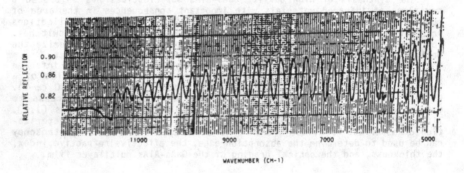

Fig. 1 Relative Reflection Spectrum for Superlattice

REFLECTANCE ANALYSIS

The average value of the reflectance interference pattern can be used to estimate the refractive index of the surface film being investigated. Suzuki and Okamota [5] used this method to estimate the refractive indices of a set of superlattice films. Applying this method to the pattern in Figure 1 yields a value of 3.145 for the refractive index at a wavelength of 1.0 micrometers which is lower than the index, 3.227, of the equivalent AlGaAs alloy film (x = 0.523) at the same wavelength. While this procedure is valid in principle, accurate values for the absolute reflectance from such samples are difficult to obtain and are very dependent upon mirror alignment, wafer position, and back surface finish of the wafer. Each of these factors can alter the absolute reflectance and lead to systematic errors. Consequently, considerable caution must be exercised when applying this procedure.

The first step in the present analysis was to calculate the reflectance pattern of the superlattice using a multilayer reflectivity program and to assume bulk optical properties for the constituent layers. The following dielectric function was used in this calculation [6,7]:

$$n^2 = A + B/(l* l - C) - D* l* l \tag{1}$$

where l is the radiation wavelength in micrometers. The parameters for GaAs are: A = 10.906, B = 0.975, C = 0.2797, D = .0025, and for AlAs: A = 7.986, B = 0.975, C = 0.0395, D = 0.0059. The results of this simulation indicated that the fringe amplitudes should be nearly constant over the entire spectral range and that the average reflectance should be higher than the experimental data. Next, a comparison was made between the reflectance pattern for the superlattice and that for the equivalent, homogeneous, AlGaAs alloy film (x = 0.523). Using Equation 1 with parameters A = 9.379, B = 0.975, C = 0.0 20, D = 0.0043, a reflectance pattern for the alloy was produced which showed only minor variations in comparison to that of the superlattice. This result can be expected since the wavelength of radiation in this spectral region is much larger that the thicknesses of the indiviual layers. The dielectric function of the superlattice should then be approximately the weighted average of the constituent films, especially when there are many layers.

Several possible explanations exist for the damping of the experimental fringe pattern including detector drift, lack of beam coherence, and absorption in the film. However, reflectance measurements of single AlGaAs layers did not show this behavior and bandedge absorption effects should not appear so far in the infrared. Another possibility is that of optical non-uniformity in the film. Hubler et al. [8] have shown that a graded optical boundary would lead to damping effects in the reflectance patterns of ion-implanted specimens. Such a graded boundary at the substrate-superlattice interface was incorporated into the multilayer reflectivity program and a close replication of the experimental data was achieved.

The next stage of this work consisted in modeling the superlattice as a uniform optical film with a graded optical boundary, having Gaussian half-width σ, on top of a GaAs substrate. In this model the dielectric function of the superlattice was described by Equation 1 with unknown parameters, A, B, C, and D. A non-linear least squares analysis, that had been previously developed to determine optical properties of ion-implanted GaAs and Si [8], was then applied. In this analysis the parameters A, B, C, D, σ were allowed to vary along with the thickness of the superlattice, tsl, and an

260

absolute reflectance scale factor F. Results of this analysis are given in
Figure 2. Here, crosses represent experimental data points and the solid
line curve is the best fit solution. The best fit parameters for the superlattice are:

$$A = 9.325 \qquad B = 0.601 \qquad C = 0.305 \qquad D = 0.0025$$

$$\sigma = 700 \text{ A} \qquad tsl = 5.65 \text{ um} \qquad F = 1.064$$

(2)

This solution indicates that there is a transition region, about 1500 A wide,
between the substrate and the superlattice. While such a transition may be
due to an interdiffusion of the first few epilayers into the substrate, the
exact mechanism giving rise to the graded optical boundary is not known at
this time. A scale factor F greater than unity implies that the average experimental reflectance was too low. Based on these parameters, a new value
of 3.192 was estimated for the effective refractive index of the superlattice
at 1.0 micrometers. This value is also lower than that of the corresponding
alloy.

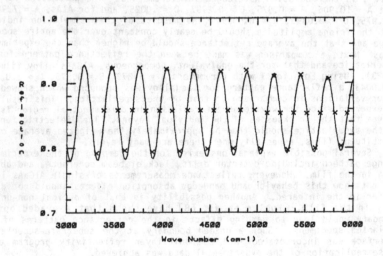

Fig. 2 Best Fit Simulation to Reflection Data

SUMMARY

In the current investigation, the non-destructive technique of infrared reflectance spectroscopy has been used to characterize the optical properties of a GaAs-AlAs superlattice. Through this technique, the absorption edge, the effective refractive index, the thickness, and the optical uniformity of the superlattice were determined. Some of these results were easier to obtain than others. The optical uniformity can be assessed from a simple inspection of the fringe pattern. The absorption edge can be easily detected if it lies within the spectral range of reflectance measurements. The thickness of the superlattice can be estimated provided the probing radiation is below the bandgap and an approximate value of the refractive index is available. Other characteristics such as values for the refractive index and descriptions of the optical non-uniformity region require a more complicated analysis. In the present case, a non-linear least squares routine was used to analyze the experimental data and quantitative estimates were achieved.

In principle, infrared reflectance can be used to characterize other infrared materials. Consideration must be given to the bandgap of these materials, to their absorption bands, and to the effect of free carriers. Care must also be given to the preparation and handling of samples under investigation and a general expression for the dielectric function is needed. However, with appropriate efforts, a wide range of information concerning the optical properties of the film can be obtained.

REFERENCES

1. A. H. Kachare, W. G. Spitzer, F. K. Euler, and A. Kahan, J. Appl. Phys. 45, 2938 (1974).
2. A. H. Kachare, W. G. Spitzer, J. E. Fredrickson, and F. K. Euler, J. Appl. Phys. 47, 5374 (1976).
3. W. D. Laidig, D. K. Blanks, and J. F. Schetzina, J. Appl. Phys. 56, 1791 (1984).
4. N. Holonyak, W. D. Laidig, M. Camras, J. J. Coleman, and P. D. Dapkus, Appl. Phys. Lett. 39, 102 (1981).
5. Y. Suzuki and H. Okamoto, J. Electron. Mater. 12, 397 (1983).
6. J. T. Boyd, IEEE J. Quantum Electron., QE-8, 788 (1972).
7. H. C. Casey, D. D. Sell, and M. B. Panish, Appl. Phys. Lett. 24, 63 (1974).
8. G. K. Hubler, P. R. Malmberg, C. N. Waddell, W. G. Spitzer, and J. E. Fredrickson, Rad. Effects 60, 35 (1982).

CHARACTERIZATION OF ULTRA HIGH PURITY SILICON EPITAXY USING PHOTOLUMINSCENCE SPECTROSCOPY

J. E. Huffman*, M. L. W. Thewalt** and A. G. Steele**
 *Rockwell International Science Center, 3370 Miraloma Avenue, Anaheim, CA
  92803
**Department of Physics, Simon Fraser University, Burnaby, B.C., Canada
  V5A 1S6

ABSTRACT

High purity epitaxial silicon samples, grown on indium doped and on ultrahigh resistivity silicon substrates, were analyzed for impurity content using photoluminescence spectroscopy (PL) and spreading resistance analysis (SRA). Calibrated SRA indicated typical net carrier concentrations of $< 3 \times 10^{12} cm^{-3}$ in the epitaxial layers, and about $7 \times 10^{11} cm^{-3}$ in the substrates. Impurities were identified by collecting highly resolved, very clean no-phonon and TO-phonon replica PL spectra at liquid helium temperatures. Spectra were taken on the substrate material alone and on substrates with epitaxy. Ga, As, Al, B and P contamination was evident in the epitaxy. Correlation of SRA and PL results on samples with various levels of contamination at the epitaxy substrate interface identified Al as the main interfacial impurity.

INTRODUCTION

Ultrahigh purity (UHP) epitaxial Si is receiving increasing attention for microwave, CMOS, and electro-optical device applications. While growth of layers having less than 10^{13} electrically active impurities per cm^3 is becoming routine, a technique for identifying the impurity species in these relatively thin layers has not been established. Spreading resistance analysis (SRA) measures only the net carrier concentration as a function of depth into the layer. Secondary ion mass spectrometry (SIMS) is capable of both elemental analysis and depth profiling, but lacks the sensitivity needed for this application. Other techniques (cryogenic Hall effect, capacitance-voltage, etc.) are not capable of elemental analysis when several species are present.

Photoluminescence spectroscopy (PL), however, has demonstrated the required sensitivity and selectivity to identify multiple impurities at concentrations well below $10^{10} cm^{-3}$ in bulk Si.[1] While some previous work has dealt with the application of PL to the characterization of epitaxial Si, none has dealt with UHP layers.[2] We report the first application of PL, together with SRA depth profiling, to the study of such layers. As, P and Ga are found to be the main, reproducible contaminants in the epitaxial layer, and a process-related interfacial contaminant layer is shown to be due to Al.

EXPERIMENTAL DETAILS

UHP silicon epitaxial layers were grown by chemical vapor deposition (CVD) in a horizontal flow RF heated reactor, operating at atmospheric pressure. Using UHP silane (SiH_4) in purified hydrogen carrier gas, a substrate temperature of 1020°C and silane partial pressure of 10^{-3} atm. resulted in a layer growth rate of 0.6 µm/min. Measured temperatures were corrected for reflection and absorption of the reactor system and the emissivity of silicon. The substrate material used in several runs was obtained from a 1-in. diameter boule of UHP vacuum float zoned Si oriented along <111>. Substrates of nominal 0.4 mm thickness were cut from this boule, and four-point probe and SRA analyses calibrated to NBS n- and p-type standards indicated net p-type carrier concentrations varying from 5×10^{11} to 7×10^{11} cm^{-3} along the length of the boule section used. Current SRA standards exist only for carrier concentrations $> 10^{13}$ cm^{-3}; consequently, measurements below that range are extrapolated. All substrates received the same polishing and cleaning[3] steps immediately prior to growth.

All PL spectra were taken with 250 mW of 514.5 nm excitation in a 3 mm diameter beam. The samples were immersed in liquid He between 4.2 and 1.8 K, and the luminescence was dispersed by a 3/4 m double spectrometer and detected by a Varian VPM159A3 photomultiplier tube in the photon counting mode.

RESULTS AND DISCUSSION

The three samples grown on UHP Si substrates differ only in the *in situ* cleaning used prior to epitaxial layer growth. SRA profiles of these samples are shown in Figure 1. Sample (c) received only a 1190°C, 10 min. bake, while samples (b) and (a) received *in situ* HCl etches removing <0.5 and 2 μm, respectively, of Si prior to growth. Relatively thick (35 μm) Si layers were then grown in order to increase the epilayer PL compared to that of the substrate. Stacking fault densities determined by chemical etch testing ranged from 100 to 1000 cm^{-2}, with a moderate density of point defects. X-ray rocking curves using the (331) reflection indicated good crystalline quality in the epitaxial layers, with peak widths varying from the same as, to less than twice, that obtained from high quality bulk Si. For samples (a) and (b), both the epilayer and substrate were p-type, while for (c) the

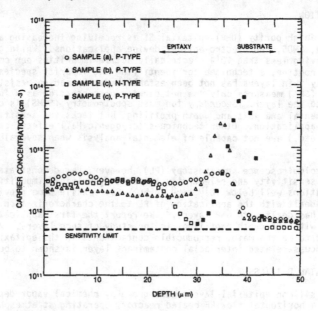

Figure 1   Net concentration of electrically active impurities vs depth for samples (a)-(c). The interface between the epitaxy and substrate is at 35μm. Samples (a) and (b) were determined to be p-type, whereas (c) had an n-type epilayer and substrate with a p-type interfacial layer. The flat, dashed curve in this figure indicates a net carrier concentration below the limit of the SRA measurement, and applies to the substrate region for sample (c).

epilayer and substrate were n-type. Sample (b) and particularly sample (c), have a high conductivity layer at a depth of 35 μm. This indicates the incorporation of a contaminant on the substrate surface, which can be removed by *in situ* HCl etching before growth. From Seebeck effect measurements, incorporated into the SRA system, the high conductivity layers in samples (b) and (c) were determined to be p-type.

Figure 2 shows high resolution PL spectra of samples (a) to (c) and the substrate material (d) taken in the no-phonon spectral region. The substrate PL shown in Figure 2(d) reveals bound excitons (BE) associated with B, P and a trace of Al. The B^2 and * features are due to bound multiexciton complexes (BMEC).[4] The substrate-plus-epitaxy PL shown in Figure 2(a) to (c) show additional lines associated with Ga and As. The bound exciton species were determined primarily on the basis of the observed transition energies. The various no-phonon to TO-phonon (NP/TO) line intensity ratios were used to verify these identifications, based on the observed relations:[4]

$$B'_{TO} \gg B'_{NP}, \quad P'_{TO} \simeq P'_{NP}, \quad As'_{TO} \simeq \frac{1}{2} As'_{NP}$$

$$Ga'_{TO} \simeq Ga'_{NP} \quad \text{and} \quad Al'_{TO} \simeq 2 \times Al'_{NP} \tag{1}$$

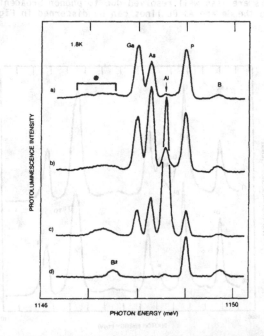

Figure 2   The no-phonon PL spectra of the three epitaxial samples (a) - (c) and the substrate material (d) used in these growth runs. All spectra have been normalized to equal FE TO-phonon replica intensity (not shown) to allow for accurate BE intensity comparisons. B, P, Al, As, and Ga label the BE lines associated with these impurities. B^2 is the two exciton B BMEC, while the band label * in (a) - (c) is a superposition of such BMEC lines for B, Al, and Ga.

where, for example, B'_{TO} represents the line intensity of the boron BE TO
phonon replica. The NP/TO ratio does not vary from sample to sample for a
given impurity. All four spectra in Figure 2 have been scaled so as to
equalize the intensity of the intrinsic free exciton (FE) PL lines. Thus, the
relative peak heights of a given BE line between the samples directly give the
relative average concentrations of that impurity in the region being probed by
PL. The B and P peak heights are seen to be relatively constant in all four
spectra, indicating that the concentration of these two impurities in the
epilayer is not much different than in the substrate material.

All three epilayer spectra show strong Ga and As lines, indicating that
these are the dominant electrically active impurities present in the epitaxial
layer. A very strong Al peak was observed for sample (c), together with a
moderate Al peak for (b) and almost no signal for (a). This correlates well
with the p-type interfacial contaminant seen in Figure 1.

In Figure 3 we show the phonon-replica spectra of the best epitaxial
sample (a) and the substrate (d). Note that the B BE is much stronger in the
TO-phonon replica than in the no-phonon region. The relative BE peak heights
for different impurities in the TO-phonon replica are much more closely
related to the actual relative concentrations of the impurities than are the
no-phonon peak heights, but accurate calibrations are only known[1,2] for B
and P. The peaks are less well resolved due to phonon broadening, but
shoulders due to the Ga and As BE lines can be discerned in Figure 3(a).

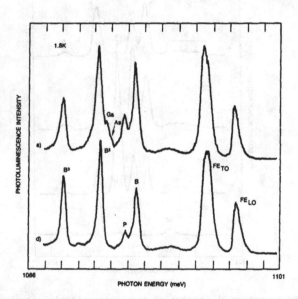

Figure 3   The optical phonon assisted PL spectra of the best epitaxial
sample (a) is compared to that of the substrate material (d). Both
TO and LO phonon replicas are observed for the FE, but the BE and
BMEC couple predominantly to TO-phonons. Phonon broadening results
in lower resolution than in Figure 2, but shoulders due to Ga and As
BE can still be resolved.

Experiments using layers grown on specially doped substrates have shown
that a considerable portion of the FE as well as B and P BE PL originates from
the substrate even with these relatively large epitaxial layer thicknesses.
This results from large FE diffusion lengths in the UHP epitaxial layer, and
makes it difficult to arrive at absolute impurity concentrations. To
circumvent this weakness of PL analysis, UHP layers were grown on indium doped
silicon substrates. Indium doping of the substrate quenches the PL of both
free excitons and bound excitons associated with shallow impurities in the
substrate, while the PL associated with indium occurs outside the spectral
range of interest. In Figure 4 we show the SRA of one such sample. The PL
spectra obtained on the indium-doped substrate material (not shown) indicated
no PL other than that associated with indium. In Figure 5 we show the PL of
that epitaxial sample in the NP and optical-phonon regions. The inset in
Figure 5 shows a high resolution scan of the NP spectral region. Here, the
FE_{TO} and As, Ga, P and B BE NP and TO replica are due to the epitaxy only,
allowing direct evaluation of impurity type and relative abundance in the
layer.

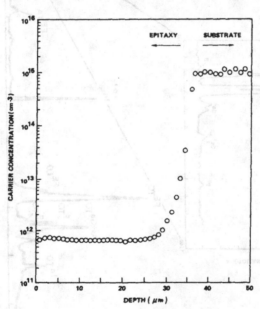

Figure 4   Net concentration of electrically active impurities vs depth for UHP
epitaxy on a Si:In substrate. Resistivity values for the epitaxy
were at or above 20KΩcm. Plotted is the estimated hole
concentration corresponding to that resistivity.

SUMMARY

We have shown that PL has sufficient sensitivity and selectivity to
characterize shallow donor and acceptor impurities in UHP epitaxial Si
layers. As, P and Ga were identified as the primary, reproducible
contaminants in our epitaxial layers, apparently arising from reactor
background contamination. Al was identified as a process-related interfacial
contaminant. Further, work is underway in order to provide accurate, absolute
concentrations from the PL spectra alone, as well as to use PL to observe
lattice defects resulting from unoptimized epitaxial growth.

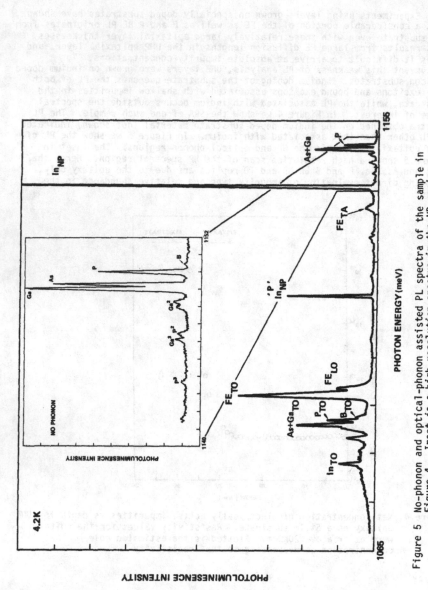

Figure 5 No-phonon and optical-phonon assisted PL spectra of the sample in Figure 4. Inset is a high resolution spectra in the NP region indicating B, P, As and Ga contamination in the epitaxy.

ACKNOWLEDGEMENTS

M.L.W.T. and A.G.S. were supported by the Natural Sciences and Engineering Research Council of Canada. J.E.H. thanks A. Crouse for preparation of the substrate material, H. Glass for X-ray measurements, and J. Quetsch for his support of this work.

REFERENCES

1. M. Tajima, T. Masui, T. Abe and T. Iizuka, Semiconductor Silicon 1981, (Electrochemical Society Inc., Princeton N.J. 1981) pp. 72-89.

2. M. Tajima and M. Nomura, Japan, J. App. Phys. 20, L697 (1981).

3. W. Kern and D. Puotinen, RCA Review 31, 187 (June, 1970).

4. For a review of BE and BMEC in Si, and further references, see: M. L. W. Thewalt in Excitons, edited by E. I. Rashba and M. D. Sturge (North Holland, Amsterdam, 1982) pp. 393-458.

LOW TEMPERATURE PHOTOLUMINESCENCE STUDY OF DOPED CdTe AND CdMnTe FILMS GROWN BY PHOTOASSISTED MOLECULAR BEAM EPITAXY

N. C. GILES, R. N. BICKNELL, AND J. F. SCHETZINA
Department of Physics, North Carolina State University, Raleigh,
North Carolina, 27695-8202

ABSTRACT

N-type and p-type (100) CdTe films have been grown on (100) CdTe substrates by photoassisted molecular beam epitaxy, using indium and antimony as n-type and p-type dopants, respectively. The application of this growth technique to substitutionally dope another II-VI material is demonstrated by the successful n-type doping of (100) CdMnTe films with indium. Modulation-doped superlattices consisting of barrier layers of CdMnTe:In alternating with CdTe have also been grown. The point defect nature of these in situ doped films and multilayers is studied with low temperature (1.6-5 K) photolumin-escence and excitation photoluminescence measurements. The introduction of the dopant atoms using this new growth technique produces immediate changes in the photoluminescence spectra of the epilayers. Photoluminescence studies of the superlattices show the effects of quantum well confinement and band filling due to free carriers.

INTRODUCTION

One of the principal problems that has hindered the use of II-VI com-pounds in device applications has been the tendency for these materials to self-compensate. As a consequence, the intentional introduction of either n-type or p-type dopants usually results in low activation of the impurity, yielding high resistivity as-grown material. The recent development of a thin film growth technique, compatible with molecular beam epitaxy (MBE), in which control of the electrical properties of layers of both conductivity types is demonstrated has been reported [1,2]. This new doping technique which was developed at North Carolina State University, photoassisted MBE, allows intro-duction of selected dopant species during the actual film growth. This proce-dure differs from conventional MBE in that the substrate is illuminated during the growth of the epilayers. We find that this process produces immediate changes in both the electrical and optical properties of the as-grown films. In fact, only those films grown under illumination are conducting. A descrip-tion of the growth details, and structural and electrical properties of the doped CdTe and CdMnTe films can be found in the literature [3,4].

In this paper, we report the results of a low temperature (1.6-5 K) photoluminescence (PL) study of the as-grown, electrically active, in situ doped CdTe and CdMnTe epilayers, and n-type CdMnTe-CdTe modulation-doped superlattices (SLs). Indium and antimony were used as dopant atoms for the n-type and p-type layers, respectively. The characteristic low temperature PL spectra normally associated with as-grown CdTe containing these dopants, which is often high resistivity, is dominated by defect band emission (1.4-1.45 eV). In contrast, the PL spectra from conducting epilayers grown by photoassisted MBE are dominated by shaop, near-edge radiative transitions. We are thus able to use the PL characterization technique to gain information about the point defect structure in these conducting films. In addition, high resolution excitation PL measurements on the films grown by photoassisted MBE have been performed, using a tunable photo-pumped dye laser. The excitation PL tech-nique allows an identification of the processes feeding the radiative recom-bination channels. As a consequence, it is possible to distinguish PL lines due to excitonic recombinations from those involving impurity level-to-band transitions in the epilayers. In the modulation-doped SLs, the excitation PL measurements are used to determine the positions of quantum well energy levels.

EXPERIMENT

A description of the luminescence apparatus used in this study is given in an earlier report [5]. The samples are mounted on a copper sample holder and cooled to liquid helium temperatures in a Janis Research Products Super-Varitemp devar. The chopped focused excitation beam, from either a He-Ne or Ar^+ laser, and the emitted luminescence were directed to and from the sample in a near-backscattering geometry. Detection of the luminescence signal is made with a photomultiplier tube with S-1 response and a lock-in amplifier. The excitation PL measurements were obtained using a photo-pumped dye laser. The range of the dye used (Exciton LDS 751) is 720-835 nm, and is continuously tunable with a multiplate birefringent filter. The wavelength setting of the spectrometer and the wavelength output of the dye is under computer control.

RESULTS AND DISCUSSION

CdTe:In

The low temperature PL spectra commonly observed for bulk In-doped CdTe, which is often high resistivity as-grown, is dominated by defect band emission at about 1.4 eV [6,7]. Similar results have been observed in PL spectra at 77 K for CdTe:In films grown by conventional MBE [8]. This broad emission band in CdTe:In has been identified as being related to donor-neutral vacancy complexes [9,10]. These complexes are believed to form when the dopant atoms are incorporated into the II-VI lattice displaced from their appropriate tetrahe-drally-bonded lattice sites. The CdTe:In films grown by conventional MBE in our lab are insulating as-grown, and exhibit low temperature PL spectra which are typically dominated by emission at about 1.4 eV, similar to the PL spectra reported for bulk CdTe:In. In sharp contrast to this, we have observed spec-tra from conducting n-type films, grown by photoassisted MBE, in which the dominate PL emission peaks are located in the near-edge region (1.580-1.595 eV), and the defect band emission (1.4 eV) is noticeably absent [1,5]. Thus, the incident photon beam may provide the energy required to surmount surface potential barriers which limit the incorporation of dopant atoms at substitu-tional lattice sites, resulting in a reduction of donor-neutral vacancy com-plexes and the absence of PL emission at 1.4 eV. A summary of our PL study of CdTe:In films is provided here; additional details can be found in the literature [5].

The edge emission region for high quality insulating In-doped CdTe films at 1.6 K is dominated by sharp (FWHM=0.5-0.8 meV) acceptor-bound exciton (A^0, X) recombination lines occurring at about 1.5892 eV. In addition, PL recom-bination at about 1.584-1.585 eV is observed. The exact position of the peak differs slightly from film to film. Excitation PL measurements identify the peak at about 1.584 eV as donor-valence band (D^0,h) recombination [5], with a donor ionization energy of about 21-22 meV. We suggest this deep donor level is created when the In atom does not substitutionally replace the Cd atom in the lattice at a tetrahedrally-bonded site, but instead is slightly displaced from the optimum site. Thus, a donor energy level could be created that is deeper than the hydrogenic donor level expected for indium (14.1 meV [11]).

In sharp contrast to the PL spectra for insulating films grown by con-ventional MBE, the near-edge PL emission for conducting films grown by photo-assisted MBE is dominated by donor-associated recombinations. For every n-type film, activation was accompanied by the appearance and enhancement of two PL transitions at about 1.580 eV and 1.592 eV, which are identified as

(D^0,h) recombinations. In addition, the exact positions of the two peaks depend on the carrier concentrations in the films, as may be expected for activated donor level transitions in n-type material. The ionization energies E_D for the two activated donor levels are 14.8 meV and 26.4 meV, for the peaks at 1.592 eV and 1.580 eV, respectively. One level is due to the incorporation of In (14.8 meV); the other level (26.4 meV) we believe is related to the presence of halogen impurities (chlorine, bromine), which may be introduced into the films from the source material or from the substrates [3,5]. If our interpretation of the line at 1.580 eV is correct, this suggests that the photoassisted MBE growth process may be used to activate a variety of impurity atoms that are present at the film growth surface. The (A^0,X) transition at 1.589 eV and the (D^0,h) line at 1.584 eV, associated with the off-site incorporation of indium into the lattice, are low in amplitude. The photoassisted MBE process thus appears to reduce the concentrations of deep off-site indium donors and the accompanying compensating acceptors. As a consequence, the CdTe:In films are highly activated [1,3] and conducting.

CdTe:Sb

In contrast to the PL spectra (1.6 K) for n-type films which show donor-associated transitions in the near-edge region, the low temperature PL spectra from p-type films are dominated by acceptor-related emission lines [2,5]. The deep level defect band normally seen in high resistivity doped CdTe at 1.4-1.45 eV is absent. The main recombination peak, at 1.6 K, occurs at about 1.5896 eV and is identified as (A^0,X) recombination associated with the Sb acceptor level. In an earlier report, the ionization energy E_A of the acceptor level was estimated to be about 68 meV [5]. We have recently observed PL emission at about 1.541 eV (FWHM=3 meV) from a conducting CdTe:Sb film which we believe, on the basis of excitation PL measurements, is due to conduction band-acceptor level (e,A^0) recombination. The acceptor level is almost certainly due to the Sb. In agreement with our earlier report, we find that the Sb acceptor level, for this film, is about 65 meV above the valence band.

CdMnTe:In

In order to demonstrate that the photoassisted MBE growth technique can be applied to other materials, we have prepared n-type conducting layers of the dilute magnetic semiconductor $Cd_{1-x}Mn_xTe$ [4,13]. We are unaware of previous reports of PL studies of n-type CdMnTe grown by conventional bulk or thin film techniques. A low temperature (1.6 K) PL spectrum for an In-doped $Cd_{0.85}Mn_{0.15}Te$ epilayer is shown in Fig. 1. The PL spectrum is dominated by a single bright peak at 1.8051 eV (FWHM=15.3 meV), believed to be bound exciton recombination. The expected alloy broadening contribution to the PL linewidth of bound exciton lines, for this Mn concentration, is of the order of 12 meV [12]; therefore, broadening contributions from the introduction of dopants appear to be small. Low amplitude (A^0,X) emission from the CdTe buffer layer is observed at 1.5894 eV.

A low temperature (1.6 K) PL spectrum for a $Cd_{0.84}Mn_{0.16}Te$:In epilayer is shown in Fig. 2. The PL spectrum for this film is dominated by a single emission peak at 1.8215 eV (FWHM=15.9 meV). Note the shift to higher energy of the PL emission with increasing Mn concentration. Again, broadening contributions related to dopant incorporation appear to be small. Longitudinal phonon emission of the main PL peak is seen as a small shoulder at 1.798 eV. No deep level defect band emission was observed in the PL spectra from the conducting CdMnTe:In films, attesting to the high quality of the epilayers.

CdMnTe:In-CdTe Superlattices

The growth and electrical properties of n-type modulation doped CdMnTe-CdTe SLs grown by photoassisted MBE have been reported [4,13]. A low temperature PL spectrum for a $Cd_{0.81}Mn_{0.19}Te$:In-CdTe SL is shown in Fig. 3. The SL consists of 50 double layers of CdTe (L_w=155 A) alternating with CdMnTe (L_b=155 A). The PL spectrum is dominated by a sharp peak at 1.6014 eV due to recombination of excitons in the CdTe quantum wells of the SL. This peak is shifted to a slightly higher energy compared with free exciton luminescence in bulk CdTe (1.5964 eV) because of the quantum size effect associated with carrier confinement in superlattices. The peak at 1.5931 eV is believed to be due to (D^0,h) transitions in the CdTe quantum wells [13]. The sharp feature in the PL spectrum at 1.5891 eV, along with the phonon replica at 1.5683 eV, is attributed to (A^0,X) recombination coming from the CdTe substrate and buffer layer. The PL emission tail observed at higher energies than the main peak is believed to arise from band filling due to free carriers.

Excitation PL measurements were performed to determine the positions of quantum well energy levels that are not seen in the standard PL spectra of superlattice structures. The excitation PL spectrum (1.6 K) shown in Fig. 4, for the SL described above, was taken with the spectrometer set to 1.6015 eV, close to the energy peak of the main PL line. Several resonances in luminescence intensity are observed. Based on our interpretation of excitation PL data reported earlier [14] for undoped CdMnTe-CdTe SLs, in which strain-induced shifts of the heavy- and light-hole bands were accounted for, an identification of the resonances in Fig. 4 can be made. Although the Kronig-Penney model predicts five quantized conduction bands in this superlattice structure, only the n=1 and n=2 energy bands (E_1=1.6181 eV, E_2=1.6525 eV) are within the range of the experiment. The resonances at 1.606 eV and 1.635 eV are associated with the recombinations of n=1 heavy-hole excitons, $E_{1,hh}$, and n=1 light-hole excitons, $E_{1,lh}$, respectively. The feature at 1.614 eV is close to the calculated energy of the n=1 conduction band state, so is believed to be due to excited states of the n=1 heavy-hole exciton. A similar resonance feature has been observed in an undoped CdMnTe-CdTe SL [14]. The n=2 heavy-hole exciton, $E_{2,hh}$, should occur at about 1.640 eV, coincident with the $E_{1,lh}$ resonance feature.

A PL spectrum (1.6 K) for an n-type $Cd_{0.81}Mn_{0.19}Te$-CdTe SL with $L_w=L_b=74$ A is shown in Fig. 5. The SL emission peak now occurs at 1.6324 eV, due to the reduced well thickness. As before, low amplitude emission peaks from the CdTe substrate and buffer layer are observed in the spectrum. The tailing of the main PL peak to high energies is, again, due to band filling with free carriers. A Kronig-Penney model calculation predicts three quantized conduction band energy states in this superlattice structure. An excitation PL spectrum (1.6 K) for this sample, with E_s=1.6320 eV, close to the SL emission peak, is shown in Fig. 6. We identify the resonances at 1.638 eV and 1.656 eV as being associated with the $E_{1,hh}$ and $E_{1,lh}$ transitions, respectively. The excitonic transitions associated with the n=2 quantized conduction band energy (1.746 eV) are outside the range of the experiment.

SUMMARY

In summary, a low temperature PL and excitation PL study of electrically active, in situ doped CdTe and CdMnTe films, and n-type modulation-doped CdMnTe-CdTe SL grown by photoassisted MBE is reported. The PL spectra from the conducting CdTe films contrast sharply with the spectra normally associated with as-grown doped CdTe. We observe spectra at 1.6 K which are dominated by sharp, near-edge PL transitions. The defect band typically seen at about 1.4 - 1.45 eV in doped CdTe is low in amplitude, or completely absent in the luminescence spectra.

The near-edge PL emission region has been studied to gain information about the point defect structure in these materials. A deep In donor level is identified in the insulating CdTe:In epilayers grown by conventional MBE as being accompanied by the creation of a compensating acceptor level. Two activated donor levels are identified from the PL spectra of n-type CdTe:In films grown by photoassisted MBE, with ionization energies of 14.8 meV (In), and 26.4 meV (Cl, Br), respectively. The PL spectra from p-type CdTe:Sb films are dominated by sharp (A^0,X) recombination. The ionization energy of the activated antimony level is estimated to be about 65-68 meV.

The PL spectra from In-doped CdMnTe films grown by photoassisted MBE are dominated by relatively narrow, bright emission lines believed to be bound exciton recombination. Deep level defect band emission is not observed, attesting to the high quality of the epilayers.

The PL emission peaks from n-type CdMnTe-CdTe SL show the effects of quantum well confinement. Tailing of the SL emission peaks to higher energies is believed to result from band filling due to free carriers. Heavy-hole and light-hole exciton transitions are identified from excitation PL measurements.

ACKNOWLEDGEMENTS

The authors wish to acknowledge the several sources of support for this research effort: the Army Research Office (through contract DAAG29-84-K-0039), the National Science Foundation (through grant DMR83-13036), and the Defense Advanced Research Projects Agency (through DARPA/ARO joint contract DAAL03-86-K-0146).

REFERENCES

1. R. N. Bicknell, N. C. Giles, and J. F. Schetzina, Appl. Phys. Lett. 49, 1095 (1986).
2. R. N. Bicknell, N. C. Giles, and J. F. Schetzina, scheduled to appear in the December 22, 1986 issue of Appl. Phys. Lett. (manuscript L-8403).
3. R. N. Bicknell, N. C. Giles, and J. F. Schetzina, Proc. of the 1986 U.S. Workshop on the Physics and Chemistry of MCT, to be published in the Nov. 1987 issue of J. Vac. Sci. and Technol. A.
4. R. N. Bicknell, N. C. Giles, and J. F. Schetzina, Proc. of the 1986 MRS Symposium on "Materials for Infrared Detectors and Sources", Boston, Mass.
5. N. C. Giles, R. N. Bicknell, and J. F. Schetzina, Proc. of the 1986 U.S. Workshop on the Physics and Chemistry of MCT, to be published in the Nov. 1987 issue of J. Vac. Sci. and Technol. A.
6. C. E. Barnes and K. Zanio, J. Appl. Phys. 46, 3959 (1975).
7. C. B. Norris and K. Zanio, J. Appl. Phys. 53, 6347 (1982).
8. K. Sugiyama, Jpn. J. Appl. Phys. 46, 665 (1982).
9. Y. Marfaing, Revue de Phys. Appl. 12, 211 (1977).
10. N. V. Agrinskaya, E. N. Arkad'eva, and O. A. Matveev, Sov. Phys. Semicond. 5, 767 (1971).
11. J. L. Pautrat, J. M. Francou, N. Magnea, E. Molva, and K. Saminadayar, J. of Crystal Growth 72, 194 (1985).
12. P. Becla, D. Kaiser, N. C. Giles, Y. Lansari, and J. F. Schetzina, accepted for publication in J. of Appl. Phys.
13. R. N. Bicknell, N. C. Giles, and J. F. Schetzina, submitted to Appl. Phys. Lett.
14. D. K. Blanks, R. N. Bicknell, N. C. Giles-Taylor, J. F. Schetzina, A. Petrou, and J. Warnock, J. Vac. Sci. Technol. A 4, 2120 (1986).

276

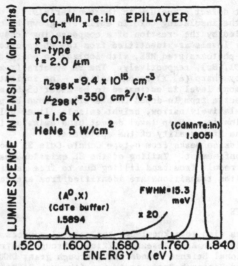

Fig. 1. Photoluminescence spectrum (1.6 K) for an n-type $Cd_{0.85}Mn_{0.15}Te$:In epilayer.

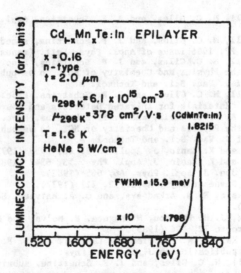

Fig. 2. Photoluminescence spectrum (1.6 K) for an n-type $Cd_{0.84}Mn_{0.16}Te$:In epilayer.

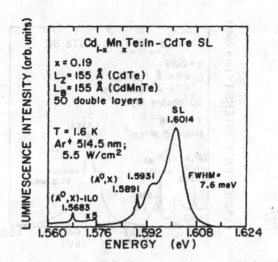

Fig. 3. Photoluminescence spectrum (1.6 K) for an n-type modulation doped $Cd_{0.81}Mn_{0.19}Te$-CdTe superlattice with a period of 310 A.

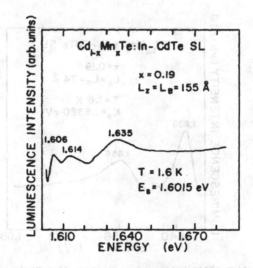

Fig. 4. Excitation photoluminescence spectrum (1.6 K) for an n-type modulation doped $Cd_{0.81}Mn_{0.19}Te$-CdTe superlattice with a period of 310 A.

278

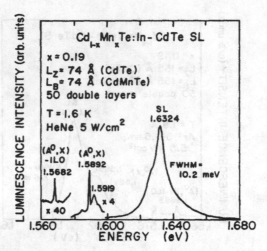

Fig. 5. Photoluminescence spectrum (1.6 K) for an n-type modulation doped $Cd_{0.81}Mn_{0.19}Te$-CdTe superlattice with a period of 148 A.

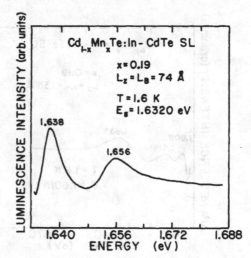

Fig. 6. Excitation photoluminescence spectrum (1.6 K) for an n-type modulation doped $Cd_{0.81}Mn_{0.19}Te$-CdTe superlattice with a period of 148 A.

EFFECTS OF BORON IMPLANTATION ON SILICON DIOXIDE PASSIVATED HgCdTe

R. C. BOWMAN, JR.[*], J. MARKS[*], R. G. DOWNING[**], J. F. KNUDSEN[*], AND G. A. TO[*]
The Aerospace Corporation, Laboratory Operations, P. O. Box 92957 Los Angeles, CA 90009
[**]Center for Analytical Chemistry, National Bureau of Standards, Gaithersburg, MD 20899

ABSTRACT

The influence of boron ion implants on the optical and physical properties of photochemically deposited SiO_2 films on $Hg_{0.7}Cd_{0.3}Te$ and silicon has been investigated. The distributions of the boron atoms between the SiO_2 film and substrate have been determined by a non-destructive neutron depth profiling method. The implants produce an apparent densification of the SiO_2 films, which is accompanied by an increase in refractive index and changes in the infrared vibrational spectra for these films.

INTRODUCTION

Due to the inherent instability of the HgCdTe surface [1], various methods have been used to grow thin surface films that can passivate the material during the numerous processing steps required to fabricate infrared sensor devices [2]. It was recently shown [3] that photochemically (PC) deposited SiO_2 films can give greatly improved electrical properties for the photodiodes when compared to HgCdTe materials that were passivated with native or anodic oxides. However, the reproducibility of high quality PC-SiO_2 passivated photodiodes has not been very good and there are also problems with the adhesion of these films. Consequently, modifications to the original Hg sensitized deposition process [3] are being investigated [4] to produce improved PC - SiO_2 films on HgCdTe.

Ion implantation is commonly used [1,2,5-7] to form the diode junctions in HgCdTe photovoltaic devices. Implants through passivation films can produce significant differences in the electrical properties of the HgCdTe layer when compared to direct implants into uncoated samples. For example, an enhanced donor concentration was recently observed from Hall measurements [8] when boron ions were implanted through nominal 100 nm PC-SiO_2 films as compared to an identical implant into a Br_2 - methanol etched surface. This difference was correlated [8] with boron depth profiles that showed the boron concentration peak to lie immediately below the SiO_2-HgCdTe interface in the passivated sample; whereas, this maximum occured 100 nm deeper in the etched-only sample. Similar shifts in the boron profiles were also found from several other PC-SiO_2 passivated HgCdTe samples [8] for different boron implant energies. During these earlier studies, the ellipsometry data that had been obtained before and after the boron implants implied significant changes in the film thickness (d) and refractive index (n) were produced by the boron implants. This paper

---

[**]Certain commercial equipment, instruments, or materials are identified in this paper in order to adequately specify the experimental procedure. Such identification does not imply recommendation or endorsement by National Bureau of Standards, nor does it imply that the materials or equipment identified are necessarily the best available for the purpose.

presents subsequent experimental results on the behavior of boron implants through SiO_2 passivated $Hg_{0.7}Cd_{0.3}Te$ and silicon single crystals. Both the distribution of the boron between the SiO_2 and substrate layers and changes in the SiO_2 films have been examined. Anneals up to $400^{\circ}C$, which are probably at the upper limits for conventional processing[2,5-7] of HgCdTe sensor devices, do not significantly alter the implant modified properties of the SiO_2 films.

EXPERIMENTAL DETAILS

The nominal $Hg_{0.7}Cd_{0.3}Te$ samples were approximately 15-20 μm thick liquid-phase epitaxial layers grown on <111>-CdTe substrates. These epilayers were initially p-type prior to the boron implants, which have been shown [8] to produce degenerate n-type layers. The silicon samples were phosphorus doped n-type wafers with polished <111>-faces.

The silicon dioxide films were deposited by three different photochemical (PC) reactions:

(A) $\quad SiH_4 + N_2O + Hg \xrightarrow{h\nu(253.7\ nm)} SiO_2 + by\text{-}products$

(B) $\quad Si_2H_6 + NO_2 \xrightarrow{h\nu(253.7\ nm)} SiO_2 + by\text{-}products$

(C) $\quad SiH_4 + NO_2 \xrightarrow{h\nu(104.0\ nm)} SiO_2 + by\text{-}products$

These reactions were carried out in a low pressure chemical vapor deposition reactor where the sample is placed on a heated stage. All the present PC-depositions were done with the samples heated to $100^{\circ}C$. Method A, which is often denoted [3] as the Photox^{TM} process, involves a Hg catalyzed production of atomic oxygen to react with silane. Because the amount of oxygen generated depends on the Hg concentration and the photon flux, the compositions of the SiO_x films produced by Method A were found to vary as reflected by differences in their n values between 1.46 and 1.98. Method B uses a photo-enhanced reaction between disilane and nitrogen dioxide to form silicon dioxide films with refractive indices of the stoichiometric composition, but these films were less dense and had some adhesion problems. Furthermore, the infrared spectra for the Method B SiO_2 films indicated the presence of Si-H and other molecular species. Finally, Method C used the reaction between silane and nitrogen dioxide in the presence of vacuum UV radiation from windowless microwave excited argon discharge lamp. The PC-SiO_2 films grown by Method C had indices of refraction between 1.45 and 1.46 and no evidence of the Si-H stretch mode in their IR spectra.

A model 400 MPR-Veeco/AI ion implanter has been used to implant either $^{11}B^+$ or $^{10}B^+$ ions into electrically grounded samples. The beam current was sufficiently low to minimize heating effects during implantation. The implant energies are summarized in Table I were the total boron implant dose was 2×10^{15} B^+/cm^2 in each case.

The properties of the SiO_2 films were monitored by ellipsometry measurements with the 632.8 nm line of a He-Ne laser and room temperature IR spectra obtained with a Nicolet model MX-1 Fourier transform spectrometer. The distributions of the implanted boron (^{10}B) atoms were obtained by the non-destructive neutron depth profiling (NDP) method [9,10] at the National Bureau of Standards 20 MW research reactor. Descriptions of this profiling procedure as well as special demands for NDP analysis in

Table I. Properties of epitaxial $Hg_{0.7}Cd_{0.3}Te$ and <111>-Silicon samples that have been implanted with boron ions to a total dose of 2×10^{15} ions/cm^2. The SiO$_2$ film thicknesses (d) and indices of refraction (n) determined by ellipsometry. The boron peak depth (R_p) is from semiconductor surface or to interface with PC-SiO$_x$.

Sample I.D. Number	Material	Treatment	d (nm)	R_p (nm)	n	SiO$_x$ IR Peaks (cm^{-1}) I	II	III
NDP-6	$Hg_{0.67}Cd_{0.33}Te$	250 keV ^{10}B Implant	0	-	-	-	-	-
NDP-7	$Hg_{0.67}Cd_{0.33}Te$	SiO$_x$ (Method A)	107	-	1.63	-	-	-
"	"	250 keV ^{10}B Implant	96	325	1.84	1021	-	441
"	"	400°C Anneal	94	-	1.82	1036	-	-
NDP-A	$Hg_{0.67}Cd_{0.33}Te$	100 keV ^{10}B Implant	0	212	-	-	-	-
"	"	SiO$_x$ (Method C)	115	217	1.46	1068	-	~460
NDP-B	$Hg_{0.67}Cd_{0.33}Te$	SiO$_x$ (Method B)	175	-	1.45	-	-	-
"	"	100 keV ^{10}B Implant	149	94	1.53	1047	807	435
"	"	400°C Anneal	146	-	1.53	-	-	-
NDP-C	Silicon	100 keV ^{10}B Implant	0	326	-	-	-	-
"	"	SiO$_x$ (Method C)	96	326	1.45	1067	809	~440
NDP-D	Silicon	SiO$_x$ (Method B)	227	-	1.44	1068	871	449
"	"	100 keV ^{10}B Implant	186	151	1.52	1045	823	425
"	"	400°C Anneal	187	-	1.49	1041	823	423
IM-2	Silicon	SiO$_x$ (Method C)	189	-	1.46	1074	921	~460
"	"	100 keV ^{11}B Implant	173	-	1.46	1065	810	~450

HgCdTe have been reported previously [8]. Because of the "pulse pile-up" effects caused by the energetic electrons emitted by the neutron-activated cadmium isotopes, the boron NDP curves have been asymmetrically broadened beyond normal total system resolution. Hence, the current boron profiles are somewhat distorted but the main conclusions of this study are not seriously affected. Analysis procedures to accurately correct the boron NDP profiles for these pulse pile-up contributions are being developed.

RESULTS AND DISCUSSION

The NDP boron (^{10}B) distributions in Fig. 1 illustrate the influence of ion implant energies. When boron ions are implanted into etched-only surfaces, the profile peaks (R_p) vary between 146 nm and 420 nm below the surface when the implant energies were increased from 40/100 keV to 250 keV, respectively. These results are in reasonable agreement with recent projected range parameter calculations and NDP measurements in bulk $Hg_{0.8}Cd_{0.2}Te$ [9]. The boron profiles in $Hg_{0.7}Cd_{0.3}Te$ for two implant energies that were done with and without the PC-SiO_2 passivation films were compared in Figs. 2 and 3. The NDP data do not show significant differences in the profiles or major discontinuities in the boron contents at the $Hg_{0.7}Cd_{0.3}Te$-SiO_2 interfaces. However, the R_p results in Table I show that passivation films cause the boron peak to occur less deep (which is approximately the SiO_2 film thickness) in the $Hg_{0.7}Cd_{0.3}Te$ layer. In addition, the absolute boron contents at the HgCdTe-SiO_2 interfaces are greater than at the surfaces of the etched-only sample which gives the enhanced carrier contents that were previously described [8]. The NDP results in Fig. 4 show the boron distributions to be very similar for 100 keV boron implants through PC-SiO_2 films on silicon although the latter NDP curves are sharper and the R_p values are shifted deeper compared to the profiles for $Hg_{0.7}Cd_{0.3}Te$ results in Fig. 3.

The changes in film thickness (d) and index of refraction (n) for PC-SiO_2 films on $Hg_{0.7}Cd_{0.3}Te$ and Si that occur following the boron implants are summarized in Table I. Initially, these changes were thought to be artifacts in the ellipsometry analyses due to alterations in the substrate optical properties during the implants. However, the result obtained from PC-SiO_2 films grown on previously implanted $Hg_{0.7}Cd_{0.3}Te$ and Si substrates have indicated only minor contributions can be attributed to implant modification of the substrate properties. Consequently, these boron implants have caused apparent compactions of the PC-SiO_2 films (which are much greater in those films grown by Methods A and B) along with corresponding increases in the refractive index. The similar behavior that was observed from ion-implanted fused silica [11] and laser-irradiated thermally-grown SiO_2 on Si wafers [12] has been associated with changes in glass structure (i.e., variations in Si-O bonds and ring dimensions [12]). The boron implants also altered the IR spectra of the PC-SiO_2 films as shown in Fig. 5 for a Method B film on Si where three IR peaks have been identified. The IR peaks I and III have been assigned [13,14] to the Si-O bond-stretching and bond-bending vibrational modes, respectively, in amorphous SiO_2 while peak II may represent either a Si-motion mode [13] or a Si-N stretch [15] due to incidental nitrogen incorporation in the deposited films. As shown in Fig. 5, the boron implants cause substantial broadening of the Si-O stretching modes as well as systematic negative shifts in the three IR peak positions, which are summarized in Table I along with the d and n values for various PC-SiO_2 films after different treatments. The less thoroughly characterized (due to experimental difficulties) IR peaks for PC-SiO_2 films on the $Hg_{0.7}Cd_{0.3}Te$ samples are also given in Table I and imply similar shifts to lower frequencies after the boron implants. Although these changes in the IR spectra could be caused by implant induced modifications in the SiO_2 structure, small

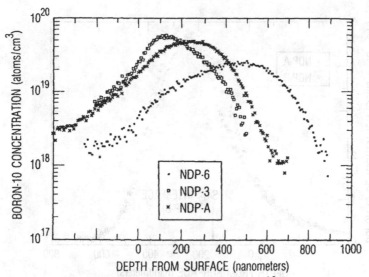

Fig. 1. Neutron Depth Profiles (NDP) of ^{10}B distributions in epitaxial $Hg_{0.7}Cd_{0.3}Te$ for different energy implants into chemically etched only surfaces.

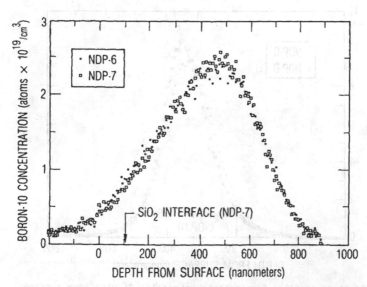

Fig. 2. Comparison of ^{10}B profiles for 250 keV boron implants into SiO_2 passivated and bare surface $Hg_{0.7}Cd_{0.3}Te$ samples (NDP 7 and NDP 6, respectively.) The arrow denotes position of SiO_2-$Hg_{0.7}Cd_{0.3}Te$ interface.

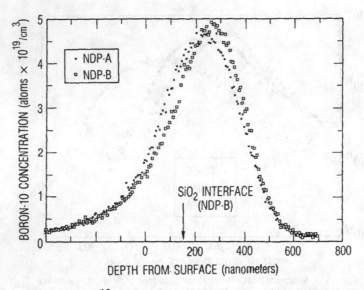

Fig. 3. Comparison of [10]B profiles for 100 keV boron implants into $Hg_{0.7}Cd_{0.3}Te$ samples with a SiO_2 film (NDP-A) and without a passivation film (NDP-B). The arrow denotes the SiO_2-$Hg_{0.7}Cd_{0.3}Te$ interface.

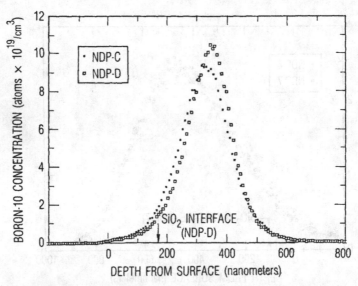

Fig. 4. NDP boron profiles for implants into <111>-Si with bare surface (NDP-C) and PC-SiO_2 film (NDP-D) where arrow denotes Si-SiO_2 interface.

285

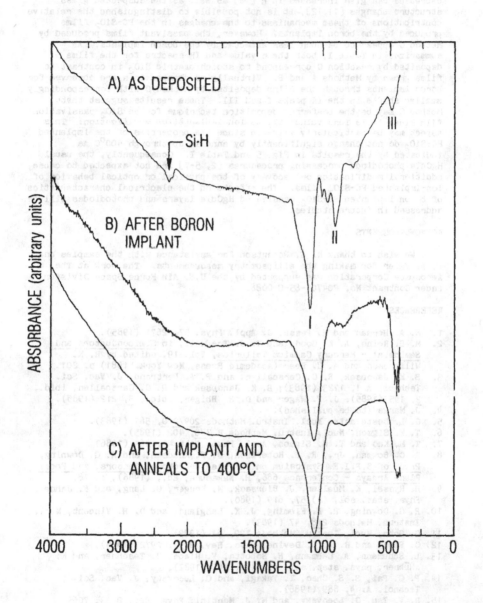

Fig. 5. Infrared spectra for PC-SiO₂ on a silicon substrate (NDP-D).

decreases (i.e., about 10% or so) in the SiO_x stoichiometries would also lead to negative shifts in the IR peaks [13,14]. Because stoichiometry decreases can give increases in n [16] as well as the suspected glass structure changes [11,12], it is not possible to distinguished the relative contributions of these mechanisms to the changes in the $PC-SiO_2$ films produced by the boron implants. However, the passivant films produced by Method C appear to be much less affected by the boron implants. As summarized in Table I, both the n value and IR spectra for the films deposited by reaction C correspond to stoichiometric SiO_2 in contrast to films grown by Methods A and B. Virtually no changes in n are observed for boron implants through the films deposited by Method C with correspondingly smaller shifts in the IR peaks I and III. These results suggest that Method C may be the preferred deposition technique for an SiO_2 passivation film since it is less vulnerable to ion implantation modification. This aspect may be particularly valuable since the properties of the implanted $PC-SiO_2$ do not change significantly by anneals up through $400^{\circ}C$ as indicated by the results in Fig. 5 and Table I. Consequently, the usual HgCdTe photodiode processing procedures [2, 5-7] are not expected to cause additional modification or recovery of the physical or optical behavior of ion-implanted $PC-SiO_x$ films. The effects on the electrical characteristics of boron implanted $PC-SiO_2$ passivated HgCdTe layers and photodiodes will be addressed in future studies.

ACKNOWLEDGEMENTS

We wish to thank R. E. Robertson for assistance with the samples and R. A. Egler for making the ellipsometry measurements. The work at The Aerospace Corporation was supported by the U.S. Air Force Space Division under Contract No. FO4701-85-C-0086.

REFERENCES

1. M. A. Herman and M. Pessa, J. Appl. Phys. 57, 2671 (1985).
2. M. B. Reine, A. K. Good, and T. J. Tredwell, in Semiconductors and Semimetal - Mercury Cadmium Telluride, Vol. 19, edited by R. K. Willardson and A. C. Beer (Academic Press, New York, 1981) p. 201.
3. B. K. Janousek, R. C. Carscallen, and P. A. Bertrand, J. Vac. Sci. Technol. A 1, 1723 (1983); B. K. Janousek and R. C. Carscallen, ibid., 3, 195 (1985); J. F. Wager and D. R. Rhiger, ibid., 3, 212 (1985).
4. J. Marks (to be published).
5. G. L. Destefanis, Nucl. Instru. Methods 209/210, 567 (1983).
6. T. W. Sigmon, Nucl. Instru. Methods B 7/8, 402 (1985).
7. T. M. Kao and T. W. Sigmon, Appl. Phys. Lett 49, 464 (1986).
8. R. C. Bowman, Jr., R. E. Robertson, J. F. Knudsen, and R. G. Downing, Proc. of S.P.I.E. Symposium on Infrared Detectors, Sensors, and Focal Plane Arrays - Conference 686, H. Nakamura, Ed., (1986), p. 18.
9. H. Ryssel, K. Mueller, J. Biersack, W. Kruger, G. Lang, and F. Jahnel, Phys. stat. sol. (a) 57, 619 (1980).
10. R. G. Downing, R. F. Fleming, J. K. Langland, and D. H. Vincent, Nucl. Instrum. Methods 218, 47 (1983).
11. G. Götz, Nucl. Instru. Methods 199, 61 (1982).
12. C. Fiori and R. A. B. Devine, Phys. Rev. B 33, 2972 (1986).
13. L. Schumann, A. Lehmann, H. Sobotta, V. Riede, U. Teschner, and K. Hübner, phys. stat. sol. (b) 110, K69 (1982).
14. P. G. Pai, S. S. Chao, Y. Takagi, and G. Lucovsky, J. Vac. Sci. Technol. A. 4, 689 (1986).
15. D. V. Tsu, G. Locovsky, and M. J. Mantini, Phys. Rev. B 33, 7069 (1986).
16. H. N. Rogers and D. R. Rhiger, Air Force Contractor Final Report AFWAL-TR-84-4056.

ON THE AUGER RECOMBINATION PROCESS IN P-TYPE LPE HgCdTe

J.S. CHEN, J. BAJAJ, W.E. TENNANT, D.S. LO, M. BROWN* AND G. BOSTRUP
Rockwell International Science Center, 1049 Camino Dos Rios, Thousand
Oaks, CA 91360
* Present Address: Grumman Corporate Research Center, Bethpage, NY

ABSTRACT

Minority carrier lifetime measurements at 77K were carried out in p-type liquid phase epitaxial (LPE) $Hg_{1-x}Cd_xTe/CdTe$ (x = 0.22) using the photoconductive decay technique. Lifetimes of 20 to 7000 ns were obtained in samples with hole concentrations, p_0, in the range 10^{14} to 10^{16} cm^{-3}. The hole concentrations were determined by analyzing the Hall data using a double-layer model. It was found that the minority carrier lifetime is inversely proportional to $p_0^{1.86}$. This result demonstrates that the Auger mechanism may be the dominant recombination process in p-type LPE $Hg_{0.78}Cd_{0.22}Te/$ CdTe. The temperature dependence of minority carrier lifetime was also measured between 10 and 200K for several samples.

INTRODUCTION

Even though the minority carrier lifetime and recombination mechanisms in HgCdTe have been extensively investigated experimentally and theoreti-cally, most of the studies were focused on bulk HgCdTe [1-15]. In addition, the conclusions of various reports differ drastically. For example, for n-type bulk HgCdTe with x ≈ 0.22, the measured minority carrier lifetimes have been interpreted by different groups in terms of various recombination proc-esses such as Auger [1,4,5,10,11], Shockly-Read [3,4,5,9,12], radiative [1], and even combination of Auger and radiative [1,5]. The number of published results on the recombination mechanism for p-type bulk HgCdTe is much less than that for n-type. However, different conclusions were also arrived at about the dominant recombination process in p-type bulk HgCdTe [2,5-8,14].

The minority carrier lifetime in LPE-grown HgCdTe on CdTe substrates may be shorter than in bulk materials. In addition to bulk recombination, a nearby recombining surface or interface will reduce the lifetime of carriers within a diffusion length (usually comparable to the layer thickness). Little work has been reported on the study of minority carrier lifetime and recombination in LPE HgCdTe [14,16].

This paper presents a systematic study of the minority carrier lifetime in p-type LPE $Hg_{1-x}Cd_xTe$ with x = 0.22. The recombination process in these p-type LPE layers was investigated by measuring the minority carrier life-time as a function of temperature and hole concentration. Hall effect mea-surements were carried out on all the layers to determine the hole concen-trations. For low-hole concentration samples that showed negative Hall coefficients and low mobilities, a two-layer model was used to derive the hole concentration.

The three possible bulk recombination mechanisms that limit the minor-ity carrier lifetime in HgCdTe are radiative, Auger and Shockley-Read mech-anisms. The details of these recombination processes have been described by many authors [1,2,4-8]. The expressions for the lifetimes limited by these three recombination processes are reproduced as follows [2]:

$$\tau_R = n_i^2/[G_R(n_0 + p_0 + \Delta n)] \tag{1}$$

$$\tau_A = 2n_i^2 \tau_A^i / \{(n_0 + p_0 + \Delta n)](n_0 + \Delta n) + \beta(p_0 + \Delta n)]\} \qquad (2)$$

$$\tau_{SR} = [\tau_{p_0}(n_0 + n_1 + \Delta n) + \tau_{n_0}(p_0 + p_1 + \Delta n)]/(n_0 + p_0 + \Delta n) \qquad (3)$$

where τ_R, τ_A and τ_{SR} represent radiative, Auger and Shockly-Read lifetimes, respectively.

The relevant symbols used in the lifetime equations are

n_i = intrinsic carrier concentration

G_R = thermal equilibrium generation rate

n_0 = equilibrium electron concentration

p_0 = equilibrium hole concentration

Δn = injected excess carrier concentration

τ_A^i = intrinsic Auger lifetime

β = hole-hole collision term

τ_{po} = shortest time constant for hole capture

τ_{no} = shortest time constant for electron capture

P_1 = $N_v \exp[(E_t-E_v)/kT]$

n_1 = $N_c \exp[(R_c-E_t)/kT]$

whre E_c and E_v are energies at conduction and valence-band edge, respectively, E_t is the energy of Shockley-Read center, and N_c and N_v are the densities of state for conduction and valence band.

EXPERIMENTAL

All samples used were LPE $Hg_{0.78}Cd_{0.22}Te$ grown on CdTe substrates from a Te-rich solution. Samples with different hole concentrations were prepared by annealing the LPE layers at different temperatures under Hg-saturated conditions. The LPE samples of typical dimensions 5 x 5 mm^2 were etched in Br-ethylene glycol solution prior to minority carrier lifetime measurements. After mounting in a cryogenic dewar, the LPE samples were masked such that only the central 1 x 1 mm^2 areas of the samples were exposed.

The photoconductive decay technique using short (20 ns) pulses from a GaAs laser was used to determine the minority carrier lifetime. Different laser intensities were used to verify that the injected carrier concentration was always kept low. The photoconductive response was recorded on a Tektronix Model 7854 waveform processing oscilloscope. The stored decay trace was plotted on a semilog scale to check if the decay was exponential.

The Hall effect and conductivity measurements were performed at 77K using the Van der Pauw method. The magnetic field, usually 2 kG, is applied perpendicular to the sample. In some cases, 7 kG was applied to check the magnetic field dependence of the Hall coefficient when the sample was found to have mixed conduction. The variable temperature Hall effect measurements were also carried out on some LPE layers between 5 and 300K.

RESULTS AND DISCUSSION

Variable Temperature Lifetime Measurements

Results of variable temperature lifetime measurements on a p-type LPE layer is shown in Fig. 1. The relevant electrical properties characterized at 77K are also listed. Three distinct regions are observed in Fig. 1. At temperatures higher than 100K, the measured minority carrier lifetime decreases with increasing temperature. This exponential temperature dependence of lifetime in the intrinsic region can be attributed to the Auger process (Eq. (2)). It is consistent with most of the published results for long-wavelength HgCdTe [1-15]. At temperatures around 100K, a maximum in lifetime is observed. For temperatures less that 100K, the minority carrier lifetime decreases slowly with decreasing temperature. This temperature dependence in the extrinsic region (T < 100K) is usually indicative of the dominant recombination process. Similar temperature-dependent lifetime measurements have been reported earlier for bulk HgCdTe by different groups, but the dominant recombination process is still debatable. Most of the studies have attributed the lifetime to a Shockley-Read recombination process. The energy level of the Shockley-Read recombination center, E_t, was usually arrived at by fitting the experimental data with the aid of Eq. (3).

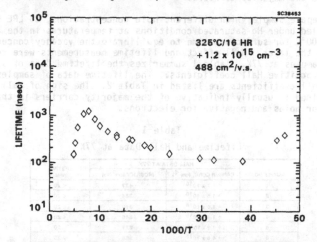

Fig. 1 Variable temperature lifetime data on LPE Hg$_{0.78}$Cd$_{0.22}$Te.

However, for p-type materials when the effects of carrier freeze-out and background radiation are considered, the Auger and radiative lifetimes will have similar temperature dependence as Shockley-Read lifetime at temperatures less than 100K [5]. Therefore, the temperature dependence of lifetime between 30 and 100K as shown in Fig. 1 cannot be used for the determination of the dominant recombination process in these p-type samples.

As indicated by Schacham et al [5], at very low temperatures a pronounced increase in Auger and radiation lifetimes will be observed in p-type samples due to carrier freeze-out. However, at low temperatures, as long as the density of Shockley-Read center is less than the majority concentration, the Shockley-Read lifetime will be constant. Equation (3) becomes

$$\tau_{SR} \simeq \tau_{no}(1+p_1/p_0) = \tau_{no}(1+\exp[(E_F-E_t)/kT]) \sim \tau_{no}$$

where E_F is the Fermi energy.

As shown in Fig. 1, at temperatures less than 30K, we found that the minority carrier lifetime increases rapidly with decreasing temperature. This indicates that the Auger or radiative mechanism instead of Shockley-Read mechanism is the dominant recombination process in p-type LPE $Hg_{0.78}Cd_{0.22}Te$.

Lifetime vs Majority Carrier Concentration

Due to the similarity in temperature dependence, the dominant lifetime-limiting process cannot be unambiguously determined simply by measuring the minority lifetime as a function of temperature. This necessitated the investigation of the recombination mechanisms by determining the dependence of lifetime on the majority carrier concentration. For p-type material, according to Eqs. (1) and (2), the dependence of lifetime on the majority carrier will be $1/p_0^2$ and $1/p_0$ for Auger and radiative processes, respectively. For the Shockley-Read process, Eq. (3), the lifetime is either proportional to $1/p_0$ when the recombination centers are related to the majority carriers or independent of p_0 when no relation exists between recombination centers and the majority carriers.

To prepare samples with different hole concentrations, 25 LPE samples were annealed under Hg-saturated conditions at temperatures in the range 250°C to 400°C for sufficient time to equilibrate the vacancy concentration throughout the layer. Hall effect and lifetime measurements were conducted on these samples at 77K. Table 1 summarizes the lifetime data of samples which show positive Hall coefficients. The lifetime data of samples having negative Hall coefficients are listed in Table 2. The sign of Hall coefficient obtained is usually indicative of the majority carriers in the sample; positive for holes and negative for electrons.

Table 1
Lifetime and Hall Data at 77K

SAMPLE NO.	HALL DATA AT 77K		LIFETIME (nsec)
	CARRIER CONC. (cm^{-3})	MOBILITY (cm^2/Vs)	
1	+ 1.7 x 10^{16}	437	< 20
2	+ 1.6 x 10^{16}	384	< 20
3	+ 1.5 x 10^{16}	310	< 20
4	+ 1.4 x 10^{16}	346	< 20
5	+ 1.1 x 10^{16}	627	< 20
6	+ 4.6 x 10^{15}	554	~ 20
7	+ 3.7 x 10^{15}	637	40
8	+ 2.3 x 10^{15}	502	56
9	+ 2.3 x 10^{15}	386	68
10	+ 2.2 x 10^{15}	575	100
11	+ 1.6 x 10^{15}	392	111
12	+ 1.2 x 10^{15}	488	330

Figure 2 illustrates the dependence of the minority carrier lifetime on the hole concentration at 77K for samples having positive Hall coefficients. The resolution of lifetime measurements using photoconductive decay technique is about 20 ns. For hole concentrations higher than 5×10^{15} cm^{-3}, the lifetimes are shorter than the system limit. For hole concentrations between 1.2×10^{15} to 5×10^{15} cm^3, the measured lifetimes vary from 20 ns to 330 ns. Least square fitting gives a slope of -1.85 with a standard deviation of 0.24. It indicates that the Auger process is probably the dominant lifetime-limiting mechanism for p-type LPE $Hg_{0.78}Cd_{0.22}Te$.

Table 2
Lifetime and Hall Data at 77K

SAMPLE NO.	HALL DATA AT 77K		LIFETIME (nsec)	CALCULATED HOLE CONCENTRATION cm^{-3}
	CARRIER CONC. (cm^{-3})	MOBILITY cm^2/Vs		
13	-3.3 x 10^{15}	712	50	2.2 x 10^{15}
14	-2.5 x 10^{15}	1466	56	3.2 x 10^{15}
15	-1.1 x 10^{15}	4398	60	3.5 x 10^{15}
16	-6.3 x 10^{14}	3658	113	1.8 x 10^{15}
17	-7.2 x 10^{14}	5268	171	1.6 x 10^{15}
18	-9.5 x 10^{14}	2837	332	2.2 x 10^{15}
19	-2.5 x 10^{14}	3804	660	7.3 x 10^{14}
20	-5.0 x 10^{14}	2659	900	1.1 x 10^{15}
21	-2.6 x 10^{14}	2603	900	5.6 x 10^{14}
22	-1.1 x 10^{15}	600	1037	6.0 x 10^{14}
23	-8.1 x 10^{13}	11506	2000	3.8 x 10^{14}
24	-2.8 x 10^{14}	2846	2500	6.5 x 10^{14}
25	-9.8 x 10^{13}	17417	7000	2.1 x 10^{14}

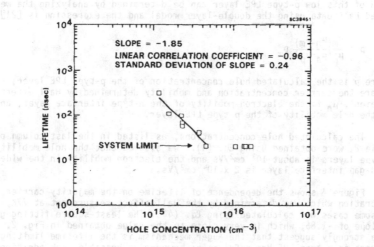

SLOPE = -1.85
LINEAR CORRELATION COEFFICIENT = -0.96
STANDARD DEVIATION OF SLOPE = 0.24

SYSTEM LIMIT ⟶

Fig. 2 Dependence of lifetime on hole concentration for LPE Hg$_{0.78}$Cd$_{0.22}$Te.

For the samples listed in Table 2, although the measured Hall coefficients are negative, the measured hole mobilities are one to two orders of
magnitude less than that of a normal n-type material, which is typically
10^5 cm^2/Vs. Figure 3 illustrates the variable temperature Hall data on a
normal n-type LPE layer and on two LPE layers which show low mobilities
after a low-temperature anneal under Hg-saturated conditions. In the literature, several models have been proposed to explain this anomalous behavior
in HgCdTe [17-20]. However, based on several experimental results [21], we
exclude the surface inversion layer model and propose that the Hall anomalies are due to a double-layer structure which is composed of a low p-type
LPE layer and an n-type LPE/substrate interface layer. The hole concentra-

292

SC86-34336

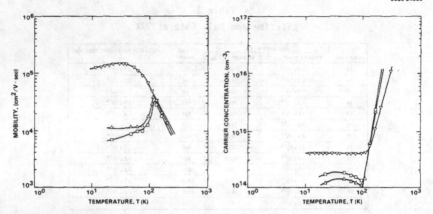

Fig. 3 Variable temperature Hall data on LPE $Hg_{0.78}Cd_{0.22}Te$. after low-
temperature anneal.

tion of this low p-type LPE layer can be determined by analyzing the measured Hall data using the double-layer model and the expression is [21]

$$p = \frac{\mu_m(\mu_n - \mu_m)}{\mu_p(\mu_n - \mu_p)} p_m \qquad (4)$$

where p is the calculated hole concentration of the p-type LPE layer, p_m and
μ_m are the carrier concentration and mobility determined by Hall effect measurement, μ_n is the electron mobility of the n-type interface layer, and μ_p
is the hole mobility of the p-type LPE layer.

The calculated hole concentrations, as listed in the last column of
Table 2, were obtained using Eq. (4). We assumed that the hole mobility in
p-type layers is about 10^3 cm^2/Vs and the electron mobility in the wide
band-gap interface layer is 2×10^4 cm^2/Vs.

Figure 4 shows the dependence of lifetime on the majority carrier concentration which was determined by the Hall effect measurement at 77K, and
in some cases was calculated using Eq. (4). The least-square fitting gives
a slope of -1.86, which is very close to the value obtained in Fig. 2. The
data strongly suggest that the Auger mechanism is the lifetime limiting-
process in p-type LPE $Hg_{0.78}Cd_{0.22}Te$; however, a possible weak additional
dependence on other mechanisms cannot be excluded.

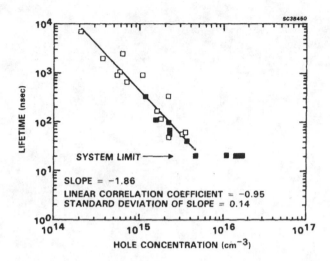

Fig. 4 Dependence of lifetime on hole concentration for LPE $Hg_{0.78}Cd_{0.22}Te$.

REFERENCES

1. I.M. Baker, R.A. Capocci, D.E. Charlton and J.T.M. Wotherspoon, Solid State Electron. 21, 1475 (1978).
2. R.G. Pratt, J. Hewett and P. Capper, J. Appl. Phys. 60, 2377 (1986).
3. J. Calas, J. Allegre and C. Fau, Phys. Stat. Sol. B 107, 275 (1981).
4. M.A. Kinch, M.J. Brau and A. Simmons, J. Appl. Phys. 44, 1649 (1973).
5. S.E. Schacham and E. Finkman, J. Appl. Phys. 57, 2001 (1985).
6. D.L. Polla, S.P. Tobin, M.B. Reine and A.K. Sood, J. Appl. Phys. 52, 5182 (1981).
7. T.N. Casselman, J. Appl. Phys. 52, 848 (1981).
8. T.N. Casselman and P.E. Petersen, Solid State Commun. 39, 1117 (1981).
9. W.A. Radford, J.F. Shanley and O.L. Doyle, J. Vac. Sci. Tech. A1, 1700 (1983).
10. G. Nimtz, G. Bauer, R. Dornhaus and K.H. Muller, Phys. Rev. B10, 3302 (1974).
11. A.V. Voitsekhovskii and Yu.V Lilenko, Sov. Phys. Semicond. 15, 845 (1981).
12. D. Polla and C.E. Jones, Solid State Commun. 36, 809 (1980).
13. G. Nimtz and K.H. Muller, Phys. Stat. Sol. A22, K215 (1974).
14. J. Bajaj, S.H. Shin, J.G. Pasko and M. Khoshnevisan, J. Vac. Sci. Tech. A1, 1749 (1983).
15. R.G. Pratt, J. Hewett, P. Capper, C.L. Jones and M.J. Quelch, J. Appl. Phys. 54, 5152 (1983).
16. R.D. Graft, F.F. Carlson, J.H. Dinan, P.R. Boyd an R.E. Longshore, Extended Abstracts, 1982 U.S. Workshop on the Physics and Chemistry of Mercury Cadmium Telluride.
17. M.C. Chen, S.G. Parker and D.F. Weirauch, J. Appl. Phys. 58, 3150 (1985).
18. L.F. Lou and W.H. Frye, J. Appl. Phys. 56, 2253 (1984).
19. C.T. Elliot and I.L. Spain, Solid State Commun. 8, 2063 (1970).
20. W. Scott and R.J. Hager, J. Appl. Phys. 42, 803 (1971).
21. J.S. Chen and W.E. Tennant, to be published.

Characterization of HgCdTe Epilayers and HgTe-CdTe Superlattice Structures Grown by Molecular Beam Epitaxy

T.H. Myers[*], R.W. Yanka[*], J.P. Karins[*], K.A. Harris[**], J.W. Cook[**], and J.F. Schetzina[**]
[*]Electronics Laboratory, General Electric Company Syracuse, New York 13221
[**]North Carolina State University, Raleigh, North Carolina 27695

ABSTRACT

The results of an in-depth study of HgCdTe epilayers and HgTe-CdTe superlattice structures are summarized. In particular, both spectral and transient photoconductance measurements have been made on samples which were also characterized by x-ray diffraction, Hall, IR transmission, IR photoluminescence, and transmission electron microscopy measurements. Selected HgCdTe samples grown at General Electric during this investigation exhibit some of the best structural and electrical properties reported to date for MBE-grown HgCdTe. Sharp photoconductance spectra have been obtained for HgTe-CdTe superlattice structures grown at North Carolina State University. This study is part of an on-going collaborative effort between General Electric's Electronics Laboratory and North Carolina State University.

INTRODUCTION

The growth of HgCdTe epilayers and HgTe-CdTe superlattice structures by molecular beam epitaxy (MBE) is of considerable interest due to the potential use of such materials in infrared detector applications. Superlattice structures, in particular, have been the subject of much attention in recent years because of the interesting properties which they exhibit. For example, the electrical and optical properties of such structures can be modified through proper choice of layer thicknesses and materials -- a form of "band gap engineering". The MBE technique, with its slow, controlled deposition rates and low growth temperatures, offers the best possibility for synthesizing Hg-based layered structures with sharp interfaces such as HgCdTe heterojunctions and HgTe-CdTe superlattices.

In this paper, the results of an in-depth study of HgCdTe epilayers and HgTe-CdTe superlattice structures are summarized. In particular, both spectral and transient photoconductance measurements have been made on samples which were also characterized by x-ray diffraction, Hall, IR transmission, IR photoluminescence, and transmission electron microscopy measurements. This study is part of an on-going collaborative effort between General Electric's Electronics Laboratory and North Carolina State University.

EXPERIMENTAL DETAILS

The epilayers and superlattices were prepared at General Electric's Electronics Laboratory (GE) and at North Carolina State University (NCSU) utilizing MBE systems designed and built specifically for growing Hg-based films and multilayer structures. The system has been extensively described elsewhere [1]. $Hg_{1-x}Cd_xTe$ epilayers were grown on $(100)Cd_{0.96}Zn_{0.04}Te$, $(111)Cd_{0.96}Zn_{0.04}Te$ and $(100)CdTe$ substrates. The substrates were chemimechanically polished on both sides and etched in a Br-methanol solution prior to loading into the MBE system. Preheat temperatures ranging from 350 to 400°C were used for the $Cd_{0.96}Zn_{0.04}Te$ and CdTe substrates. Immediately prior to the growth of a Hg-based film on CdTe, a CdTe buffer layer was deposited onto the substrate. Buffer layer thicknesses ranged from 2-10 μm and were grown at substrate temperatures T_s = 275-300°C. For growth of layers containing Hg, substrate temperatures of T_s = 170-220°C were employed.

The infrared transmittance of selected films was measured using a Perkin-Elmer 983-G double beam ratio-recording spectrophotometer at NCSU or a Nicolet 60SX FTIR spectrophotometer at GE. To correct for substrate effects, the transmittance of each polished substrate was recorded prior to its use for film growth and stored as a baseline correction file.

Van dér Paaw Hall effect measurements were completed on selected epilayers and superlattices over the temperature range 20K - 300K using an applied magnetic field of 2-3 kG. In these experiments, rectangular samples with In-soldered contacts located at the center of each of the four edges were employed.

An Oriel model 77250 grating monochromator and a chopped blackbody source were used to perform spectral photoconductance measurements. The signal was detected using a PAR model 116 preamplifier and a PAR HR-8 lock-in amplifier. A BNC model 6020 light pulse generator utilizing a GaAs laser was used to generate light pulses for transient photoconductance measurements. The transient response was measured using a Tektronix 7854 storage oscilloscope with the waveform calculator option. Lifetimes down to a few nanoseconds can be measured using this system.

RESULTS AND DISCUSSION

Both n-type and p-type epitaxial $Hg_{1-x}Cd_xTe$ films have been grown with x-values ranging from zero (HgTe) to one (CdTe). The x-value of the HgCdTe epilayers has been varied systematically by changing the Cd to Te vapor flux ratio from growth to growth. Hall mobilities in the range of 3.3-3.5x10^5 cm²/V·s have been obtained for several x=0.16 films grown at GE and NCSU with substrate temperatures of 190 °C. Typical 80K mobilities for for x = 0.18-0.20 films grown on CdZnTe substrates are in the range of 1-2x10^5 cm²/V·s. In general, carrier concentrations are intrinsic down to below 100K and are in good agreement with theory. At 77K, carrier concentrations as low as 2.7×10^{14} cm^{-3} have been observed for x=0.2 HgCdTe films. This indicates that contamination from unintentional impurities in the MBE system is not a problem. IR transmittance measurements have also been completed on all samples. A typical IR transmittance measurement exhibits a sharp cut-off illustrating the sharp absorption edge for the epilayers. The x-values for the layers obtained from these measurements are in good agreement with those obtained from the Hall effect data.

Several HgCdTe epilayers grown at GE are of exceptional structural quality[2]. Figure 1 shows a double crystal x-ray diffraction pattern obtained from a 7 μm thick HgCdTe epilayer grown on a 1cm x 1cm (111) CdZnTe substrate. This peak, taken from a 2mm^2 region, has a full width half maximum (FWHM) of 26 arcseconds. To our knowledge, this is the smallest FWHM ever reported for HgCdTe grown by any technique. While all

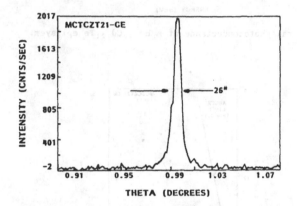

Figure 1. Double crystal x-ray diffraction rocking curve for high structural quality HgCdTe.

regions of the film were not this perfect, an average FWHM of 65 arcseconds was obtained for the sample. The central portion of this sample exhibited FWHMs of ~40-50 arcseconds with larger values being observed near the edges of the sample. The larger values near the edge may have occurred due to a cooling effect from the molybdenum mask holding the substrate to the molybdenum transfer block.

Figure 2 shows a spectral photoconductance which is typical of x = 0.2 HgCdTe epilayers measured in this study. The very abrupt turn-on of photoconductance with increasing energy is indicative of high quality material of uniform composition. The temperature dependence of the minority carrier lifetime as determined by transient photoconductance for an n-type $Hg_{0.79}Cd_{0.21}Te$ epilayer grown at GE is shown in Figure 3. This sample exhibits Auger-recombination limited behavior down to ~125K. The lifetime then peaks at 3.2 μsecond and slowly decreases with further decrease in temperature. This value, 3.2 μsecond, is the longest minority carrier lifetime reported for MBE-grown HgCdTe that we are aware of. Values

298

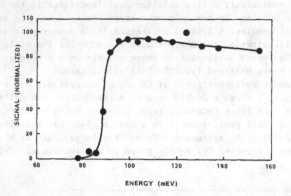

Figure 2. Spectral photoconductance of a $Hg_{.79}Cd_{.21}Te$ epilayer.

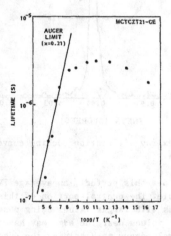

Figure 3 Minority carrier lifetime for n-type $Hg_{0.79}Cd_{0.21}Te$

of ~1 μsecond have been observed for several HgCdTe epilayers grown both at NCSU and GE.

HgTe-CdTe superlattices have been grown on both (111) $Cd_{0.96}Zn_{0.04}Te$ and (100) CdTe substrates. The IR transmittance of a superlattice consisting of 250 HgTe-CdTe double layers on a (100) CdTe substrate is shown in Figure 4. Note that the absorption edge is shifted toward the visible relative to HgTe. This 1.9 μm thick superlattice exhibited n-type conductivity with a carrier concentration ranging from 4×10^{16} cm^{-3} at room temperature down to 2×10^{15} cm^{-3} at 12 K as determined by Hall measurements. The mobility increases from 6×10^3 cm^2/V·s at 300 K to a plateau of about 4×10^4 cm^2/V·s at temperatures below 50 K.

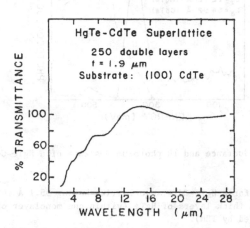

Figure 4. IR transmittance of a HgTe–CdTe superlattice grown on (100)CdTe.

This sample actually consists of two stacked superlattices consisting of an initial 70 pairs of 56Å HgTe/64Å CdTe followed by 180 pairs of 22Å HgTe/54Å CdTe. A previous TEM study [3] of this sample performed at Purdue by N. Otsuka determined the second set of layer thicknesses to be $L_B = 54\pm2$ Å (CdTe) and $L_Z = 22\pm2$ Å (HgTe). No misfit dislocations were found at the interfaces of the superlattice, but a few threading dislocations from the substrate were present with a density of approximately 1×10^4 cm^{-2}. Six orders of distinct satellite spots were observed in electron diffraction patterns obtained for this sample indicating a nearly perfect periodicity.

Figure 5 shows the spectral variation in photoconductance measured for this sample. The onset of photoconductance begins at about 326 meV (3.8 μm) and reaches a maximum at about 475 meV (2.6 μm). Also shown is an IR photoluminescence spectrum obtained for this sample at Hughes Research Laboratories (Malibu) [3]. The IR photoluminescence spectrum contains one broad luminescence peak centered at 357 meV (3.47 μm) with a full width at half intensity of 64 meV. There is excellent agreement between the position of the photoluminescence peak and the onset of photoconductivity. A correlation can also be seen between the width of the photoluminescence peak and the "softness" of the increase in the photoconductance signal with increasing energy. This softness can be attributed to the two sets of HgTe-CdTe layers. Thus it can be seen that photoconductance and IR photoluminescence yield the same type of information. Both techniques are in reasonable agreement with theory [3] that predicts a band gap of 373 meV

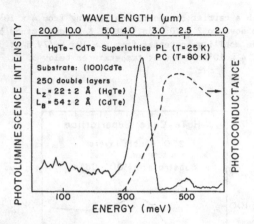

Figure 5. Photoconductance and IR photoluminescence of a HgTe-CdTe superlattice.

(3.32 μm) at 25 K for a HgTe-CdTe superlattice having 55.1 Å thick layers of CdTe and 25.8 Å thick layers of HgTe (within one monolayer of the layer thicknesses measured by TEM).

A transient photoconductance measurement was made on this sample to determine excess carrier lifetime at 80 K. The value obtained was 13 milliseconds! This measurement indicates an excess carrier decay time much longer than that obtained in the equivalent alloy. No variation in decay time was observed for a variation in excitation energy of 100. We believe that this long decay time may be due to a junction effect between the superlattice structure and the substrate similar to the persistent photoconductivity effect found in many III-V heterostructures [4]. This result is preliminary and we are in the process of a more complete investigation of this effect.

The IR transmittance of a 3.9 μm thick superlattice is shown in Figure 6. TEM vertical cross sections of this superlattice show well-formed repetitive layers of thickness L_B = 58±3 Å (CdTe) and L_Z = 96±3 Å (HgTe). Interdiffusion effects at the interfaces appear to be minimal. The spectral variation of photoconductance measured for this superlattice is shown in Figure 7. There is reasonable agreement between the onset of photoconductance and the absorption illustrated by the IR transmittance. Of particular interest is the abruptness of the onset of the photoconductance signal. The sharp cut-on indicates a fairly rapid increase of absorption with increasing photon energy in this sample. This result is favorable for the use of superlattices in the construction of IR detectors. Transient photoconductance measurements were also made on this sample. An excess carrier decay time of about 2 milliseconds was measured at 78 K.

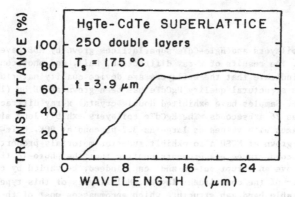

Figure 6. IR transmittance of a HgTe-CdTe superlattice on (111)CdZnTe.

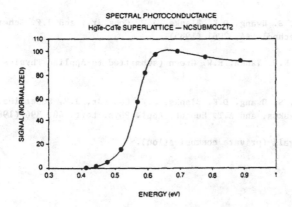

Figure 7. Photoconductance of a HgTe-CdTe superlattice on (100)CdZnTe.

SUMMARY

HgCdTe epilayers and HgTe-CdTe superlattices grown by MBE have been characterized. the results of x-ray diffraction, Hall, and photoconductance measurements indicate that the epilayers are device quality material. Extremely high structural quality HgCdTe has been grown at GE on (111) $Cd_{0.94}Zn_{0.04}Te$. Samples have exhibited double crystal x-ray diffraction FWHMs as low as 26 arcseconds. The HgCdTe epilayers exhibit long minority carrier lifetimes with values as large as 3.2 µsecond at 90K. HgTe-CdTe superlattices grown at NCSU also exhibit superior materials properties. Spectral photoconductance measurements confirm that the photoelectric response can have an abrupt cut-on and can, indeed, be varied by changing the thicknesses of the constituent layers. Multilayers of this type can provide a variable band gap structure which encompasses most of the IR spectral region.

REFERENCES

1. K.A. Harris, S. Hwang, D.K. Blanks, J.W. Cook,Jr., and J.F. Schetzina, J. Vac. Sci. Technol. A4, 2061 (1985).

2. T.H. Myers, R.W. Yanka, R.W. Green (submitted to Applied Physics Letters).

3. K.A. Harris, S. Hwang, D.K. Blanks, J.W. Cook,Jr, J.F. Schetzina, N. Otsuka, J.P. Baukus, and A.T. Hunter, Appl. Phys. Lett. 48, 396 (1986).

4. J.M. Ballingall (private communication).

SURFACE RECOMBINATION MEASUREMENTS IN HgCdTe
BY OPTICAL MODULATION FREQUENCY RESPONSE*

J.A. MROCZKOWSKI, E. LESONDAK AND D. RESLER
Honeywell Electro-Optics Division, Lexington, Massachusetts 02173

ABSTRACT

The frequency response in the modulation of the excess electron concentration, produced by a modulated photogenerating pump beam, is used to determine bulk lifetime and the surface or interface recombination velocity. The depth-wise integrated excess electron concentration is contactlessly monitored by the proportional absorption/transmission modulation of a second probe beam. Using this approach over the 20 kHz to 3 MHz frequency range recombination velocities up to 10^4 cm/sec have been measured in n-type epitaxial HgCdTe films.

INTRODUCTION

The measurement of excess carrier recombination parameters is most frequently performed by monitoring some form of electrical transient induced by the carriers such as photoconductivity. An alternate approach, which is particularly suited to narrow bandgap HgCdTe with low electron mass, is to monitor the change in the absorbtance of the material. The absorbtance changes respond immediately to excess carrier modulations as do resistivity changes in photoconductivity measurements. The absorption changes can be monitored as transmission variations in a probe beam. This paper reports the use of the Optically Modulated Absorbtion (OMA) technique, previously reported for bulk or effective lifetime measurements [1,2] to surface or interface recombination measurements.

RESPONSE MODEL

The response model is based on the configuration shown in Figure 1(a). Figures 1(b), (c) illustrate the temporal behavior of the incident pump and probe beams, and Figure 1(d) shows the impact of excess carriers on the transmission of the probe beam.

The OMA approach adopted in this work used a probe beam with energy slightly greater than the bandgap. Photogenerated excess electrons in the conduction band reduce the density of final states available for direct gap transitions causing the transmission just above the bandgap energy to increase significantly.

Frequency Response

In principle, the recombination parameters can be determined from the temporal transmission relaxation response as shown in Figure 1(d). The instantaneous modulated transmission $\Delta I(t)$ at time t is proportional to the total number of excess electrons $\Delta n(t)$ the probe beam encounters in passing through the epitaxial film. Instead of sampling $\Delta I(t)$ at different times the transmission modulation frequency response $\Delta I(\omega)$ was mea-

---
* This work was supported under NVEOC contract
DAAK-83-C-0184. The funding organization was
AMMRC, Watertown, Massachusetts.

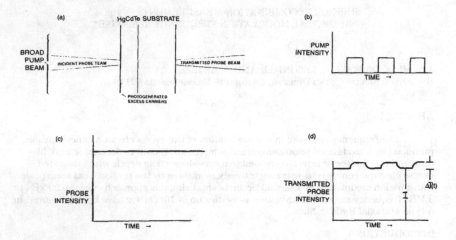

Figures 1(a) - (d). Schematic Illustration of Experimental Approach

sured. In this case $\Delta I(\omega)$ is proportional to the peak amplitude of the fundamental component in the periodic $\Delta I(t)$ modulation. This approach was simpler to implement; it also enabled measurements at much lower photogeneration levels since the measurement bandwidth was only 20 Hz.

The general solution to $\Delta n(t)$ involving all possible sources of recombination is quite complex since this involves the frontside epitaxial surface recombination velocity S_1, recombination in the bulk material characterized by a lifetime τ, and the epitaxial/substrate interface recombination velocity S_2. The special use of interest in this work, which enabled surface or interface recombination to be evaluated, required the effective diffusion length of excess carriers L to be much less than the epitaxial thickness. Under these conditions the normalized frequency response $S(\omega)$ can be expressed as [3,4]

$$S(\omega) = \frac{1}{\sqrt{[a^2 + b^2] \ [(a + R_i)^2 + b^2]}}$$ (1)

where $a^2, b^2 = \dfrac{(1 + \omega^2\tau^2)^{1/2} \pm 1}{2}$, $R_i = S_i \, \tau \, L_o$

and S_i represents the surface or interface ($i = 1$ or 2) recombination velocity, ω is the modulation frequency and L_o is the low frequency diffusion length. In the limiting case where $R_i \ll 1$

$$S(\omega) = \frac{1}{\sqrt{1 + \omega^2 \tau^2}}, \quad (2)$$

and the lifetime can be determined from a closed form analysis of $S(\omega)$.

When the surface or interface contributes significantly to the total recombination where $R_i >> 1$, the response at lower frequencies where $R_i >> \omega\tau$, becomes

$$S(\omega) = \frac{1}{\sqrt[1/4]{1 + \omega^2 \tau^2}} \quad (3)$$

The effective diffusion length L is related to the low frequency diffusion length L_o by $L = L_o/(a + ib)$. The physical significance of this relationship can be recognized at high frequencies when the rise or fall in the excess electron concentration $\Delta n(x,t)$ near the surface is not in phase with the response in the $\Delta n(x,t)$ distributions deeper in the film. This situation results in the temporal modulation depth decreasing more rapidly as a function of distance x into the film. When $\omega\tau >> 1$ the amplitude of the shorter effective diffusion length is $L = L_o/(\omega\tau)^{1/2}$.

When L is greater than the epitaxial thickness, the frequency response approaches the simple pole form given by equation (2). The recombination measurement then represents an effective lifetime in which the contributions of S_1, S_2 and τ become more difficult to separate.

Determination of τ and S

A frequency response analysis using equation (1) provides values of τ and the compound parameter $R_i = S_i \tau/L_o$. S_i can then be determined using the relationship between the diffusion length L_o and the lifetime τ [5].

$$L_o = \left((kT/e)\, \mu_a \tau \right)^{1/2} \quad (4)$$

where μ_a is the ambipolar diffusion mobility.

In n-type HgCdTe the diffusion of excess carriers is limited by the hole mobility μ_H, which at the measurement temperatures of 90 K is approximately 500 cm²/V-s. Using these values in equation (4) gives

$$L_o = 20\, \tau^{1/2}\, (\mu s)\ \ \mu m \quad (5)$$

S_i can then be determined from

$$S_i = 2 \times 10^3\, R_i\, \tau^{1/2}\, (\mu s)\ \ cm/s \quad (6)$$

EXPERIMENTAL

The experimental set-up is shown in Figure 2. The pump beam is a 3.4 micron CW He-Ne laser modulated by an acusto-optic modulator (AOM). The probe beam is a filtered blackbody source with the bandpass filter selected so that only radiation near the material bandgap is transmitted. Most of the modulated transmitted probe beam signal is contributed by radiation whose energy is just above the bandgap. A second bandpass filter is occasionally used behind the epitaxial sample to block any pump radiation not completely absorbed by the sample. The resulting signal induced by the probe beam incident on a wide bandwidth HgCdTe diode detector is amplified and measured by a variable frequency programmable AC voltmeter (HP3586C).

The frequency response measurements and data analyses were performed as follows. First, the combined response of the modulator, detector and preamplifier were measured as a function of frequency with the 3.4 micron pump beam directly incident on the test station detector. The epitaxial sample was then introduced and the modulated probe intensity was measured at the same frequencies. The system response was then removed from the probe modulation data, to obtain the true sample response. The data, normalized to the maximum value was fit to a simple pole rolloff. If required a more complex fit including surface recombination effects was applied.

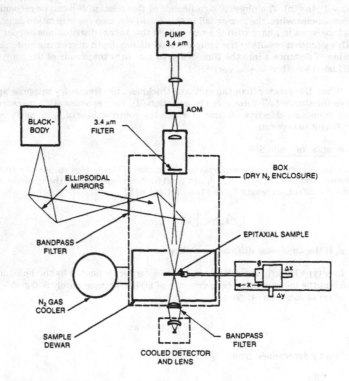

Figure 2.

307

RESULTS AND DISCUSSION

More than 40 HgCdTe epitaxial films were evaluated during the course of developing the OMA lifetime technique. These samples spanned a range of thickness, and n and p-type carrier concentrations. Since the diffusion lengths are generally much longer in p-type HgCdTe only the thick n-type films qualified for possible surface recombination analysis with equation 1.

Figures 3(a),(b) show OMA data for an n-type film LW23 with frontside photogeneration indicating a bulk lifetime of 630 nsec, and relatively insignificant surface recombination. When this film was rotated 180 degrees to provide photogeneration near the film/substrate interface the best data fit was obtained with τ = 900 nsec and R_2 = 3 giving S_2 = 5.7 x 10³ cm/sec at the interface. The dotted line in Figure 3(b) shows the inadequacy of the initial exploratory result using the simple pole fit first which assumes R_2 = 0.

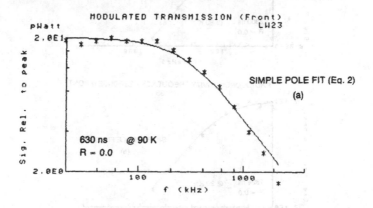

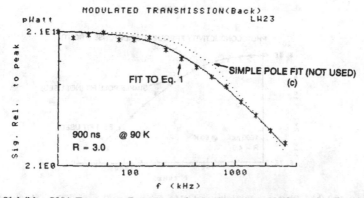

Figures 3(a),(b). OMA Frequency Responses obtained on an n-type film using Frontside (a), and Backside Interface (b) Photogeneration

308

Another thick n-type film CM138 was selected to demonstrate frontside recombination effects on the measurements and to determine correlation between photoconductivity and OMA frequency response data. Electrical leads were attached to a fragment of this film and all but a small portion of the front surface between the leads was masked off. The film electrical and optical frequency responses with frontside photogeneration were measured and are shown in Figures 4(a),(b).

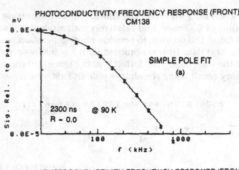

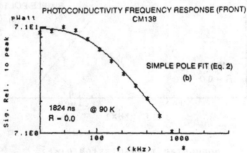

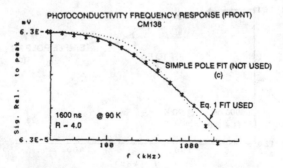

Figures 4(a),(b). Photoconductivity and OMA Frequency Response Data obtained on an N-type film showing little surface recombination. Figure 4(c) shows data after the surface was damaged.

The front surface of the film was then damaged by abrading it with a cotton swab and 0.3 micron alumina lapping grit. The frequency response measurements were then repeated. The photoconductivity frequency response shown in Figure 4(c) now clearly shows the influence of increased surface recombination. The surface recombination velocity S_1 was determined to be 10^4 cm/sec. A similar OMA frequency response could not be obtained since the signal was reduced excessively by surface damage.

Several limitations were found with the use of acusto-optic pump beam modulation. The major limitations were that the system response was not perfectly reproducible, and the nature of the modulation technique does not result in very uniform spatial frequency response at high frequencies. These problems have not been observed with electro-optic modulation, therefore an EOM system has recently replaced the AOM system. The much faster response and wider bandwidth (35 MHz) of the EOM system will significantly increase the response measurement sensitivity. This in turn will improve resolution in the analysis of all the recombination parameters; also it will be possible to evaluate a wider range of n and p-type materials.

SUMMARY

Initial results from a limited data base indicate that both the bulk lifetime and the surface or interface recombination velocity can be determined from frequency response measurements provided the diffusion length is less than the sample thickness. When the diffusion length becomes comparable with the epitaxial thickness a more complicated response analysis[4], which has not yet been developed, is required to separate out the various recombination contributions.

REFERENCES

1. O.L. Doyle, J.A. Mroczkowski and J.F. Shanley, "Electron Lifetime Measurements in HgCdTe", IRIS Specialty Group Meeting, Seattle, WA, August 1984.

2. O.L. Doyle, J.A. Mroczkowski and J.F. Shanley, "Photoabsorbtance and Electron Lifetime Measurements in HgCdTe", J. Vac. Sci. Technol. A3(1), 259 (1985).

3. J.A. Mroczkowski, "Contractless Electron Lifetime Measurements in HgCdTe Epitaxial Films", IRIS Specialty Group Meeting on Infrared Materials, SRI International, June 10, 11 (1986).

4. J.A. Mroczkowski, "HgCdTe Materials Screening Test Station", U.S. Army CECOM Report No. DAAK70-83-C-0184, 1986.

5. R.A. Smith, *Semiconductors*, 1st ed. (Cambridge University Press, 1959), Chapter 8.

A STUDY OF THE MBE HgTe GROWTH PROCESS

ROLAND J.KOESTNER AND H.F.SCHAAKE
Texas Instruments, Inc., Central Research laboratories, Dallas, Tx 75265

ABSTRACT

The MBE growth of HgTe on CdTe is examined over a hundred-fold range in Hg/Te2 flux ratio and over four separate substrate surface orientations [(111)Te, (111)Te-4 degrees, (112)Te and (001)]. The 77K Hall mobility of the (112)Te and (001) oriented HgTe layers approaches the best bulk values reported to date, although our (111)Te and (111)Te-4 deg oriented HgTe films yield much lower values. The growth process is shown to be very far from thermodynamic equilibrium at our optimal substrate temperature and calculated equivalent Hg beam pressure. Important clues to help understand the kinetics governing the HgTe growth process are uncovered by studying the defects that form under Hg- or Te-rich conditions with cross-sectional transmission electron microscopy (XTEM). Since multilayered structures are an important application for MBE growth of Hg-based semiconductors , we have also examined the interfacial roughness present in HgTe-CdTe superlattices (SL) as a function of growth orientation.

RESULTS AND DISCUSSION

As a measure of the quality of HgTe layers grown on CdTe substrates with varying surface orientations, Hall transport measurements by the Van der Pauw method was employed with a 1000 Gauss magnetic field strength. The temperature dependent Hall mobility of a 2.1 um HgTe layer on a CdTe(112)Te substrate grown under an optimal Hg/Te2 flux ratio is illustrated in Figure 1. The measured mobility curve approaches the best bulk values reported to date [1]. These bulk HgTe samples as in our MBE grown layer show a ~ 100,000 cm2/V-s mobility at 77K and a step increase in the mobility near 40K due to the cross-over from LO phonon to ionized impurity scattering. Our optimized HgTe film, however, plateaus at 120,000 cm2/V-s rather than increase to ~400,000 cm2/V-s at 10K. To compare this Hall mobility for HgTe(112)Te to other oriented HgTe layers, a range of Hg/Te2 flux ratios had to be tested due to the reported variation of the Hg sticking coefficient with growth orientation [2].

The optimal HgTe(112)Te layer was deposited at 225C with a calcu-

lated equivalent Hg beam pressure of 1 x 10-3 torr; the Te2 flux was kept
constant throughout this study and yielded an elemental deposition rate
of ~2 um/hr on Si substrates held at 100C. All the CdTe substrates prior
to HgTe overgrowth in a Riber 2300P MBE chamber had been chemimechanically
polished in 0.5% and 0.2% Br2/MeOH solutions (unless otherwise indicated
in the Figures), static etched in a 0.2% Br2/MeOH solution and finally
vacuum annealed near 300C to remove excess Te. To span the Te rich growth
conditions, the Hg cell temperature was reduced in steps by a total of
47C; and the calculated equivalent Hg beam pressure was incrementally
reduced from 1 x 10-3 to 1x 10-4 torr. To cover the Hg rich growth
conditions, the substrate temperature was reduced in steps by a total of
60C. The Hg cell temperature was not raised from the optimal (112)Te
growth conditions since the Hg charge lifetime would become prohibitively
short. The lower substrate temperature increased the Hg adsorption
lifetime which can be viewed to first order as an increase in the
effective Hg beam flux at a constant growth temperature.

Although the Hg desorption
energy has not been determined,
we will show in a later section
that it is significantly larger
than the sublimation energy for
elemental Hg. The effective Hg
beam pressure at our substrate
surface is then calculated to be
at least 1 x 10-2 torr. By
varying both the Hg cell and
substrate temperatures, at least
a hundred-fold range in Hg beam
flux has been sampled at a
constant Te2 flux.

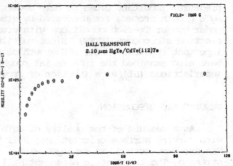

Figure 2 shows the (112)Te
oriented HgTe layer mobilities
that were measured at 77K for
this hundred-fold range in Hg
flux. The mobilities are peaked
over a 10-20C range in

Figure 1. Hall mobility for
2.10 um HgTe on CdTe(112)Te.

ΔTstoichiometry which measures the change in Hg cell(Te rich) and
substrate (Hg rich) temperatures from the optimal value. (The optimal
mobility of 60,000 cm2/V-s for the same film described in Figure 1 earlier
is low due to an error in the magnetic field calibration; however, the
relative mobilities among the separate films in Figure 2 are accurate.)
This small range in ΔTstoichiometry corresponds to approximately a factor
of 2 in Hg beam flux for optimal (112)Te oriented HgTe growth.

Figure 3 illustrates a similar 77K Hall mobility plot for (112)Te
and (001) oriented HgTe layers grown side-by-side in the MBE chamber.
Although there are only four separate growth runs with different Hg

fluxes, it appears that the optimal growth conditions for (112)Te and (001) HgTe layers are very similar. This implies that the Hg sticking coefficient for these two orientations is also very similar.

In contrast, Figure 4 shows a 77K Hall mobility plot for (111)Te and (111)Te-4° oriented HgTe layers grown over nearly the same range of Hg flux conditions as in Figures 2 and 3. There is no peaked mobility over the range in Hg flux, even though another group reported [2] a change in Hg sticking coefficient by only a factor of 4-5 between the (111)Te and (001) orientations. Indeed, preliminary XTEM micrographs indicate that excess Hg and Te conditions were achieved at each end of our Hg flux range. Under our growth conditions, the (111)Te and (111)Te-4° oriented

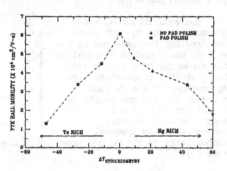

Figure 2. 77K Hall mobilities of HgTe(112)Te layers grown on CdTe.

HgTe layers show significantly lower mobilities than the (001) and (112)Te orientations. The defect mechanism underlying this dramatic sensitivity to growth orientation will be reported after a more complete XTEM analysis is performed.

Thermodynamically vs. Kinetically Limited Growth

At thermodynamic equilibrium, the range in Hg overpressure that leads to single crystalline HgCdTe (x=0.2) material has been determined for the low substrate temperatures employed during MBE growth [3]. This result is summarized in Figure 5 where the Hg and Te phase boundaries are plotted as a function of reciprocal temperature. Over the 225-650C temperature range investigated, HgCdTe (x=0.2) exists in the single crystalline state only with an excess Te concentration (or equivalently, a finite Hg vacancy

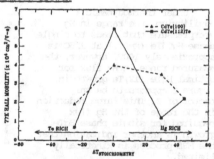

Figure 3. 77K Hall mobilities of HgTe(112)Te + (001) layers on CdTe.

314

concentration). At 225C, this vacancy concentration is measured to span from 10*13 to 10*16 cm-3. A higher metal vacancy concentration will lead to second phase Te precipitation and a lower vacancy concentration will give second phase Hg.

The equilibrium Hg vapor pressure at the Hg rich phase boundary for temperatures as low as 225C was found to be equal to the elemental Hg pressure (40 torr) at that temperature. The Hg pressure at the Te rich boundary at 225C can easily be calculated to be 6 x 10-2 torr since the metal vacancy concentration is inversely proportional to the Hg overpressure [3].

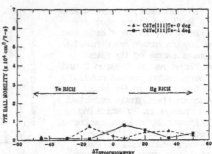

Figure 4. 77K Hall mobilities of HgTe(111)Te-0° + 4° layers on CdTe.

During MBE HgTe growth, a 1 x 10-3 torr Hg beam pressure was supplied to the substrate surface at 225C. This Hg pressure is far below the necessary amount (by a factor of ~60) to produce single phase HgTe. In the (112)Te oriented HgTe layers grown under these conditions, we find no evidence for second phase Te. In addition, the high mobilities measured for these films argue against the possibility of a ~10*18 cm-3 metal vacancy concentration since these vacancies should appear as doubly ionized scattering centers at 77K. This suggests that the MBE growth of HgTe is kinetically rather than thermodynamically controlled.

Under thermodynamic equilibrium, the range in Hg overpressure that leads to single phase HgCdTe (x=0.2) at 225C is approximately 1000; however, the measured range in Hg flux for optimal HgTe(112)Te growth in Figure 2 appears to be only factor of 2. This large reduction in the range of the Hg flux required for single phase HgTe growth is another indication that the MBE process is kinetically limited.

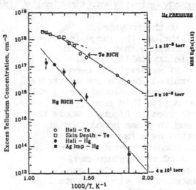

The presence of a Te2 overpressure during MBE growth has not yet been considered.

Figure 5. Single crystalline phase field for HgCdTe (x=.2).

This overpressure is much higher than the equilibrium vapor pressure of Te2 over HgCdTe (x=.2) at 225C and actually drives the deposition process. However, the stoichiometry of MBE grown HgTe is controlled by the substrate temperature and the Hg overpressure. The possible error bars in our calculated Hg flux has also not been considered yet. The Hg beam pressure was measured from the cell temperature and dimensions; the major source of error in this calculation is the cell temperature measurement. However, the Hg cell would have to be 120C higher than measured to achieve the necessary equilibrium pressure for single phase HgTe growth at the Te rich boundary.

Second Phase Hg and Te

XTEM micrographs of HgTe(112)Te layers grown under excess Hg or Te2 fluxes have been obtained and provide important clues into the actual growth process. Second phase precipitation of a low pressure, hexagonal Te has been observed in Te rich CdTe growth [4]; this phase of Te is also found in our Te rich HgTe(112)Te layers. The defect mechanism operating on the Hg rich layers was however not anticipated. The only defect present close to the HgTe-CdTe interface is a set of narrow (50-100A) void columns that run along the growth direction as is seen in a weak beam XTEM micrograph illustrated in Figure 6. Further from the interface, a large number of inclined twins originate at the void column boundary and

Figure 6. Weak beam XTEM of a HgTe(112)Te layer grown under excess Hg flux.

eventually lead to polycrystalline growth. There is no evidence for antiphase domain formation.

Figure 7 illustrates a likely growth process that would lead to the observed void columns. A thin Hg island forms on the growth surface and adsorbed Te2 reacts with Hg at the periphery of the island. No Hg condensation can occur since the Hg beam pressure (1 x 10-3 torr) is far less than the equilibrium vapor pressure for elemental Hg (40 torr) at 225C. A Hg desorption energy at the growth surface that is much larger than the elemental sublimation energy is the likely cause for the thin film present in this model. In a related study [5], the Cd desorption energy on a CdTe (001) surface was determined to be 8 times larger than the elemental sublimation energy by measuring the residence times of Cd with RHEED as a function of substrate temperature.

Te2 attack occurs at the periphery of the Hg island leading to the void column directly above the island as observed. The adsorbed Te2 probably has too short a residence time on the Hg island to allow any reaction to take place. The HgTe layers beside the void column begin to actually grow in a single crystalline manner since the excess Hg may be transported to the void columns during growth.

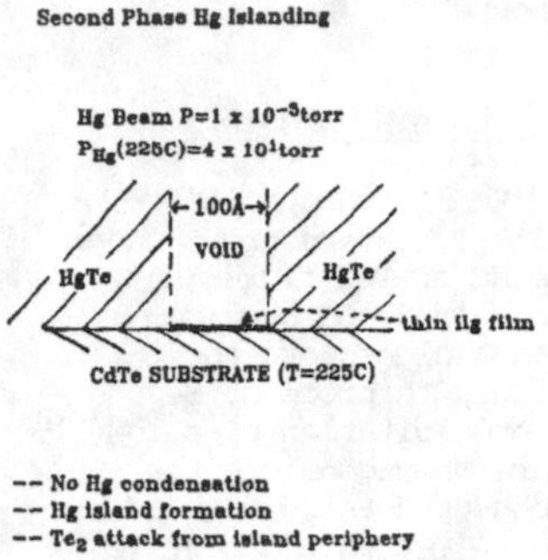

Superlattice Smoothing

Since a major application of the MBE growth method is multilayered structures, the issue of interfacial roughness at the boundary between layers is critical. In 1984, the interface roughness in AlGaAs/GaAs quantum wells (QW) was examined [6] with XTEM, cathodoluminescence (CL) and photoluminescence (PL). It was found by CL that impurities (non-radiative centers) floated on only the AlGaAs growth front, while simultaneous XTEM micrographs indicated that the AlGaAs growth front did not smooth any initial surface roughness as was observed for GaAs. In fact, PL measurements on other AlGaAs/GaAs QW's suggested thick AlGaAs cladding layers actually led to an increased surface roughness. The impurities that float on the AlGaAs layer surface probably pin the lateral motion of steps and hinder the efficient removal of inclined microfacets.

Figure 8 shows a bright-field XTEM micrograph of a 103A HgTe~57A CdTe SL grown on a CdTe (112)Te surface. The CdTe (HgTe) layers are imaged as white (dark) areas. The CdTe buffer surface exhibits a (111) oriented microfacet approximately 400-500A wide; the microfacet is inclined by the expected ~19 degrees from the (112) normal. In only five SL periods, the inclination of the microfacet is reduced considerably due to the lateral motion of steps along the microfacet surface. Similar XTEM micrographs were observed for (001) oriented SL films.

The SL smoothing occurs primarily from the lateral motion of steps in only the HgTe layers. This is best observed in Figure 9 with a bright field micrograph of a 22A HgTe-12A CdTe SL (with 10% error in layer dimensions) on CdTe(112)Te. The CdTe buffer at the bottom and the individual

Second Phase Hg Islanding

Hg Beam P=1 x 10⁻⁹torr
P_{Hg}(225C)=4 x 10¹torr

←100Å→

VOID

HgTe   HgTe

thin Hg film

CdTe SUBSTRATE (T=225C)

-- No Hg condensation
-- Hg island formation
-- Te₂ attack from island periphery

Figure 7. Model for MBE HgTe growth under excess Hg flux.

CdTe layers are imaged in white; the CdTe layers have a very uniform
thickness and very closely follow the surface roughness present in the
underlying HgTe layers. The HgTe layers, on the other hand, have a highly
non-uniform thickness due to the rapid lateral motion of steps as will be
discussed in the next section.

Facetted Growth

The interfacial
roughness in the SL
in Figure 9
propagates through
the entire film
thickness. A (001)
oriented SL grown
side-by-side with
the SL in Figure 9
did not show this
propagation of
interfacial
roughness nor did
any of the (112)Te
oriented SL with
thicker individual
layers.

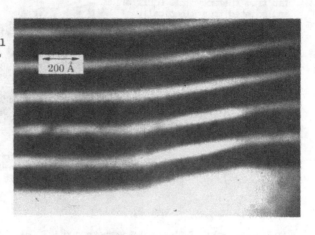

The probable
cause for this
effect is the rapid
variation in the

Figure 8. Bright-field XTEM of a 103A
HgTe-57A CdTe SL on CdTe(112)Te.

dynamic HgTe growth rate as the growth orientation varies slightly from
the (111) direction. The large twin that runs diagonally across the center
of Figure 9 fixes the (111) direction in the SL; it is inclined by the
expected ~19° from the CdTe(112)Te buffer surface. The HgTe layer thickness
appears to be very small along the (111) oriented microfacets, but shows a
pronounced bulge along the (115) oriented microfacets that are also
inclined by ~19° from the (112) direction. The transient or dynamic growth
rate along the (111) direction is expected to be low due to the difficulty
in nucleating each bilayer; however, the dynamic growth rate of the (115)
oriented microfacets is much higher since adatoms can nucleate much more
easily at the step edges. This facetted growth should then occur only for
the SL films that are oriented near the (111) direction.

SUMMARY

A dramatic sensitivity to substrate surface orientation has been
observed for our MBE HgTe growth. Important clues to help understand

318

the HgTe growth process were
uncovered by examining the defects
that occur under Hg rich growth
conditions. The MBE growth
conditions were found to be very
far from thermodynamic equilibrium
and pointed to a kinetically
limited process. Interfacial
roughness in HgTe-CdTe SL films
was reduced due to the rapid
lateral motion of steps on the
HgTe layer surface.

CdTe Buffer

Figure 9. Bright-field XTEM
of a 22A HgTe-12A CdTe SL
on CdTe(112)Te.

REFERENCES

1) J.J.Dubowski, T.Dietl, W.Szymanska and R.R.Galazka, J.Phys.Chem.
   Solids 42, 351(1981)
2) S.Sivananthan, X.Chu, J.Reno and J.P.Faurie, J.Appl.Phys. 60,
   1359 (1986).
3) H.F.Schaake, J.Electron Mat. 14, 513 (1985).
4) N.G.Chew, A.G.Cullis and G.M.Williams, Appl.Phys.Lett. 45, 1090
   (1984).
5) J.D.Benson, B.K.Wagner, A.Torabi and C.J.Summers, Appl. Phys.
   Lett. 49, 1034 (1986).
6) P.M.Petroff, R.C.Miller, A.C.Gossard and W.Wiegmann, Appl.Phys.
   Lett. 44, 217 (1984).

Epitaxial Growth I

LIQUID-PHASE EPITAXY OF $Hg_{1-x}Cd_xTe$ FROM Hg SOLUTION: A ROUTE TO INFRARED DETECTOR STRUCTURES

Tse Tung, M.H. Kalisher, A.P. Stevens, P.E. Herning
Santa Barbara Research Center, Goleta, CA 93117

ABSTRACT

Over the past few years, liquid-phase epitaxy (LPE) has become an established growth technique for the synthesis of HgCdTe. This paper reviews one of the most successful LPE technologies developed for HgCdTe, specifically, "infinite-melt" vertical LPE (VLPE) from Hg-rich solutions.

Despite the very high Hg vapor pressure ($\geq$ 10 atm) and the extremely low solubility of Cd in the Hg solution ($< 10^{-3}$ mol%), this approach was believed to offer the best long-term prospect for growth of HgCdTe suitable for various device structures. Since the initial demonstration of LPE growth of HgCdTe layers from Hg solution in experiments conducted at SBRC in 1978, the VLPE technology has advanced to the point where epitaxial HgCdTe can now be grown for photoconductive (PC) and photovoltaic (PV) as well as monolithic metal-insulator-semiconductor (MIS) and high-frequency laser-detector devices with state-of-the-art performance in the entire 2-12 μm spectral region.

A historical perspective and the current status of VLPE technology are reported. Particular emphasis is placed on the important role of the thermodynamic parameters (phase diagram) and on control of stoichiometry (defect chemistry) and impurity doping (distribution coefficient) for growth of HgCdTe layers from Hg solution. Critical material characteristics, such as transport properties, minority-carrier lifetime, morphology and crystal structure, are also discussed. Finally, a comparison with the LPE technology using Te solutions, which has been the mainstay of the remainder of the IR community, is presented.

INTRODUCTION

$Hg_{1-x}Cd_xTe$ is a ternary solid solution of interest primarily because HgTe and CdTe are miscible in all proportions and because its bandgap varies approximately linearly with composition from -0.3 eV for the semimetal HgTe to 1.6 eV for the wide-bandgap semiconductor CdTe. The narrow homogeneity ranges of the binary compounds, HgTe and CdTe, are reflected in the behavior of the ternary solid solution in that the Te-to-metal ratio is restricted to values close to 1.0, namely, y ~ 0.5, in a more exact nomenclature $(Hg_{1-x}Cd_x)_yTe_{1-y}$. This material, which is an intrinsic semiconductor for composition, x, greater than 0.15, can be tailored to cover a wide spectral region from the visible to beyond 30 μm. In the last 25 years, HgCdTe has been firmly established as the material of choice for IR detectors [1,2,3].

Historically, crystal growth has always been a major problem. This is mainly because a relatively high Hg pressure is present during growth, which makes it difficult to control the exact stoichiometry (y) and composition (x) of the grown material, and sometimes even with the threat of explosion. Low-temperature growth processes, such as MBE [4] and MOCVD [5], as well as high-temperature processes, including bulk growth and LPE, are all plagued by this universal problem. Proper Hg vapor pressure must be furnished and controlled during growth.

HgCdTe LPE layers can now be grown on CdTe or CdZnTe substrates with the required composition, morphology, and electrical and structural properties. The quality of the LPE layers is generally superior to that of bulk material due to lower growth temperature and more favorable growth thermodynamics. Availability of a near-perfect substrate, CdTe (CdZnTe), which matches closely to $Hg_{1-x}Cd_xTe$ of all compositions in lattice parameter, thermal expansion coefficient, and chemical constituents, contributes significantly to the rapid and impressive success of the HgCdTe LPE technology. The two major obstacles encountered in HgCdTe bulk crystal growth - compositional nonuniformity and long annealing time to control stoichiometry - are overcome by LPE. HgCdTe LPE layers also have the advantage of being suitable for making backside-illuminated detectors. In addition, LPE techniques can be adopted for growing HgCdTe multilayers with different compositions and doping levels. Multilayer structures are advantageous for the fabrication of the two classes of focal plane arrays (FPA) developed intensively in parallel: hybrid FPA's (HFPA), using HgCdTe detectors mated to silicon charge-coupled devices (CCD), and monolithic FPA's, comprising detectors and CCD's built in one semiconductor (HgCdTe).

In this article we review an LPE growth technique, specifically, "infinite-melt" vertical LPE (VLPE) from Hg solutions, which has successfully overcome the high Hg pressure problem through proper system design and has produced excellent HgCdTe materials. IR device structures with state-of-the-art performance have been manufactured by VLPE. A history of the evolution of the VLPE technology is presented, followed by a detailed elaboration of the phase diagram of the Hg-Cd-Te system and its inseparable relations with LPE growth processes. Material requirements for IR detectors and material characteristics realized by the VLPE technology are reported, together with a comparison with the LPE technology using Te solutions. Particular emphasis is placed on the important role of stoichiometric native defects and impurity dopants for epitaxial HgCdTe grown from Te- and Hg-solutions. We close with a discussion of the impact of VLPE technology on HgCdTe device structures.

I. HISTORICAL PERSPECTIVE

Liquid-phase epitaxy (LPE) is a single-crystal growth process in which precipitation of material from a cooling solution occurs onto a substrate. Over the last 25 years, LPE has emerged as the major technology for the fabrication of III-V optoelectronic devices [6]. This is mainly due to the fact that LPE is a very simple technique for the growth of high-quality semiconductor compounds or alloys. From a thermodynamic point of view, LPE is the growth technique closest to equilibrium and thus most likely to form stoichiometric solids. For LPE of III-V compounds, the Group III-rich melts are generally used. This reduces the concentration of Group III vacancies, which are believed to be associated with non-radiative centers that limit lifetime and diffusion length [7]. In addition, the solvent in contact with the substrate during growth tends to retain undesirable impurities and act as a getter of impurities. This gettering of impurities allows the routine growth of III-V compounds of very high purity with excellent electrical and optical properties [8]. It is, therefore, no surprise to see LPE has already proved itself as the growth technique to dominate next-generation HgCdTe FPA technology, with both vertical open-tube LPE and variations of horizontal LPE being extensively developed in the IR industry. These solution-growth techniques, the former involving either Hg-rich [9] or Te-rich [10-11] solutions and the latter exclusively Te-rich [10-25] solutions, have received considerable attention in the past several years and have advanced significantly over their earlier stages of development.

$Hg_{1-x}Cd_xTe$ epitaxial layers have been grown on CdTe substrates from
pseudobinary [12-13,26-27], Hg- [9,12-13,28-29] and Te-rich [10-25] solutions by LPE. CdTe is often used for the substrate because its lattice mismatch with epitaxial $Hg_{1-x}Cd_xTe$ is less than 0.3%. Most efforts on LPE
growth of $Hg_{1-x}Cd_xTe$ have been pursued with Te-rich solutions, primarily
because the equilibrium Hg vapor pressure over the Te-rich solution is significantly lower than the other two cases. It has been, nonetheless, difficult to prepare homogeneous, high-quality epitaxial layers of HgCdTe by LPE
even from Te-rich solutions, mainly because of the relatively high vapor
pressure of Hg (~ 0.1 atm) at typical growth temperatures (~ 500°C). Unlike
the LPE growth of III-V compounds, the initial growth solution composition
can be drastically changed during the growth cycle due to Hg being absorbed
or evaporated from the growth solution [30]. Unless special provision is
made during the LPE growth process to keep the growth solution from varying
in composition, $Hg_{1-x}Cd_xTe$ epilayers of uniform composition simply cannot be
reproducibly grown.

Several approaches have been attempted to overcome this problem. The
first two are based upon reducing Hg loss from the melt by applying a high
argon or hydrogen overpressure [10,16] or by confining the melt in a tight-
fitting slider containing extra HgTe [12,15,18]. A sealed, close-tube
tilting technique has also been used [13,20-22]. Another approach is to
control actively the Hg vapor pressure over the growth solution by means of
a two-temperature-zone system [14,23-24]. One zone controls the solution
temperature while the other controls the Hg pressure over the solution. For
a particular growth temperature and solution composition, the three-phase
diagram, shown in Figure 7, can be used to choose a partial pressure of Hg
such that the composition of the growth solution remains constant with
time [30].

LPE growths from Hg-rich solutions were first attempted with unsatis-
factory results by Bowers et al. [12] and Konnikov et al. [28]. Sealed
quartz tubes were used to contain the high vapor pressure of Hg. The small-
size melts they used, however, made it impossible to grow epitaxial layers
of uniform composition as a result of Cd depletion during growth. Despite
the very high Hg vapor pressure (> 10 atm) and the extremely low solubility
of Cd in the Hg solution (< 10^{-3} mol%) indicated by the preliminary phase
diagram studies in 1977*, the "infinite-melt" VLPE growth from Hg solution
was selected as the best long-term prospect for growth of HgCdTe suitable
for future IR device requirements. The term "infinite melt" refers to the
use of a very large (> 2 kg) melt which is maintained at constant temperature and composition for the life of the melt. Infinite-melt VLPE was first
developed to grow thin-film GaAs and InP by S. Kamath [31] of Hughes Research Laboratories (HRL) in 1972. In 1975, the VLPE technology was transferred from HRL to SBRC. It was modified later to grow epitaxial HgCdTe.

Since the initial demonstration of LPE growth of HgCdTe layers from Hg
solution in experiments conducted at SBRC in 1978, the VLPE technology has
advanced to the point where epitaxial HgCdTe can now be grown for photoconductive (PC) and photovoltaic (PV) as well as monolithic metal-insulator-
semiconductor (MIS) and high-frequency laser-detector devices in the entire
2-12 μm spectral region. IR device structures with state-of-the-art performance, including p-on-n as well as n-on-p double-layer heterojunctions
(DLHJ), have been manufactured by VLPE. This is mainly due to the unique
characteristics inherent in the Hg solution:

* This work was supported in part by the U.S. Air Force Materials Laboratory
of Wright-Patterson Air Force Base (R.L. Hickmott, Technical Monitor).

324

1. Low liquidus temperature (< 450°C), which makes the cap-layer growth steps required for the DLHJ structure feasible

2. Ease of intentionally incorporating temperature-stable impurity dopants such as As, Sb, and In in the HgCdTe layers during growth.

Furthermore, the use of the infinite melt provides extremely uniform temperature control of the growth interface and results in excellent uniformity and reproducibility of composition and thickness.

The major VLPE HgCdTe technology milestones that have been achieved since 1977 are shown in Table I.

Table I. Technical Milestones of the Infinite-Melt VLPE Technology

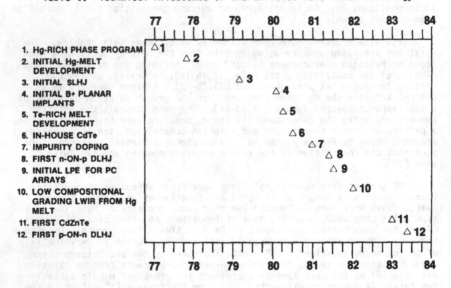

II. APPARATUS AND EXPERIMENTAL PROCEDURE

The VLPE system is shown schematically in Figure 1 [9] and pictorially in Figure 2. A computer-controlled furnace allows accurate control of the system's thermal characteristics. Because of the steep temperature gradient in the system, the vapor, which is composed predominantly of Hg, Te_2, and Cd becomes supersaturated and condenses in fine droplets that fall back into the solution to be evaporated again. At the reflux interface, the vapor is in pressure equilibrium with the H_2 gas. In operation the system H_2 overpressure is regulated. The combination of pressure-balancing H_2 gas and steep temperature gradient acts as a flexible membrane by trapping the escaping vapor. The stream of vapor progressively removes the non-condensable H_2 gas from the high-temperature region [32]. As a result, the solution is maintained in an atmosphere of volatile Hg vapor that prevents its decomposition. In addition, the more volatile impurities can also be swept out of the solution, resulting in a purification effect.

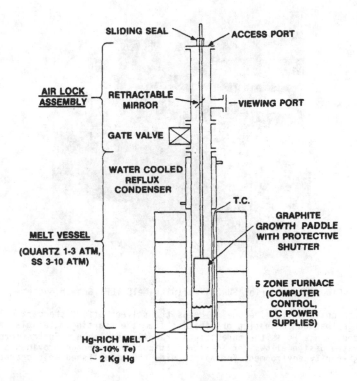

Figure 1. Schematic of Infinite-Melt VLPE Growth System

Te-rich solutions were also used for growth of HgCdTe in the VLPE systems. It was found difficult to maintain melt composition at a constant level due to lack of efficient refluxing in Te-rich solutions, where equilibrium Hg pressure is relatively low (~ 0.1 atm). Refluxing becomes more intense and more efficient at higher pressures and temperatures, from the heat and mass transfer consideration.

A graphite LPE growth paddle, with shutters to protect the substrate from Hg-vapor etching, is attached to the actuator rod. This complete growth assembly is used not only to hold substrates for growth experiments but also to hold pieces of bulk CdTe for the solubility measurements in the phase diagram study.

The infinite melts are always kept saturated and are maintained at the appropriate growth temperature between successive runs. The prepared substrates are introduced into the melt through an air lock. The air lock assembly allows access to the melt without the need to cycle the system to ambient temperature, which would reduce throughput. Most importantly, large melts weighing up to 13 kg provide extremely uniform temperature control at the growth interface and result in layers with excellent compositional and

Figure 2. Four of SBRC's Infinite-Melt VLPE Growth Systems

thickness uniformity. The use of Hg as the solvent offers the particular advantage that contamination originating from the starting materials can be minimized, since Hg is the purest metal readily available. Furthermore, the VLPE system design allows for in-situ melt purification and provides a stable high-purity environment for maintaining the established melt characteristics.

Net donor densities, a measure of the residual background impurities, on the order of 10^{14} cm^{-3} for annealed VLPE layers grown from undoped Hg-rich solutions have been achieved consistently. The use of large melts results in a near constant saturation temperature from run to run and assures excellent reproducibility of layer characteristics, since the amount of material removed from the melt during each growth run is relatively insignificant. Large melts additionally allow doping impurities to be accurately weighed for incorporation into growing layers.

III. PHASE DIAGRAM AND DEFECT CHEMISTRY

The major difficulty in the growth of homogeneous alloy crystals from dilute or stoichiometric melts results from the differences in the equilibrium compositions of the liquid and solid phases, which leads to segregation during solidification and compositional variations in the grown crystals. Knowledge of the solid-liquid phase relation is essential for proper use of solution- as well as bulk-growth processes. In addition, the solid-vapor and liquid-vapor phase relations are of practical importance in materials preparation, especially in view of the high Hg pressure in the growth process [33] and the effect of vapor of constituent components upon post-growth annealing and the consequent electrical properties of HgCdTe [34-45]. Theoretical phase-diagram calculations have proved highly useful in extrapolating and interpolating from the limited amount of experimental data and hence minimizing the laborious experimental work otherwise required for developing complete phase-diagram data. Furthermore, a reliable means of evaluating the existent experimental data and of guiding further experiments is thus obtained [46-47].

The full regular associated solution (RAS) model has been recently developed to successfully describe and predict the phase diagram of the entire Hg-Cd-Te system. Extensive experimental phase-diagram and thermodynamic data have also been critically reviewed along with the calculated results [47-56]. A significant portion of the thermodynamic data for this system consists of the partial pressures of Hg, Cd, and Te_2 for both the metal-saturated and the Te-saturated solid solution, and quantities derived from these [54-56]. All these partial pressures, especially the predominant Hg vapor pressures, are crucial for defect chemistry analysis [57].

The calculated liquidus isotherms and solid isoconcentration lines over the entire Gibbs composition triangle are shown in Figure 3 [49]. It can be immediately deduced that on the Hg-Cd side the tie lines tend to converge near the Hg corner with decreasing temperature. This includes x-values as high as 0.9999. It is clear that growth from Cd-rich solutions produces a solid that is almost pure CdTe. Any attempt to use Cd as the solvent for LPE growth of $Hg_{1-x}Cd_xTe$ with $x \leqslant 0.9$ is practically impossible. In fact, essentially pure CdTe ($x > 0.99$) can be grown even from Hg-rich solution dissolved with CdTe only (i.e., no excess Te) at temperatures as low as 250°C [58]. In this case, Hg is almost an ideal solvent for LPE growth of CdTe.

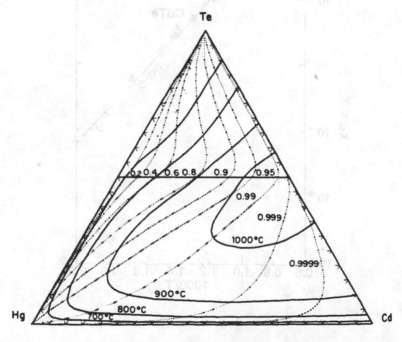

Figure 3. Calculated Liquidus Isotherms (Solid Lines) at High Temperatures and Solid-Solution Isoconcentration Lines (Dashed Lines) Plotted Across the Entire Gibbs Composition Triangle

328

The solubility of CdTe in Hg is shown in Figure 4. Vanyukov et al. [59] determined the solubility of CdTe in Hg between 350 and 580°C by the phase-separation and the dew-point techniques. The calculated results agree well with these data and with data measured in the VLPE system at SBRC by the dissolution technique [9]. The solubilities of Te in Hg and CdTe in Hg in the presence of Te were also measured at SBRC. It was found that Te is much more soluble in Hg than is CdTe and the presence of Te or CdTe depresses the solubility of the other. It was further found that excess Te must be present in the melt to grow $Hg_{1-x}Cd_xTe$ epitaxial layers with $x <$ 0.9. This is clearly illustrated in Figure 5, an expanded view of the Gibbs composition triangle near the Hg corner showing the low-temperature liquidus isotherms and the solid isoconcentration lines. There is good agreement between the calculated results shown here and the limited experimental data measured at SBRC.

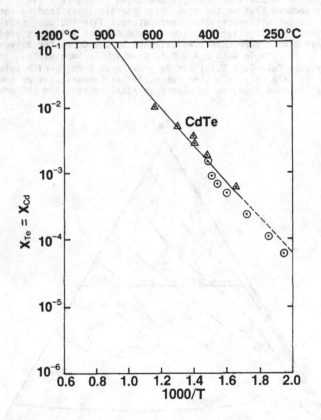

Figure 4. Calculated Curve and Experimental Points for the Solubility of CdTe in Hg. Triangles are from Vanyukov et al. [59] and circles from SBRC.

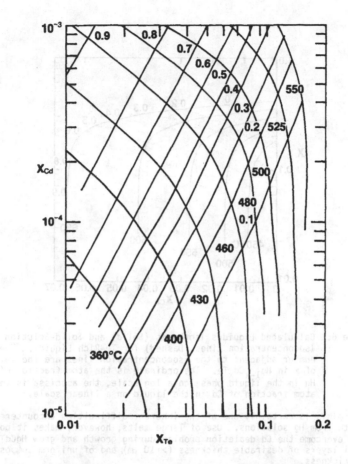

Figure 5. Calculated Liquidus Isotherms and Solid Isoconcentration Lines
for Hg-Rich Liquids. Axes are atom fractions in the liquid.
Nearly straight lines slanting up towards the right are the
solid isoconcentration lines for labeled values of x in the
formula, $Hg_{1-x}Cd_xTe$. Isotherms with labeled values of
temperatures in degree Celsius are curves bending downwards
left to right.

Calculated results for the Te corner detailed in an expanded plot are
shown in Figure 6 [49]. Along each liquidus isotherm the solubility of Cd
increases with decreasing Hg content, although with a very slow rate for
melts of low Hg content. This is slightly different from Harman's observation that along the liquidus isotherm the solubility of Cd has a maximum
with respect to changes in the Hg content [14]. Otherwise, the calculated
results agree fairly well with the experiments.

As an example to compare growth parameters for Te solutions with those
for Hg solutions, consider the growth of LWIR $Hg_{1-x}Cd_xTe$ (x = 0.2) at 500°C
by LPE from both Te and Hg solutions. According to the phase diagrams shown
in Figure 5 and 6, the atom fraction of Cd (x_{Cd}) for Te-rich solutions is
8.3×10^{-3} while x_{Cd} for Hg-rich solutions is 2.6×10^{-4}, a factor of 32

330

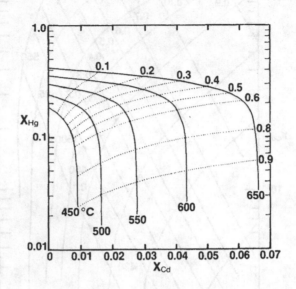

Figure 6. Calculated Liquidus Isotherms (Solid) and Solid-Solution
Isoconcentration Lines (Dashed) for Te-Rich Liquids. The
number adjacent to the isoconcentration lines are the values
of x in $Hg_{1-x}Cd_xTe$. The ordinate is the atom fraction of
Hg in the liquid phase on a log scale, the abscissa is the
atom fraction of Cd in the liquid on a linear scale.

smaller. This is certainly one of the inherent difficulties encountered in
LPE growth from Hg solutions. Use of large melts, however, makes it possible to overcome the Cd depletion problem during growth and grow HgCdTe
epitaxial layers of desirable thickness (> 10 μm) and of uniform composition
across thickness.

The ternary semiconductor alloy $Hg_{1-x}Cd_xTe$ in equilibrium with its own
vapor has three degrees of freedom, by the Gibbs phase rule. Thus, for a
given x and temperature, only one partial pressure need be fixed to fix all
of the intensive variables, including the carrier concentrations. For such
purposes it is valuable to know the partial pressure of the predominant
vapor species for both the metal-saturated and the Te-saturated solid solutions [54]. The partial pressures of Hg, Te_2 and Cd along the three-phase
curves, where the solid solution coexists with the vapor and the liquid,
have been measured for various solid solutions by the optical absorbance
technique [54-56]. Calculated curves of P_{Hg} for various solid solutions and
the measured data are shown in Figure 7 [49]. Calculated curves are also
shown for x values (0.8,0.9) for which there are not available experimental
data. The area within each three-phase curve specifies those values of P_{Hg}
and temperature for which the solid solution is a stable phase. For semiconductor compounds or alloys such as $(Hg_{1-x}Cd_x)_yTe_{1-y}$ the upper and lower
branches of the three-phase curve can be separated at a given temperature by
several orders of magnitude or more in P_{Hg}, even when the width of the narrow homogeneity range is well under 1 at % (i.e., y - 0.5 << 1) [60-62].
Similar three-phase curves also exist for partial pressures of Te_2 and Cd

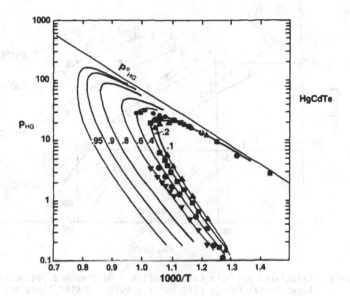

Figure 7. Partial Pressure of Hg Along the Calculated Three-Phase Curves for Various Solid Solutions. The labels are the values of x in the formula, $Hg_{1-x}Cd_xTe$. Experimental points from Schwartz et al. [54] and Tung et al. [55]

for various solid solutions [49]. It is apparent from Figure 7 that equilibrium Hg pressure over Te-rich solutions is 0.13 atm while that over Hg-rich solutions is 7.3 atm for growth of LWIR $Hg_{1-x}Cd_xTe$ (x = 0.2) at 500°C. This high Hg pressure imposes a major design challenge for an open-tube LPE technique with Hg as the solvent.

It has been observed that the concentration of carriers in $(Hg_{1-x}Cd_x)_yTe_{1-y}$ for a given Hg-to-Cd ratio (x) can be varied by equilibration at temperatures above 200°C under various Hg pressures [36-45]. In accordance with the simple defect chemistry of solids, this implies a varying concentration of native atomic point defects, which can act as donors or acceptors, and consequently a small range of stability around an atomic fraction of 0.5 for Te (i.e., y = 0.5) [39]. Undoped $Hg_{1-x}Cd_xTe$ with x < 0.4 can be made with an excess of conduction-band electrons over valence-band holes only by equilibrating the material in the low-temperature range (< 300°C) so that it is as metal-rich as possible. It is believed that this n-type conductivity is due to residual background donors in the low 10^{14} cm^{-3} range in high-purity samples [36-45, 63].

P-type conductivity as a function of equilibration temperatures and Hg pressures as shown in Figure 8, however, is believed to result from activation of the doubly-ionized metal (Hg-Cd) vacancies [39]. On the other hand, it is known that Cd-saturated CdTe is n type for temperatures as high as 900°C [35]. It can be expected that n-type $Hg_{1-x}Cd_xTe$ with 0.4 < x < 1.0 will be obtainable by metal-saturation not only at temperatures below 300°C but also at higher temperatures [39]. This has been confirmed experimentally for x = 0.7. It was observed that the net electron concentrations of

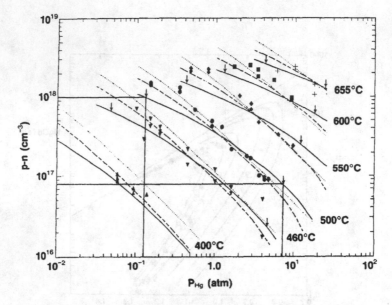

Figure 8. Calculated Net Hole Concentration - Hg Pressure Isotherms and Experimental Points [37] for x = 0.20. Dotted lines are calculated in Ref. [37] using the law of mass action. Solid lines are calculated in Ref. [39] using the defect chemistry analysis in conjunction with the band structure and assuming degeneracy. Dashed lines are calculated in Ref. [39] assuming non-degeneracy. The two arrows on each solid line indicate the range of P_{Hg} for which $Hg_{0.8}Cd_{0.2}Te$ is stable at each temperature.

bulk material (x = 0.7) increase with the annealing temperatures up to 400°C under saturated Hg vapor, characteristic of donor native defects (e.g., Te vacancies) [45].

A complete analysis of the defect chemistry requires a theoretical model capable of fitting the experimental data, such as shown in Figure 8, and thereby establishing values for the intrinsic material parameters [60-61]. These parameters include the partial pressure of Hg over the intrinsic pure material, the equilibrium constant for the ionized predominant native defects (e.g., Schottky constant), and the intrinsic carrier concentration. Once these are reliably established, the degree of self-compensation among the native defects and between these and donor or acceptor impurities can be calculated. These factors are related to ionized-impurity scattering and to doping efficiency. Such an analysis has been carried out recently for x = 0.2, 0.4, and 1.0 $Hg_{1-x}Cd_xTe$ [39]. The analysis relies upon a single equation among the net hole concentration, partial pressure, and temperature that is equivalent to a mass-action-law analysis but that already takes account of the electro-neutrality equation and so gives a solution valid from extrinsic p-type to extrinsic n-type in a single equation.

As shown in Figure 8, the net hole concentration of bulk $Hg_{0.8}Cd_{0.2}Te$ equilibrated under Te-saturated condition (P_{Hg} = 0.13 atm) at 500°C is 10^{18} cm^{-3}, while that of Hg-saturated $Hg_{0.8}Cd_{0.2}Te$ (P_{Hg} = 7.3 atm) is 8×10^{16} cm^{-3}. The net hole concentration (77K) of SBRC's LPE $Hg_{0.8}Cd_{0.2}Te$

grown from the Hg-rich solution at 515°C, is consistently $1-2 \times 10^{17}$ cm^{-3}, in agreement with that of bulk $Hg_{0.8}Cd_{0.2}Te$ equilibrated under Hg-saturated conditions [36-37,42-44]. It has been widely accepted that excess Te is accommodated in the $Hg_{1-x}Cd_xTe$ lattice through the formation of metal (Hg-Cd) vacancies acting as doubly-ionized acceptors. It is expected, therefore, that internal microprecipitation as a result of the retrograde solubility of Te in $Hg_{1-x}Cd_xTe$ should occur more readily in Te-saturated material than in Hg-saturated material [34,64]. It is also possible that electrically neutral (inactive as donors or acceptors) native defects are significant in Te-saturated material, such as is indicated to be the case for some of the Te-saturated II-VI compounds [65].

IV. MATERIAL CHARACTERISTICS

Material characteristics realized by the VLPE technology will be elaborated in five areas: impurity doping, minority-carrier lifetime, compositional uniformity, crystal quality and morphology, and multiple-layer heterojunction structures.

1. Impurity Doping

Impurity doping is one of the key elements in the development of complicated semiconductor devices. In order to produce heterostructure detectors with HgCdTe epitaxial layers it is essential to have an understanding and control of dopant behavior. Undesirable impurities may be present in the melt or in grown layers, and affect layer properties and device performance. Intentional doping is usually necessary to produce well-behaved and stable junctions for photovoltaic devices. Although it is possible to use stoichiometric annealing methods [17, 41, 66-67] to control the carriers due to native defects such as Hg vacancies, this approach does not ensure the stability and repeatability required for device structures. It is known that Hg vacancies diffuse rapidly even at relatively low temperatures [68, 69]. The Hg-vacancy concentration is subject to unintentional changes during the subsequent high-temperature processing steps.

It is generally accepted that the Group V elements that substitute in the Te sublattice are much slower diffusing acceptors than the Group I elements, which substitute in the metal sublattice [66, 70-72]. The Group V elements are, therefore, preferred to other impurity dopants for p-type HgCdTe materials used in monolithic as well as PV devices. An ideal impurity dopant should have low vapor pressure, low diffusivity, and small impurity ionization energy. It will be clear later that the Group V dopants indeed qualify as the p-type dopants of choice for HgCdTe. Furthermore, for the native-defect-dominated II-VI compounds such as HgCdTe, dramatically improved material characteristics are expected in impurity-doped materials, since the concentration of the native defects that are often found to deteriorate material characteristics can be significantly reduced by incorporation of impurity dopants through the law of mass action [65, 73]. Such material characteristics have been found to include the minority-carrier lifetime [70, 74-77], and the Hall mobility of free carriers [19, 22, 78-79].

Initial experiments in doping layers grown from the Hg solution were begun here at SBRC in late 1980 using Sb [80]. In early and mid-1981, additional experiments were performed using In and As. In all three cases it was evident that Hg-rich melts could readily be doped to produce both n- and p-type temperature-stable layers. Since then, many other dopants have been investigated as shown in Table 1 of Reference 80.

Determination of dopant concentration in the solid involves the use of Hall-effect measurements as well as corroboration with secondary-ion mass spectrometry (SIMS) concentration profiles. The correspondence in the concentration of a dopant measured by SIMS and Hall-effect or capacitance-voltage (C-V) methods is an indication of activity of that dopant.

In the case of As, Figure 9 is a plot of the acceptor concentration versus the concentration measured by SIMS. It is clear that the two measurements correlate and that the arsenic is 100% active over three orders of magnitude. In, is also 100% active as shown in Figure 10 [81]. If the desired dopant concentration is lower than the native defect concentration dictated by the growth conditions, the layer must be annealed at low temperature (< 300°C) under saturated Hg vapor. This treatment does not "activate" the dopants but fills or eliminates the Hg vacancies, allowing the dopants rather than the Hg vacancies to dominate the electrical properties of the layer.

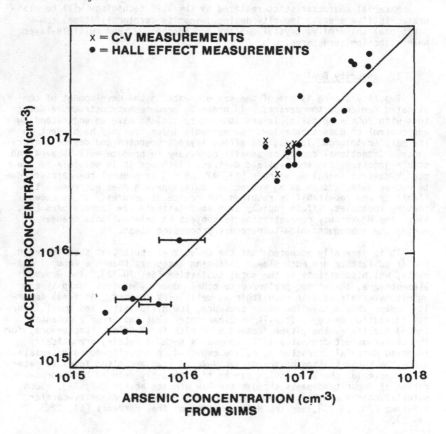

Figure 9.   Comparison of As Concentration (Measured by SIMS) and Acceptor Concentration in Intentionally Doped HgCdTe Epitaxial Layers Grown from Hg-Rich Solution

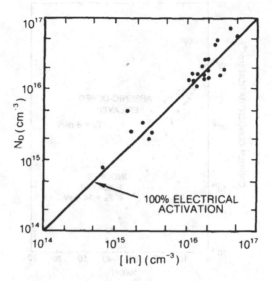

Figure 10.   Comparison of In Concentration (Measured by SIMS) and Donor
             Concentration (Measured by Hall-Effect Measurements) in
             Intentionally Doped HgCdTe Epitaxial Layers Grown From Hg-Rich
             Solution

The contribution of active p-type dopants, however, to the carrier concentration is a function of temperature. The ionization energy for As is about one-half that of Hg vacancies [82]. This can be seen in Figure 11. This means that the electrical properties are half as sensitive to temperature changes in As-doped material as they are in Hg-vacancy doped material. This agrees very well with the reported work on Sb-doped Bridgman-grown HgCdTe [72]. It was also found that the activation energy decreased as the As concentration was raised above 10^{17} cm^{-3}, and no freezeout was observed for concentrations greater than 10^{18} cm^{-3}. This is due to the breakdown of the usual localized level concept, which is applicable only when the impurity concentration of semiconductors is low [83]. In heavily doped semiconductors the impurity levels merge with the adjacent bands [84]. The best criterion for the concentration at which the ionization energy goes to zero is the impurity concentration at which metallic impurity conduction occurs. This transition has been predicted by Mott et al. [85] to occur when the ratio of the average separation of the impurity atoms to the radius of the hydrogenic impurity is about three. For p-type $Hg_{0.7}Cd_{0.3}Te$ the calculated value for the acceptor ionization energy to approach zero is 2×10^{18} cm^{-3}, in good agreement with the experiment [80]. Most n-type dopants are 100% active at all temperatures, because the ionization energies of donors are very small, even for doping concentrations as low as 10^{14} cm^{-3}. Again, this can be explained by Mott's criterion.

The elements that have been most useful as dopants for HgCdTe are those in Groups III and V. A trend that is obviously visible in Group III and partly visible in Group V is the decrease in the distribution coefficient with increasing atomic number. At present, it is not clear whether this trend is due to an increase in atomic size or to the decrease in electron affinity [80].

336

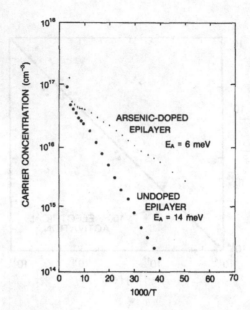

Figure 11. Temperature Dependence of Carrier Concentrations in As-Doped
and Nonstoichiometric p-Type VLPE Layers Grown from Hg-Rich
Solution

The distribution coefficients of impurity dopants vary with the actual
growth conditions and the concentration of the dopants. Among the major
growth conditions that may be varied is the Te fraction in the melt. The
dependence of distribution coefficient on this quantity can be seen in Figure 12 for the three main dopants as well as Cd, which is included for
reference and is based on calculated phase-diagram results [49]. In the
case of In, the effective distribution coefficient has been measured from a
high of 100 to a low of 8. For As, the high value is 10 and the low is
0.1. In all cases, the distribution coefficient decreases with increasing
Te content in the melt.

The distribution coefficient is also a function of the concentration of
the dopant in solution. Because each dopant has specific solubility limits
both in a given melt and a given layer, its distribution coefficient decreases as the concentration in the melt increases. Figure 6 of Reference 80 shows the dependence of the distribution coefficient on concentration in the Hg-rich melt for Ga, In, and As. In the case of both As and
Sb, the distribution coefficients from Hg-rich solutions at ~ 500°C are at
least a factor of 100 greater than those from Te-rich solutions [17] as
shown in Figure 13.

Statistical thermodynamic considerations indicate that the chemical potential of a substitutional impurity also depends upon the chemical potential of the substituted host atom [60, 61]. It should be emphasized that
the chemical potentials of the host atoms are strong functions of nonstoich-
iometry of the compound semiconductor, as indicated by the pressure-tempera-
ture phase diagram mentioned in Section III. The chemical potential of a
substitutional dopant for a fixed value of temperature and dopant concentration in the solid solution is, therefore, expected to vary significantly

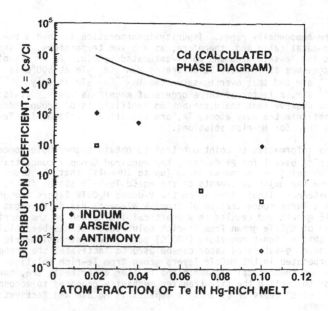

Figure 12. Distribution Coefficient (K) of In, As, and Sb in HgCdTe Grown from Hg-Rich Melt as a Function of Melt Composition. The distribution coefficient is the concentration ratio of the impurity incorporated into the epitaxial layer and that dissolved in the melt.

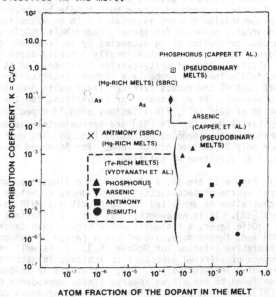

Figure 13. Distribution Coefficient of the Group V Elements in HgCdTe as a Function of the Dopant Concentration in Hg-Rich, Te-Rich [17] and Pseudobinary Melts [86]

across the homogeneity range. Impurity incorporation is then a function of the Te-to-metal ratio and, therefore, at a given temperature, it should be different for Te-saturated and metal-saturated HgCdTe. As an example, the partial pressure of Te_2 over Te-saturated $Hg_{0.8}Cd_{0.2}Te$ at 500°C is 6.4×10^{-4} atm and that over Hg-saturated $Hg_{0.8}Cd_{0.2}Te$ is only 2.2×10^{-7} atm, a factor of three orders of magnitude lower. It is not surprising to observe that the distribution coefficients of Group V dopants which substitute the host atoms, Te, are significantly lower for Te-rich solutions than for Hg-rich solutions.

It is informative to point out that to obtain higher hole concentration ($> 10^{15}$ cm^{-3}) useful for PV devices, the required Group V concentrations in the Te-rich solution become so large (up to 10 mol%) that the Group V dopants become the major components of the Hg-Cd-Te-V (V stands for P, As, Sb, or Bi) quaternary liquid from which the V-doped HgCdTe layers are grown [17]. This large concentration of Group V elements has deleterious effects on the LPE growth and results in a practical limitation on the carrier concentration of HgCdTe grown from Te-rich solutions. It has been claimed recently that a high-temperature (500°C) post-growth stoichiometric annealing under Hg-saturated vapor can be used to "activate" the Group V dopants incorporated in LPE HgCdTe layers grown from Te-rich solutions [17]. More rigorous theoretical arguments and further experiments are, however, called for to confirm the site-transfer mechanism proposed to account for the "amphoteric" behavior of Group V dopants in HgCdTe and "activation" of these dopants.

2. Minority-carrier Lifetime

The excess carrier lifetime is one of the most important material characteristics of HgCdTe, since it governs the device performance and frequency response. The lifetime in undoped p-type bulk HgCdTe has been reported to be limited by recombination centers associated with Hg vacancies. Very low lifetimes are generally observed for these undoped bulk HgCdTe which are made p type through generation of Hg vacancies by stoichiometric annealing. It was found, however, that intentionally impurity-doped HgCdTe can be obtained with relatively high minority-carrier lifetime. It is believed that in undoped HgCdTe there is probably a higher concentration of trapping deep levels that control the excess carrier lifetimes in Shockley-Read-Hall recombination processes. As an example for comparison, for the same net hole concentration of 10^{16} cm^{-3}, 77K lifetimes of gold-doped bulk [75-77] and As-doped epitaxial $Hg_{0.8}Cd_{0.2}Te$ [87] are approximately several hundred nanoseconds, while those of undoped epitaxial [88, 89] and bulk [77] p-type $Hg_{0.8}Cd_{0.2}Te$ are only 20 nanoseconds or less. The As-doped epitaxial LWIR HgCdTe were grown by VLPE from Hg solutions.

As shown in Figure 14 and 15, a clear inverse linear dependence of the lifetime measured by the photoconductivity (PC) decay technique at 77K on the doping concentration is exhibited for both LWIR and MWIR As-doped HgCdTe epitaxial layers [87]. It is noteworthy that the 77K lifetimes for As-doped MWIR (x = 0.3) HgCdTe layers are significantly higher than those for undoped bulk HgCdTe [90] and, furthermore, are within a factor of two of theoretical values of the radiative lifetime for HgCdTe [91]. It is well known that the maximum performance of infrared detectors is achieved in materials that are limited by the radiative lifetime [92, 93]. Lifetimes in In-doped MWIR HgCdTe were also found to exhibit an inverse linear dependence on the doping concentration [87], with $N_D\tau$ products similar to $N_A\tau$ products observed for the As-doped material.

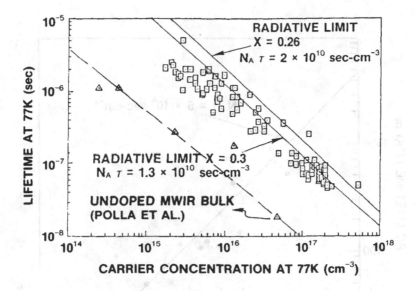

Figure 14. PC Decay Lifetime for As-Doped VLPE HgCdTe (x ~ 0.3) Layers and Undoped Bulk HgCdTe (x = 0.32) as a Function of Carrier Concentration

The status of deep level studies in HgCdTe has been reviewed recently by Jones et al. [74]. Deep-level transient spectroscopy (DLTS) and electron paramagnetic resonance (EPR) were used to characterize electrically the deep levels in both undoped and doped HgCdTe. It was found that the deep-level defects in undoped p-type HgCdTe act donor-like, with electron capture being easier than hole capture. The same deep level defect that is believed to limit the minority carrier lifetime also exists in As-doped HgCdTe. The concentration of this Hg-vacancy-related deep-level defect in As-doped HgCdTe can be reduced significantly by low-temperature stoichiometric annealing under saturated Hg vapor to remove Hg vacancies. In addition to this minor deep level, there exists in As-doped HgCdTe another deep level of significantly higher concentration. This major deep level in As-doped HgCdTe acts as if it is a multiple acceptor, with both hole and electron capture taking place in a negative center. Because of its small minority-carrier (electron) capture cross section, the major center does not control the minority-carrier lifetime. This explains why the As-doped HgCdTe shows significantly higher minority-carrier lifetime than does the Hg-vacancy dominated HgCdTe at the same net hole concentration.

3. Compositional Uniformity

Good compositional uniformity across both the thickness and diameter of HgCdTe samples is essential for the required uniform device performance, especially in the LWIR spectral region, since the bandgap energy of HgCdTe in this region is an extremely strong function of material composition [94, 95]. Based upon the theoretical analysis of the Hg-Cd-Te phase diagram [49, 96], the compositional change due to cooling (dx/dT) for the Hg-rich solution in equilibrium with LWIR HgCdTe (x = 0.2) is only a factor of two larger than that for the Te-rich solution at 500°C, dx/dT being 0.007 per °C

340

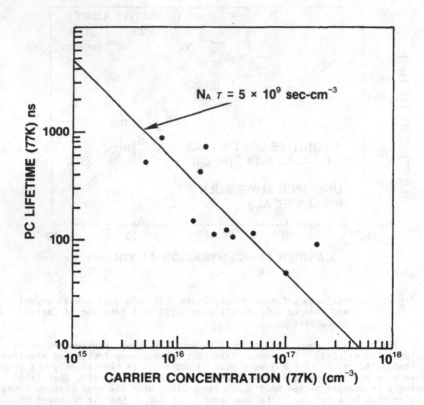

Figure 15.   PC Decay Lifetime for As-Doped $Hg_{0.8}Cd_{0.2}Te$ VLPE Layers as a Function of Carrier Concentration

and 0.003 per °C [97,98] respectively. The compositional grading through the layer, dx/dt, however, is determined not only by the solidus slope, dx/dT, but also by the deposition rate, dt/dT. In fact, $dx/dt = (dx/dT)(dt/dT)^{-1}$. Note here T is temperature in °C and t is thickness in μm.

The deposition rate depends upon various parameters, including the melt size, the degree of supercooling and mixing of the melt, the geometrical configuration of the growth system, and the phase diagram. The VLPE growth parameters have been optimized to obtain consistently a deposition rate in excess of 10 μm per °C. This deposition rate corresponds to a compositional grading of $dx/dt < 7 \times 10^{-4}$/μm for LWIR HgCdTe, comparable to that typically achieved in Te-rich-melt LPE technology. The excellent compositional uni-formity across a 4×5 cm^2 LWIR ($\bar{x} = 0.227$, $\sigma = 8 \times 10^{-4}$) VLPE layer grown from Hg-rich solution is shown in Figure 16. The layer also exhibits good uniformity in thickness ($\bar{t} = 14.8$ μm, $\sigma = 1$). Both composition and thickness of the layer were derived from fast fourier transform IR (FFTIR) transmission measurements.

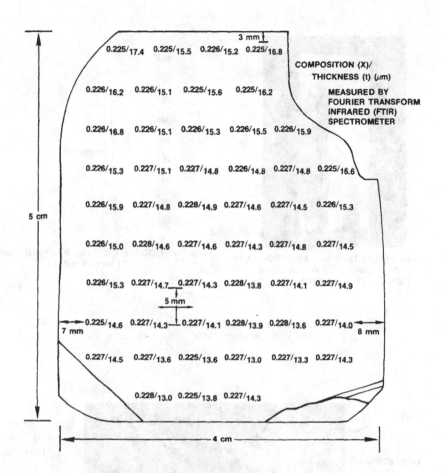

Figure 16. Composition and Thickness Mapping Across a 4×5 cm^2 LWIR VLPE Layer. Average composition ($\bar{x}$) is 0.227 with standard deviation, $\sigma = 8 \times 10^{-4}$; average thickness ($\bar{t}$) is 14.8 μm with standard deviation, $\sigma = 1$ μm.

4. Crystal Quality and Morphology

Good crystal quality of VLPE materials is demonstrated in the scanning Lang x-ray reflection topograph, along with the double-crystal rocking curve, shown in Figure 17 for a typical LWIR $Hg_{0.81}Cd_{0.19}Te$ layer on a CdTe substrate. The double-crystal rocking curve was taken with an InSb (111) reference crystal set for CuKα (333) reflection. The reflection topograph exhibits the same characteristic features (sub-grain boundaries with uniform morphology within the individual grains) as those seen in the topograph of the substrate, and the full width at half maximum (FWHM) is ~ 64 arc sec. The corresponding FWHM in the substrate is ~ 32 arc sec. The interfacial region between layer and substrate is shown in a cross-sectional transmission electron microscopy (TEM) micrograph (Figure 18). Dislocations and a Te precipitate are visible in the photograph. The Te precipitate lies in

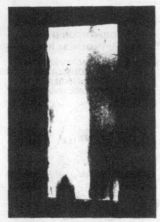

**LANG REFLECTION X—RAY TOPOGRAPH
WITH CuK$_\alpha$ (333) REFLECTION
FOR LAYER II-408.1 (Hg$_{0.81}$Cd$_{0.19}$Te)**

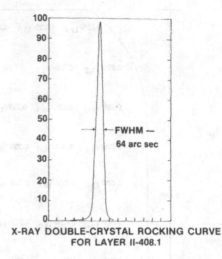

**X-RAY DOUBLE-CRYSTAL ROCKING CURVE
FOR LAYER II-408.1**

Figure 17. X-Ray Topograph and Double-Crystal Rocking Curve for
HgCdTe VLPE Layer on CdTe Substrate

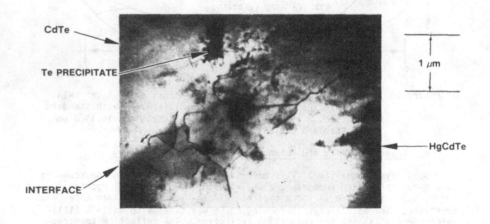

Figure 18. Bright Field (110) Foil Orientation XTEM Micrograph of LPE
HgCdTe Grown in CdTe. A high dislocation density is visible
near Te precipitate in substrate. Dislocation propagation
into first two microns of LPE layer is shown.

the CdTe substrate near the interface. Associated with the precipitate are
a strain field and numerous dislocations. Many dislocations are generated
at the interface in close proximity to the precipitate. The interfacial
dislocations generated by lattice mismatch and the strain field associated
with the Te precipitate propagate into the HgCdTe LPE layer as shown in Figure 18. Some of the dislocations are "bent-over" into the interface and act
as misfit dislocations, which can effectively block threading dislocations
[99]. The blocking structure shown in Figure 18 is incomplete due to the
interaction of the blocking dislocations with other dislocations and with
impurity atoms. The dislocation density decreases from $1 \times 10^9/cm^2$ in the
near interfacial region to $< 1 \times 10^8 cm^{-2}$ at a distance of ~ 1.5 μm from the
interface. Decreased dislocation density with increased distance from the
interface is generally seen. Fewer interfacial dislocations were found in
LPE HgCdTe layers grown on lattice-matched CdZnTe substrates [100-102].

This layer and a few other VLPE HgCdTe layers were further examined by
the x-ray precession technique. It was found that the twin faults present
in epitaxial CdTe, HgCdTe and HgMnTe with {111} orientation grown by MBE and
MOCVD are absent in the VLPE materials [103]. There is experimental evidence to suggest that the twinning is a nucleation phenomenon intrinsic to
II-VI compounds with {111} orientation [104-105]. The surface deposition
techniques such as MBE and MOCVD are, therefore, more susceptible to this
problem.

The ease of decanting the Hg-rich melt after layer growth in the Hg-
melt VLPE process results in smooth and specular surface morphology if a
precisely oriented, lattice-matched CdZnTe substrate is used. Figure 19
shows six MWIR n-on-p double layers used for lot device processing. The
layers were grown on lattice-matched $Cd_{0.96}Zn_{0.04}Te$ substrates.

Figure 19. MWIR Double Layers Used for Lot Device Processing

344

5. Multiple-layer Heterojunction Structures

An important application of the VLPE technology involves the fabrication of double-layer heterojunction (DLHJ) structures [106-108]. The trend for developing such architecture is in the direction of vapor-phase epitaxial (VPE) techniques such as MOCVD [5, 105] and MBE [4, 109]. LPE will, however, continue to play a major role in development of such structures for some time. Because of the dedicated nature of the VLPE furnace systems and their large melts, only one type of layer can be grown from a given melt. By "type", we mean a given dopant and level and an approximate growth temperature/composition range. While there is some flexibility in the dopant concentration (i.e., it can be increased but not easily decreased) and the grown layer composition, these are not usually modified once established for a given requirement. This means that the substrate must be transferred from melt to melt for each layer of the multilayered structure and additional melt systems are required to grow the multilayered structure.

Both n-on-p and p-on-n double-layer heterostructures have been grown and successfully fabricated into state-of-the-art infrared focal plane arrays. The dopant profile of an n-on-p DLHJ structure analyzed by SIMS is shown in Figure 20. Here, an indium-doped cap layer has been grown on an

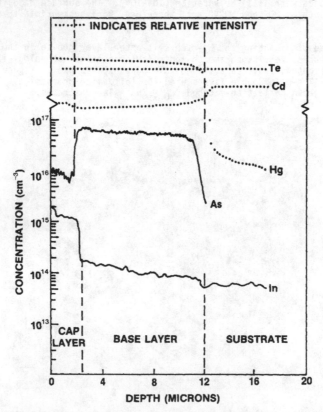

Figure 20. SIMS Profiles of As and In Concentrations in an n-on-p DLHJ Structure

arsenic-doped base layer. The cap-layer growth was performed at 375°C for two hours, but there was no significant diffusion of either As or In. It should be noted that the sensitivity of SIMS for detecting As is relatively low ($\sim 10^{16}$ cm^{-3}), so the 10^{16} cm^{-3} As level shown in the n-type cap layer is merely the measurement noise. The compositional profile of a typical LWIR DLHJ measured by the photoreflectance technique is shown in Figure 21. The HgCdTe HJ was formed by VLPE growth of a cap layer of n-type wide-bandgap $Hg_{0.7}Cd_{0.3}Te$ onto a p-type narrow-bandgap $Hg_{0.77}Cd_{0.23}Te$ base layer on CdZnTe substrate. The HJ is graded due to interdiffusion between Hg and Cd during VLPE growth [106, 110], which is usually carried out at temperatures between 350 and 450°C.

V. IMPACT OF THE VLPE TECHNOLOGY ON HgCdTe DEVICE STRUCTURES

Materials of uniform composition and thickness, high purity, minimum concentration of deep levels and defects, good crystal quality and surface morphology, and proper control of doping are essential for the successful fabrication of any IR device structure. To be specific, n-type HgCdTe of $\sim 10^{14}$ cm^{-3} net donor concentration is required for PC devices; both n- and p-type HgCdTe with low carrier concentration, slightly above or below 10^{15} cm^{-3} at operating temperatures, are recommended for use in monolithic MIS devices and CO_2 laser detectors. The excellent material characteristics of VLPE HgCdTe epilayers reviewed in the previous section have made it possible to meet the stringent material requirements for the fabrication of PC, monolithic MIS, and CO_2 laser detectors with state-of-the-art performance.

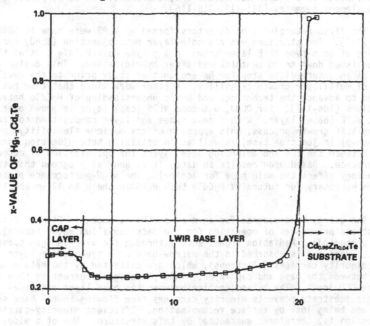

Figure 21. Compositional Profile of a Typical LWIR DLHJ Measured by Photoreflectance

The application of backside-illuminated HFPA technology to IR detection has emerged in the past few years as the most attractive approach for both scanning and staring modes in either tactical or strategic application. Given here is a brief summary of the structure and performance of the HgCdTe junction photodiodes for use with silicon CCD in hybrid mosaic focal plane arrays for direct detection in the 2-12 µm spectral region.

A wide variety of techniques has been used to form p-n junctions in HgCdTe. The most technologically significant of these techniques for high performance IR detectors appear to be ion implantation and heterojunction growth. For the implanted homojunction structure, p-type HgCdTe material, either bulk or epilayer, is usually used [111-113]. The requirement of p-type material reflects the fact that the n-type damage in HgCdTe arising from the implantation process dominates any activation of the implanted dopant species. The ion-implanted devices exhibit both diffusion and depletion-layer lifetimes that are significantly lower than the starting material lifetimes that are discussed in Part 2 of Section IV. On the other hand, DLHJ's fabricated by the relatively benign heterojunction growth process consistently exhibit less degradation of the starting material lifetime, and specific devices exhibit device diffusion lifetimes that approach the starting material lifetime [82]. Extensive efforts have been made in developing satisfactory post-implant annealing techniques to effectively remove the implantation damage and maintain a high-quality junction [111-113]. There is, however, strong experimental evidence pointing to the formation of antisites and two-dimensional stacking faults in the epitaxial layers as a result of the post-implant annealing [114]. Ion-implantation, nevertheless, can offer the advantage of a planar technology and, therefore, can be very useful for small element geometry [111-113, 115-116].

The first heterojunction detectors formed by VLPE were made at SBRC in 1979 [117]. The structure was a single-layer heterojunction (SLHJ) consisting of an n-type VLPE layer grown on a p-type bulk HgCdTe crystal that was polished down to 20 µm thickness after hybridization. This design was used as an intermediate step for heterojunction study prior to the development of multilayer growth capability. Further work since that time has continued to advance the technology and basic understanding of HgCdTe heterojunctions [106-107]. For DLHJ, a second VLPE (cap) layer is grown over the first VLPE (base) layer. With dopant types and layer composition controlled by the VLPE growth process, this approach offers maximum flexibility (p-on-n or n-on-p) in junction type, as well as in utilizing heterojunction formation between the cap and absorbing base layers for optimization of detector performance. Based upon results to date, it is generally agreed that this technology offers the main hope for achieving the high-performance photodiodes necessary for future PV HgCdTe LWIR HFPA in the 8 to 12 µm spectral region.

The energy band diagram for an n-on-p DLHJ is shown in Figure 22 to illustrate principles of operation for the backside-illuminated heterojunction photodiode. Radiation signal enters through the wide-bandgap substrate (e.g., CdTe) and is absorbed in the narrow-bandgap p-type HgCdTe layer creating minority carriers (electrons), which are collected by the heterojunction between the base and cap layer. This photosignal current is then fed to the CCD input. The heterojunction between the base layer and wide-bandgap substrate prevents minority carriers from flowing to the back surface and being lost by surface recombination. Efficient minority-carrier collection is, therefore, guaranteed by this structure. Use of a wide-bandgap n-type cap layer minimizes the dark current associated with the diffusion of the minority holes from the cap layer [113]. Moreover, if the location of the p-n junction is pushed slightly into the wide-bandgap side of the heterojunction, both generation-recombination (g-r) current originating in the junction depletion region and tunneling current can be

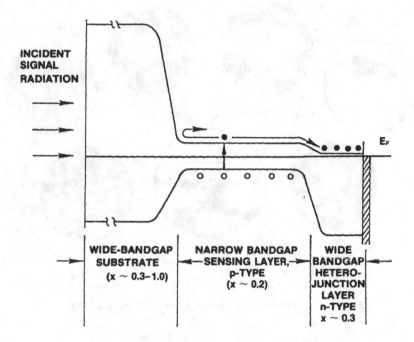

Figure 22. Energy Band Diagram for the n-on-p DLHJ $Hg_{1-x}Cd_xTe$ Detector

substantially reduced [106-107]. Alternatively, a p-on-n DLHJ can be fabricated with the same advantages by VLPE growth of a wide-bandgap p-type cap
layer over a narrow-bandgap n-type sensing layer. P-on-n DLHJ's of excellent device performance have been fabricated successfully at SBRC since 1983
[118].

The fabrication of an HFPA from epitaxial HgCdTe consists of two steps:
(1) detector array processing, and (2) hybridization to a CCD readout or to
a fanout test chip. The processing sequence for the HFPA made of DLHJ is
illustrated in Figure 23.

The zero-bias dynamic resistance-area (R_0A) product of a p-n junction
photodiode is an important parameter because it sets the upper limit on
achievable spectral detectivity. High R_0A values are desired to permit efficient transfer of signal current into the CCD and minimizing loss of spectral detectivity due to CCD noise. As an example to demonstrate the excellent capabilities of the VLPE technology for producing device-quality materials used in a variety of photodiode structures, the temperature dependence
of R_0A of six arrays made of n-on-p heterojunctions or homojunctions
hybridized to fanouts with cutoff λc in the SWIR, MWIR, and LWIR regions is
shown in Figure 24 [119]. The intrinsic carrier concentration n_i dependence
on temperature is also shown. Note that the SWIR array is g-r limited from
room temperature down to 120K. The MWIR devices at T < 150K are also
dominated by g-r current. The transition from the diffusion to the g-r
limited case is clearly seen in the LWIR arrays. A summary of the device
structures and selected values for R_0A is compiled in Table II [119].

348

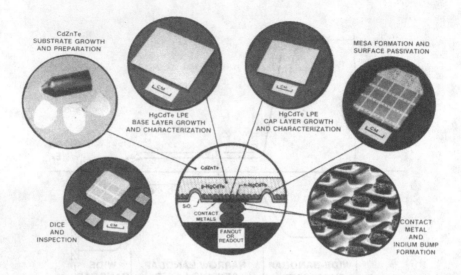

Figure 23. Fabrication Sequence for HFPA Made of DLHJ

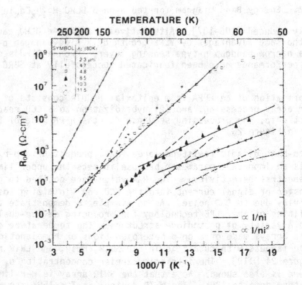

Figure 24. Temperature Dependence of R_0A for Six Detector Hybrids. Covering the 2-14 μm IR Spectrum. Note the intrinsic carrier concentration dependency.

Table II.   Summary of Array Performance Presented in Figure 24

λ_c AT 80K (μm)	STRUCTURE	R_oA (Ω-cm^2)		
		T = 60K	T = 80K	T = 200K
2.3	SLHJ	—	—	8×10^4
4.7	PLANAR IMPLANTED	—	1×10^7	8
4.8	DLHJ	3×10^7	3×10^6	1
8.5	SLHJ	4×10^4	6×10^2	—
10.3	SLHJ	3×10^3	2×10^2	—
11.5	DLHJ	2×10^2	10	—

VI.   SUMMARY AND CONCLUDING REMARKS

Infinite-melt vertical LPE (VLPE) from Hg-rich solutions has been firmly established as one of the most successful LPE technologies for preparation of epitaxial HgCdTe of excellent material characteristics. VLPE materials are being grown routinely for PC and PV as well as monolithic MIS and high-frequency laser detector devices with state-of-the-art performance in the entire 2-12 μm spectral region. Photodiode structures with excellent performance, including p-on-n as well as n-on-p double-layer heterojunctions (DLHJ), have been manufactured by VLPE. This is mainly due to the unique characteristics inherent in the Hg-rich solution:

1.   Low liquidus temperature (< 450°C), which makes the cap-layer growth step required for the DLHJ structure feasible

2.   Ease of intentionally incorporating temperature-stable impurity dopants, such as As, Sb, and In in the HgCdTe layers during growth.

It has been generally recognized that the DLHJ is the key element to further improvement in PV HgCdTe detector performance for LWIR applications.

Knowledge of fundamental material properties for the Hg-Cd-Te system has been established and utilized for better understanding and control of the VLPE growth process throughout the past ten years. We have reviewed three of the most important areas: phase diagram, defect chemistry, and impurity doping. We have also found that impurity-doped material exhibits dramatically improved material characteristics, including the minority-carrier lifetime, and the Hall mobility of free carriers. This is probably due to reduction of the concentration of the native defects by incorporation of impurity dopants through the law of mass action. Native defects are present ubiquitously in HgCdTe and are often found to deteriorate material characteristics. Further work is needed to elucidate the origin and nature of various native defects and the complex interaction between native defects and impurity dopants from both a statistical thermodynamic and material characterization point of view [60, 61, 120]. The VLPE technology may offer the best vehicle for conducting this important study, not only because of maturity of the technology, but also because thermodynamic equilibrium and some unique characteristics such as impurity doping inherent in the Hg-rich solution can be readily and simultaneously achieved during growth.

350

ACKNOWLEDGMENTS

The authors wish to thank all their colleagues at SBRC who directly or indirectly have contributed to the VLPE HgCdTe technology. The current status of the VLPE technology reviewed in this paper attests to their efforts in the past ten years. In particular, we thank Walt Konkel, George Whiteman, Ron Risser, Richard Herald, Trish Whitney, Dave Voros, Joan Chia, Ed Westen, and Roger Cole for their technical expertise in the VLPE area; Mitra Sen, and Walt Konkel in the substrate area; Betty Zuck for her meticulous skill in preparing the cross-sectional TEM sample, and Jay James for his contribution to the cross-sectional TEM work. We are also grateful to Peter Bratt, and Colin Jones for many stimulating discussions. Special thanks are due Ralph Ruth who reviewed the manuscript and made many suggestions for its improvement. The support of various government agencies is also gratefully acknowledged, in particular the U.S. Air Force Materials Laboratory of Wright-Patterson Air Force Base, and the Naval Research Laboratory.

REFERENCES

1. D. Long and J.L. Schmit, in Semiconductors and Semimetals, Vol. 5, edited by R.K. Willardson and A.C. Beer, (Academic Press, New York, 1970).

2. R. Dornhous and J. Nimtz, Solid State Physics, Springer Tracts in Modern Physics, Vol. 98, (Springer-Verlag, Berlin, 1983).

3. Semiconductors and Semimetals, Vol. 18, edited by R.K. Willardson and A.C. Beer, (Academic Press, New York, 1982).

4. M. Boukerche, P.S. Wijewarnasuriya, J. Reno, I.K. Sou, and J.P. Faurie, J. Vac Sci., Technol. A4 (4), 2072 (1986).

5. J.B. Mullin, S.J.C Irvine, and J. Giess, A. Royle, J. Crystal Growth 72, 1 (1985).

6. J.J. Hsieh, in Handbook on Semiconductors, Vol. 3, edited by S.P. Keller (North Holland, Amsterdam, 1980) Chap. 6.

7. A.S. Jordan, A.R. von Neida, R. Caruso, and C. Kim, J. Electrochem Soc. 121, 153 (1974).

8. G.B. Stringfellow, Rep. Prog. Phys., Vol. 45, 469 (1982).

9. P. Herning, J. Electron. Mater. 13, 1 (1984).

10. C.C. Wang, S.H. Shin, M. Chu, M. Lanir, and A.H.B. Vanderwyck, J. Electrochem. Soc. 127, 175 (1980).

11. C.A. Castro and R. Korenstein, Proc. Soc. Photo-Opt. Instrum. Eng. 317, 262 (1981).

12. J.E. Bowers, J.L. Schmit, C.J. Speerschneider, and R.B. Maciolek, IEEE Trans. Electron Devices ED-27, 24 (1980).

13. K.E. Mironov, V.K. Ogorodnikov, V.D. Rozumnyi, and V.I. Ivanov-Omskii, Phys. Stat. Sol. (a) 78, 125 (1983).

14. T.C. Harman, J. Electron. Mater. 9, 945 (1980).

15. Y. Nemirovosky, S. Margalit, E. Finkman, Y. Shacham-Diamand, and I. Kidron, J. Electron. Mater. 11, 133 (1982).

16. D.D. Edwall, E.R. Gertner, and W.E. Tennant, J. Appl. Phys. 55, 1453 (1984).

17. H.R. Vydyanath, J.A. Ellsworth, and C.M. Devaney, J. Electron. Mater., 16 (1), 13 (1987).

18. J.C. Tranchart, B. Latorre, C. Foucher, and Y. Le Gouge, J. Crystal Growth 72, 468 (1985).

19. K. Nakahama, R. Ohkator, K. Nishitani, and T. Murotani, J. Electron. Mater. 13, 67 (1984).

20. J.A. Mroczkowski and H.R. Vydyanath, J. Electrochem. Soc. 128, 655 (1981).

21. E. Janik, M. Ferah, R. Legros, R. Triboulet, T. Brossa, and Y. Riant, J. Crystal Growth 72, 133 (1985).

22. M. Yoshikawa, S. Ueda, K. Maruyama, and H. Takigawa, J. Vac. Sci. Technol. A3 (1), 153 (1985).

23. D. Amingual, G.L. Destefanis, S. Guillot, J.L. Ouvrier-Buffet, S. Paltrier, and D. Zenatti, 1986 Innsbuck SPIE Proceeding (to be published).

24. H. Ruda, P. Becla, J. Lagowski, and H.C. Gatos, J. Electrochem. Soc. 130 (1), 228 (1983).

25. M. Astles, G. Blackmore, N. Gordon, and D.R. Wight, J. Crystal Growth 72, 61 (1985).

26. C.J. Speerschnieder and R.B. Maciolek, Honeywell Technical Report, Contract DAHC60-70-C-008 (1973).

27. V.I. Ivanov-Omskii, K.E. Mironov, and V.K. Ogorodnikov, Phys. Stat. Sol. (a) 58, 543 (1980).

28. S.G. Konnikov, V.K. Ogorodnikov, and P.G. Sydorchuk, Phys. Stat. Sol. (a) 27, 43 (1975).

29. J.G. Fleming and D.A. Stevenson, J. Electrochem. Soc. (to be published).

30. T.C. Harman, J. Electron. Mater. 10, 1069 (1981).

31. G.S. Kamath, J. Evan, and R.C. Knechtli, IEEE Trans. Electron Devices 24, 473 (1977).

32. J. Steininger, J. Crystal Growth 37, 107 (1977).

33. J. Steiniger, J. Electron. Mater. 5, 299 (1976).

34. A.J. Strauss and R.F. Brebrick, J. Phys. Chem. Solids 31, 2293 (1970).

35. F.T.J. Smith, Met. Trans. 1, 617 (1970).

36. J.L. Schmit and E.L. Stelzer, J. Electron. Mater. 7, 65 (1978).

37. H.R. Vydyanath, J. Electrochem. Soc. 128, 2609(I); 2619(II); 2625(III) (1981).

38. C-H. Su, P-K. Liao, T. Tung, and R.F. Brebrick, J. Electron. Mater. 11, 931 (1982).

39. C-H. Su, P-K. Liao, and R.F. Brebrick, J. Electron. Mater. 12 (5), 771 (1983).

40. R.F. Brebrick and J.P. Schwartz, J. Electron. Mater. 9, 485 (1980).

41. M. Chu, J. Appl. Phys. 51, 5876 (1980).

42. H.F. Schaake, J. Electron. Mater. 14, 513 (1985).

43. J. Yang, Z. Yu, and D. Tang, J. Crystal Growth 72, 275 (1985).

44. C.L. Jones, M.J.T Quelch, P. Capper, and J.J. Gosney, J. Appl. Phys. $\underline{53}$ (12), 9080 (1982).

45. J. Zhang and J.C. Thuillier, Phys. Stat. Sol. $\underline{(a)\ 77}$, 649 (1983).

46. Y.A. Chang and J.F. Smith eds. of Calculations of Phase Diagrams and Thermochemistry of Alloy Phases (The Metallurgical Soc. of AIME, Warrendale, PA, 1979).

47. R.F. Brebrick, C-H. Su, and P-K. Liao in Semiconductors and Semimetals, Vol. 19, edited by R.K. Willardson and A.C. Beer, (Academic Press, New York, 1983).

48. T. Tung, L. Golonka, and R.F. Brebrick, J. Electrochem. Soc. $\underline{128}$, 1601 (1981).

49. T. Tung, C-H. Su, P-K. Liao, and R.F. Brebrick, J. Vac. Sci. Technol. $\underline{21}$, 117 (1982).

50. T. Tung, Ph.D thesis, Marquette University, 1983.

51. C-H. Su, J. Crystal Growth $\underline{78}$, 51 (1986).

52. P.J. Meschter, K.E. Owens, and T. Tung, J. Electron. Mater. $\underline{14}$ (1), 33 (1985)

53. C-H. Su, P-K Liao, T. Tung, and R.F. Brebrick, High Temp. Sci. $\underline{14}$, 181, (1981).

54. J.P. Schwartz, T. Tung, and R.F. Brebrick, J. Electrochem. Soc. $\underline{128}$, 438 (1981).

55. T. Tung, L. Golonka, and R.F. Brebrick, J. Electrochem. Soc. $\underline{128}$, 451 (1981).

56. C-H. Su, P-K. Liao, and R.F. Brebrick, J. Electrochem. Soc. $\underline{132}$, 942 (1985).

57. R.F. Brebrick, J. Phys. Chem. Solids, $\underline{40}$, 177 (1979).

58. M.H. Kalisher (unpublished).

59. A.V. Vanyukov, I.I. Krotov, and A.I. Ermakov, Inorg. Mater. 13, 667 (1977).

60. R.F. Brebrick, in Progress in Solid State Chemistry, Vol. 3, edited by H. Reiss, Pergamon Press, Oxford (1966).

61. R.F. Brebrick, in Treatise on Solid State Chemistry, Vol. 2, edited by N.B. Hannay, Plenum Press, New York (1975).

62. R.F. Brebrick and A.J. Strauss, J. Phys. Chem. Solids, 26, 989 (1965).

63. R.A. Reynolds, M.J. Brau, H. Kraus, and R.T. Bate, in The Physics of Semimetals and Narrow Gap Semiconductors, edited by D.L. Carter and R.T. Bate, Pergamon Press, Oxford (1971).

64. P.L. Anderson, H.F. Schaake, and J.H. Tregilgas, J. Vac. Sci. Technol., $\underline{21}$ (1), 125 (1982).

65. R.F. Brebrick (private communication).

354

66. D.D. Edwall, E.R. Gertner, and W.E. Tennant, J. Electron. Mater. 14 (3), 245 (1985).

67. O. Caporaletti and W.F.H. Micklethwaite, Phys. Lett. 89A, 151 (1982).

68. J.S. Chen, F.A. Kroger, and W.L. Ahlgren, Extended Abstracts of the 1984 U.S. Workshop on the Physics and Chemistry of HgCdTe.

69. M. Brown and A.F.W. Willoughby, J. Crystal Growth 59, 27 (1982).

70. M.A. Kinch, J. Vac. Sci. Technol., 21 (1), 215 (1982).

71. E.S. Johnson and J.L. Schmit, J. Electron. Mater 6 (1), 25 (1977).

72. P. Capper, J.J.G. Gosney, C.L. Jones, I. Kenworthy, and J.A. Roberts, J. Crystal Growth 71, 57 (1985).

73. Ching-Hua Su (private communication).

74. C.E. Jones, K. James, J. Merz, R. Braunstein, M. Burd, M. Eetemadi, S. Hutton, and J. Drumheller, J. Vac. Sci. Technol A3, 131 (1985).

75. P.R. Bratt, D.R. Rhiger, K.J. Riley, and J.Y. Wong, Final Report, Contract No. F33615-77-C-5270, Santa Barbara Research Center, August 1981.

76. S.E. Schacham, E. Finkman, J. Appl. Phys. 57 (6), 2001 (1985).

77. E. Finkman, Y. Nemirovsky, J. Appl. Phys. 59 (4), 1205 (1986).

78. M. Itoh, H. Takigawa, and R. Ueda, IEEE Trans. Electron Devices, ED-27, 150 (1980).

79. Unpublished SBRC data.

80. M.H. Kalisher, J. Crystal Growth 70, 365 (1984).

81. L.E. Lapidas, R.L. Whitney, and C.A. Crosson, in Applied Materials Characterization, edited by W. Katz and P. Williams, Mat. Res. Soc. Symp. Proc. vol. 48, 365 (1985).

82. W.A. Radford, C.E. Jones, E.J. Smith, and L.F. Lou, Proceedings of the IRIS Detector Specialty Group Meeting, Seattle, Washington (1984).

83. G.L. Pearson and J. Bardeen, Phys. Rev. 75, 865 (1949).

84. E.O. Kane, Phys. Rev. 131, 79 (1963).

85. N.F. Mott and W.D. Twose, Advan. Phys. 10, 107 (1961).

86. P. Capper, J. Crystal Growth 57, 280 (1982).

87. W.A. Radford, R.E. Kvaas, and S.M. Johnson, Proceedings of the IRIS Specialty Group on Infrared Materials, Menlo Park, California (1986).

88. J. Bajaj, S.H. Shin, J.G. Pasko, and M. Khoshnevisan, J. Vac. Sci. Technol. A1 (3), 1749 (1983).

89. J.S. Chen, J. Bajaj, M. Brown, and W. E. Tennant, Abstracts of the "Materials for Infrared Dettectors and Sources" Symposium, 1986 MRS Fall Meeting, Boston, Massachusetts, 1986, P. 660.

90. D.L. Polla, R.L. Aggarwal, D.A. Nelson, J.F. Shanley, and M.B. Reine, Appl. Phys. Lett. 43 (10), 941 (1983).

91. T.N. Casselman (private communication).

92. D. Long, in Topics in Applied Physics, edited by R.J. Keyes, Springer-Verlag, Berlin and New York (1977).

93. D. Long, T.J. Tredwell, and J.R. Woodfill, Joint Meeting of the IRIS Specialty Groups on Infrared Detectors and Imaging, Vol. 1, 387 (1978).

94. G.L. Hansen, J.L. Schmit, T.N. Casselman, J. Appl. Phys., 53, 7099 (1982).

95. J.H. Chu, S.C. Xu, and D.Y. Tang, Appl. Phys. Lett. 43, 1064 (1983).

96. T. Tung (unpublished).

97. Don W. Shaw, J. Crystal Growth 62, 247 (1983).

98. R.A. Wood and R.J. Hager, J. Vac. Sci. Technol. A1 (3), 1608 (1983).

99. T.W. James and R.E. Stoller, Appl. Phys. Lett. 44, 56 (1984).

100. S.L. Bell and S. Sen, IRIS Detector Speciality Group Meeting, Boulder, Colorado (1983).

101. S.L. Bell and S. Sen, J. Vac. Sci. Technol. A3 (1), 112 (1985).

102. M. Yoshikawa, K. Maruyama, T. Saito, T. Maekawa, and H. Takigawa, Extended Abstracts of the 1986 U.S. Workshop on the Physics and Chemistry of HgCdTe, P. O/D I-17.

103. R.D. Horning and J.L. Standenmann, Appl. Phys. Lett. 49 (23), 1590 (1986).

104. J.E. Hails, G.J. Russel, A.W. Brinkman, and J. Woods, J. Appl. Phys. 60 (7) 2624 (1986); J. Crystal Growth (To be published).

105. J.B. Mullin, J. Giess, S.J.C. Irvine, Abstracts of the "Materials For Infrared Detectors and Sources" Symposium, 1986 MRS Fall Meeting, Boston, Massachusetts, 1986, P. 663.

106. P.R. Bratt, J. Vac. Sci. Technol. A1 (3), 1687 (1983).

107. P.R. Bratt and T.N. Casselman, J. Vac. Sci. Technol. A3 (1), 238 (1985).

108. P. Migliorato and A.M. White, Solid State Electron. 26, 65 (1983).

109. P. Migliorato, R.F.C. Farrow, A.B. Dean, and G.M. Williams, Infrared Phys. 22, 331 (1982).

110. M.F.S. Tang and D.A. Stevenson, Extended Abstracts of the 1986 U.S. Workshop on the Physics and Chemistry of HgCdTe, P. O/D II-17.

111. L.O. Bubulac, D.S. Lo, W.E. Tennant, D.D. Edwall, and J.C. Robinson, J. Vac. Sci. Technol. A4 (4), 2169 (1986).

112. G.L. Destefanis, J. Vac. Sci. Technol, A3, 171 (1985).

356

113. A. Fraenkal, S.E. Schacham, G. Bahir, and E. Finkman, J. Appl. Phys. (to be published).

114. P.M. Raccah, J.W. Garland, Z. Zhang, Y. De, W.E. Tennant, and L.O. Bubulac, Extended Abstracts of the 1986 U.S. Workshop on the Physics and Chemistry of HgCdTe, P. V-13.

115. S.E. Botts, IEEE Trans. Electron Devices, ED-32 (8), 1584 (1985).

116. T.L. Koch, J.H. De Loo, M.H. Kalisher, and J.D. Phillips, IEEE Trans. Electron Devices ED-32 (8), 1592 (1985).

117. K.J. Riley and A.H. Lockwood, SPIE 217, 206 (1980).

118. R.L. Whitney, T.N. Casselman, and K. Kosai, Proceedings of the IRIS Detector Specialty Group Meeting(1985).

119. M. Lanir, K.J. Riley, IEEE Trans. Electron Devices, ED-29 (2), 274 (1982).

120. J.A. Van Vechten in Handbook on Semiconductors, Vol. 3, edited by S.P. Keller (North Holland, Amsterdam, 1980) Chap. 1.

LIQUID PHASE EPITAXY OF $Hg_{1-x}Cd_xTe$ FROM Te SOLUTIONS: A ROUTE TO IR DETECTOR STRUCTURES

E.R. GERTNER
Rockwell International Science Center, 1049 Camino Dos Rios, P.O. Box 1085, Thousand Oaks, CA 91360

INTRODUCTION

The intrinsic semiconductor mercury cadmium telluride ($Hg_{1-x}Cd_xTe$), a solid solution of HgTe and CdTe, has assumed an ever increasing role in the fabrication of infrared (IR) detectors because its energy gap (0-1.5 eV) can be tailored to match the specific needs of IR detection and fiber optic systems. In photovoltaic focal plane array (FPA) applications, low power consumption, as well as excellent sensitivity at elevated temperatures, have made $Hg_{1-x}Cd_xTe$ the material of choice for both the midwave IR (MWIR) and longwave IR (LWIR) region.

In this paper, we review the liquid phase epitaxial (LPE) technique from Te solution for the synthesis of $Hg_{1-x}Cd_xTe$ and demonstrate its suitability for IR detector structures, especially FPAs. Epitaxial growth techniques have traditionally produced higher quality material because their lower growth temperatures allow greater control of thickness, impurity background, and compositional uniformity. These advantages accrue to LPE. Lower growth temperatures have additional importance for $Hg_{1-x}Cd_xTe$ as they reduce and control the Hg vapor pressure and constituent interdiffusion, both of which are exponential functions of temperature. Another advantage of epitaxy for $Hg_{1-x}Cd_xTe$ is that near-lattice and thermal matching conditions exist over the entire compositional range, with only a 0.3% lattice mismatch between HgTe and CdTe. This is analogous to the $Ga_{1-x}Al_xAs$ III-V system in that it allows the synthesis of stress-free, epitaxial, multilayer structures over a range of energy band gaps, provided the layer interdiffusion can be controlled, which can be considerable for the Hg-Cd-Te system [1]. The goal of LPE is the reproducible and routine synthesis of compositionally uniform $Hg_{1-x}Cd_xTe$ (constant x value) tailored toward a particular application.

This review will give a short general background of the Te solution LPE growth. However, the emphasis will be on presenting data from both Rockwell's research, development and production phase of LPE $Hg_{1-x}Cd_xTe$ showing that this technique is eminently suitable as a material synthesis technique for FPA applications. For background and further information, the reader is referred to recent reviews [2-5].

LPE is a solution growth technique that involves the controlled precipitation of a solute dissolved in a solvent onto a single cyrstal substrate. For $Hg_{1-x}Cd_xTe$, bulk-grown CdTe single crystal wafers are suitable as substrates as they are thermodynamically compatible and nearly lattice matched. Hybrid substrates, i.e., an epitaxial layer of CdTe on Al_2O_3, GaAs, InSb, etc., have become viable alternatives to bulk CdTe. Of these, $CdTe/Al_2O_3$ is eminently suitable for LPE. Contamination of the growth solution by dissolution of the base substrate does not occur due to sapphire's chemical inertness. A clear demonstration of the advantages of this approach is seen in the high performance and high uniformity obtained for LPE HgCdTe on $CdTe/Al_2O_3$ substrates [6-9].

Lattice matching can be achieved by use of epitaxial buffer layers of CdZnTe, CdSeTe, and CdMnTe on these alternate substrates by the appropriate use of the Zn, Se or Mn fraction. Currently, no definitive device perform-

Mat. Res. Soc. Symp. Proc. Vol. 90. 1987 Materials Research Society ance data exists that substantiates the benefits of close lattice matching for $Hg_{1-x}Cd_xTe$, although smoother morphologies [10] and lower interfacial dislocation densities have been observed [11-13]. $Hg_{1-x}Cd_xTe$ epilayer dislocation densities appear to be primarily determined by the dislocation density of the substrate [13] with most of the misfit dislocations confined to the compositionally graded interfacial region [11-15].

Tellurium is an excellent solvent for the LPE of $Hg_{1-x}Cd_xTe$ because of a low Hg partial pressure (for example, at 500°C, the Hg partial pressure over Te saturated $Hg_{0.6}Cd_{0.4}Te$ is 0.1 atm, while that of Hg saturated $Hg_{0.6}Cd_{0.4}Te$ is 7 atm [16,17]), and relatively high solubility of column II elments (Hg, Cd) in liquid Te at low temperature [18,19]. Tellurium has been extensively used as a solution for LPE $Hg_{1-x}Cd_xTe$ as shown by a partial list of current users in Table 1. Device quality $Hg_{1-x}Cd_xTe$ has been grown from Te solutions at several labs [6-9,20-23].

Table 1
Current Users of Te Liquid for $Hg_{1-x}Cd_xTe$ Growth

Aerojet
CNRS/LETI
Fermionics
Fujitsu
Honeywell
Lincoln Lab - MIT
Mitsubishi
Rockwell International
SOREQ
Technion
Texas Instruments
Westinghouse

Despite the lower Hg vapor pressure over the Te solution, the partial pressure of Hg must be controlled in the growth system in order to obtain compositional uniformity, reproducibility, and stability. Control of the Hg partial pressure has been accomplished in various ways. The simplest of these is to carry out the entire growth process in a closed ampule [24]. However, LPE techniques that use a closed tube are limited to low production levels because of the necessity of sealing an ampule each time.

Several open-tube LPE techniques have been used successfully to control the Hg partial pressure. These include the addition of Hg to the system via an external source to replace the Hg lost from the growth system; the use of HgTe in the vicinity of the growth solution as a solid source of Hg; and the use of a high inert-gas overpressure to reduce the loss of Hg to managable proportions.

The reader is referred to references [18,20,25-31] for details on the various LPE growth techniques and repective HgCdTe layer characteristics. Below is a discussion of the LPE development at Rockwell from its initial research stage (1977) to the transition to production (1983).

At Rockwell, Wang and co-workers [32] used high pressure inert gas (Ar, H_2) to control the loss of Hg from the growth solution. The first back-side-illuminated high performance HgCdTe heterostructure diodes were fabricated in layers grown in this system [33]. References [32-44] document the versatility of this growth technique for the production of high performance, device-quality $Hg_{1-x}Cd_xTe$ epitaxial layers of various x values with excellent compositional uniformity over an area of $\approx$ 6 cm².

The driving force for larger (≥ 20 cm²) compositionally uniform epitaxial layers of HgCdTe for large-area, focal plane detector arrays led to the
development of a horizontal growth system derived from the original vertical system (Fig. 1). In this system, growth occurs in a semi-sealed
graphite boat under a high pressure of H_2, typically about 10 atm. Epitaxial layer growth is initiated by rolling an in situ-reacted Hg-Cd-Te melt
onto the substrate using the action of gravity, i.e., tipping, and growth
is terminated in the same fashion. This is an effective way of obtaining
melt-free $Hg_{1-x}Cd_xTe$ epilayers of large size. This method of melt removal
is more effective than a pushing melt motion because it takes advantage of
the low surface tension of the Te-rich melt. This tipping method has produced large, high quality $Hg_{1-x}Cd_xTe$ layers with compositions ranging from
x ≥ 0.15 to x = 1. Examples of the excellent compositional uniformity both
layer-to-layer (Fig. 2a) and across each layer (Fig. 2b) [45,46], obtained
from consecutive growths toward a specific composition, are shown. (PACE is
a Rockwell acronym for Producible Alternative to CdTe for Epitaxy, PACE-1
is CdTe/sapphire, PACE-2 is CdTe/GaAs).

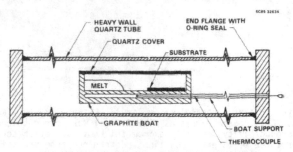

Fig. 1   Rockwell Te liquid $Hg_{1-x}Cd_xTe$ LPE apparatus.

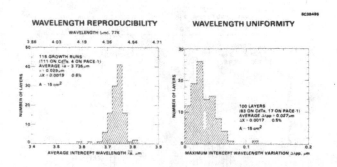

Fig. 2   Reproducibility of LPE $Hg_{1-x}Cd_xTe$. 300K target wavelength was
3.7 μm. λ_a is 300K IR absorption edge, $\Delta\lambda_{pp}$ is the maximum 300K
wavelength variation seen across the layer for three measurement
positions. (a) Run-to-run wavelength compositional reproducibility, and (b) wavelength uniformity across a layer.

Figure 3 shows the data on layer thickness reproducibility (run-to-run) (Fig. 3a) and uniformity across each layer (Fig. 3b) for the same layers of Fig. 2. Figure 4 shows similar data from the Rockwell $Hg_{1-x}Cd_xTe$ production facility. Shown in Fig. 4a is the layer-to-layer spread in composition expressed as wavelength at 300K. These layers are grown to slightly longer wavelength than the target wavelength, then "tuned" by chemical etching to the desired wavelength (Fig. 4b). As grown, 75% of the layers are within a 0.1 μm wavelength spread; after etching, 87% fall within a 0.1 μm window, which is only a 0.006 change in x value.

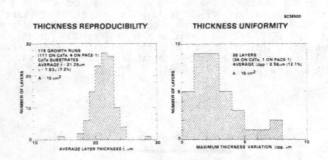

Fig. 3 Layer thickness control of Te liquid LPE $Hg_{1-x}Cd_xTe$. Target value was 20 μm. Same layers as Fig. 2. (a) Run-to-run thickness reproducibility, and (b) maximum thickness variation across layer for three measurement positions, determined from transmission interference fringes.

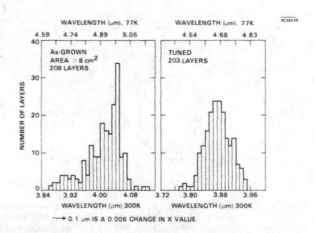

Fig. 4 Wavelength reproducibility of LPE $Hg_{1-x}Cd_xTe$ at Rockwell production facility. (a) As-grown, and (b) post-growth etched.

For MWIR $Hg_{1-x}Cd_xTe$, n/p diodes are fabricated by ion implantation [47] into the as-grown layers. Figure 5 shows the reproducibility of the 77K hole carrier concentration (Fig. 5a) and hole mobility (Fig. 5b) for the same layers shown in Figs. 2 and 3. On PACE-1 substates, both the carrier concentration and standard deviation are lower than for CdTe substrates, indicative of the higher substrate purity and uniformity of PACE-1. For LWIR $Hg_{1-x}Cd_xTe$, the as-grown p-carrier concentration (dominated by metal vacancies) is $\approx 1 \times 10^{17}$ cm-3 at 77K. Post growth anneals at lower temperatures in Hg vapor are used to reduce the vacancy concentration to levels desirable for junction formation. Figure 6 shows the hole carrier concentration and mobility reproducibility after thermal annealing of LWIR LPE $Hg_{1-x}Cd_xTe$ on CdTe substrates. Again, good control over carrier concentration and mobility are evident; indeed, a considerable part of the observed spread is due to variations in CdTe substrate purity. This is shown in Fig. 7, where we plot the 77K n-type carrier concentration vs composition. Conversion to n-type by low temperature Hg vapor annealing reveals the true impurity background by annihilating the metal vancancies. 77K electron concentration values are typically less than 1×10^{15} cm-3 and range from low 10^{14} to mid 10^{15} cm-3 with some spread in mobilities. Outdiffusion of impurities from the CdTe substrate during the LPE and annealing cycle are suspected to contribute to the observed range. Figure 8 shows typical variable temperature Hall data for n-type annealed LWIR LPE $Hg_{1-x}Cd_xTe$, indicative of classical behavior.

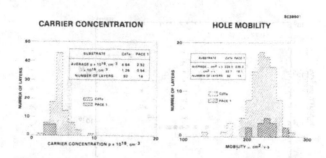

Fig. 5 Reproducibility of hole carrier concentration and mobility of MWIR LPE $Hg_{1-x}Cd_xTe$. 77K data is for layers shown in Figs. 2 and 3.

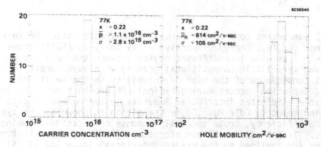

Fig. 6 77K hole carrier concentration and mobility of post growth annealed LWIR LPE $Hg_{1-x}Cd_xTe$.

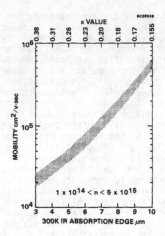

Fig. 7 77K mobility range observed for post growth n-annealed LPE
$Hg_{1-x}Cd_xTe$.

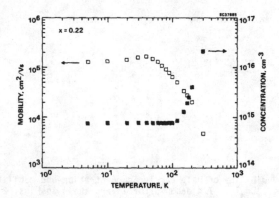

Fig. 8 Variable temperature Hall characteristics of n-annealed LWIR LPE
$Hg_{1-x}Cd_xTe$.

Table 2 summarizes some of the structural characteristics of Te-rich
LPE-grown $Hg_{1-x}Cd_xTe$. Energy dispersive x-ray analysis (EDAX) shows an
approximately 3 μm-wide interdiffused region between the epilayer and
substrate and very little compositional grading with layer depth [45] which
is a direct consequence of the high Cd solubility in the Te liquid. Insignificant Cd depletion occurs during typical LPE growth times. An in-situ
etchback cycle eliminates interfacial impurity gettering and results in low
recombination velocity CdTe-$Hg_{1-x}Cd_xTe$ transition as shown by the typical
high quantum efficiency, broad spectral response of backside illuminated
photodiodes. Transmission electron microscopy (TEM) has shown that misfit
dislocations are confined to the graded interfacial layer [12,14,15]. Dislocation etching of HgCdTe on the (111)A surface using the Polisar etch
[48] gives etch pit densities (EPD) of 1-5 × 10⁵ cm⁻² for layers on CdTe
substrates and ≈ 5 × 10⁶ cm⁻² on sapphire substrates (the higher density on

sapphire is primarily due to the lattice mismatch between CdTe and sapphire
(~ 30%)). Double crystal x-ray diffraction (using (111) reflection, CuK_α
radiation and $\approx 1 \times 4$ mm areas) indicates a 1-2 arc-min diffraction width.
$Hg_{1-x}Cd_xTe$ layers on sapphire (higher EPD) typically have the narrower
diffraction curves.

Table 2
Structural Characteristics of LPE $Hg_{1-x}Cd_xTe$ from Te Solution

Analysis	Observation
EDAX	≈ 3 μm interdiffused substrate/epitaxial inter- facial region Minor compositional grading with depth
SIMS	In-situ etch eliminates interface impurity accumulation
TEM	Dislocation network confined to graded interface region
EPD (111)A	$1-5 \times 10^5$ cm^{-2} on CdTe, $\approx 5 \times 10^6$ cm^{-2} on sapphire
X-ray FWHM	1-2 arc-min, (111) reflection, CuK_{α_1}, 1×4 mm area

In conclusion, LPE of $Hg_{1-x}Cd_xTe$ from Te-rich liquids has proven to be
a viable technique in both research and production environments, as demonstrated by the large number of current users. At Rockwell, the use of a
simple tipping arrangement and Hg pressure control give excellent control
over material characteristics, both structural and electrical, and has led
to excellent device performance over the entire composition range. Figure
9 shows device performance expressed as R_0A for ion-implanted n/p
$Hg_{1-x}Cd_xTe$ photodiodes vs temperature. This device performance equals or
exceeds that of other growth and/or junction formation techniques for
$Hg_{1-x}Cd_xTe$.

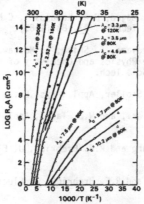

• DATA BY M. HERELD,
UNIVERSITY OF CHICAGO

Fig. 9 Device performance (R_0A) vs temperature of implanted n/p photodiodes. Diodes fabricated in Te liquid LPE grown $Hg_{1-x}Cd_xTe$
layers.

The data presented above is but a small fraction of a new technology that evolved over the past 10 years through the efforts of workers at both Rockwell's research and production facilities, who through their perseverance, dedication and inventiveness made Te liquid LPE of $Hg_{1-x}Cd_xTe$ the basis of a new IR technology.

REFERENCES

[1]   M.F.S. Fong and D.A. Stenson, 1986 U.S. Workshop on the Physics and Chemistry of HgCdTe, to be published, J. Vac. Sci. Tech.

[2]   R. Dornhaus and G. Nimtz, Springer Tracts in Modern Physics 98, 119-309 (1983).

[3]   R.K. Willardson and A.C. Beer, Mercury Cadmium Telluride 18, 1-384 (1981).

[4]   E.R. Gertner, Ann. Rev. Mater. Sci. 15, 303-28 (1985).

[5]   M.A. Herman and M. Pessa, J. Appl. Phys. 57, 2071 (1985).

[6]   W.E. Tennant, Tech. Digest, 1983 Intern. Electronic Device Meeting (IEEE, New York, 1983), p. 704.

[7]   J.P. Rode, Proc. SPIE (Intern. Soc. Opt. Eng) 443, 120 (1983).

[8]   R.A. Riedel, E.R. Gertner, D.D. Edwall and W.E. Tennant, Appl. Phys. Lett. 46, 64 (1984).

[9]   E.R. Gertner, W.E Tennant, J.D. Blackwell and J.P. Rode, J. Crystl. Gr. 72, 462 (1985).

[10]   S.L. Bell and S. Sen, J. Vac. Sci. Technol. A3, 112 (1985).

[11]   R.A. Wood, J.L. Schmit, H.K. Chung, T.J. Magee and G.R. Woolhouse, J. Vac. Sci. Technol. A3, 93 (1985).

[12]   G.R. Woolhouse, T.J. Magee, H.A. Kawajoshi, C.M. S. Leung and R.D. Ormond, J. Vac. Sci. Technol. A3, 83 (1985).

[13]   M. Yoshikawa, J. Maruyama, T. Shito, T. Maekawa and T. Takigawa, 1986 U.S. Workshop on the Physics and Chemistry of HgCdTe, to be published, J. Vac. Sci. Tech.

[14]   T.W. James and R.E. Stoller, Appl. Phys. Lett. 44, 56 (1984).

[15]   S. Wood, J. Greggi, Jr., and W.J. Takei, Appl. Phys. Lett. 46, 371 (1985).

[16]   J.P. Schwartz, T. Tung and R.F. Brebrick, J. Electrochem. Soc. 128, 439 (1981).

[17]   T. Tung, L. Golonka and R.F. Brebrick, J. Electrochem. Soc. 128, 451 (1981).

[18]   T.C. Harman, Electron. Mater. 8, 191 (1980).

[19]   T. Tung, L. Golonka and R.F. Brebrick, J. Electrochem. Soc. 128, 1601 (1981).

[20] J.E. Bowers, J.L. Schmit, C.J. Speerschneider, and R. Maciolek, IEEE Trans. Elect. Devices ED-27, 24 (1980).

[21] D. Amingual, G.L. DeStefanis, S. Guillot, J.L. Ouvrier-Buffet and S. Paltrier, to be published, Proceedings of SPIE, Innsbruck(1986).

[22] P. Nicolas, J.P. Chamomal, J. Cluzel, M. Ravetto and G. Rigaux, SPIE 686, 26 (1986).

[23] H.R. Vydyanath, S.R. Hampton, P.B. Ward, L. Fishman, J. Slawinski and T. Krueger, presented at the Detector Specialty Meeting of IRIS, Sunnyvale, CA (1986).

[24] J.A. Morczkowski and H.R. Vydyanath, J. Electrochem. Soc. 128, 655 (1981).

[25] T.C. Harman, J. Electron. Mater. 8, 191 (1979).

[26] T.C. Harman, J. Electron. Mater. 10, 1069 (1981).

[27] J.L. Schmit and J.E. Bowers, Appl. Physl. Lett 35, 457 (1979).

[28] J.L. Schmit, R.J. Hager and R.A. Wood, J. Cryst. Growth 56, 485 (1982).

[29] R.A. Wood and R.J. Hager, J. Vac. Sci. Technol. A1, 1608 (1983).

[30] Y. Nemirovsky, S. Margarlit, E. Finkman, Y. Shacham-Diamond and I. Kidron, J. Electron. Mater. 11, 133 (1982).

[31] K Nakahama, R. Ohkator, K. Nishitani and T. Murotani, J. Electron. Mater. 13, 67 (1984).

[32] C.C. Wang, S.H. Shin, M. Chu, M. Lanir and A.H.B. Vanderwyck, J. Electrochem. Soc. 127, 175 (1980).

[33] M. Lanir, C.C. Wang and A.H.B. Vanderwyck, Appl. Physl Lett. 34, 50 (1979).

[34] C.C. Wang, M. Chu, S.H. Shin, W.E. Tannant and J.T. Cheung et al, IEEE Trans. Electron. Dev. ED-27, 154 (1980).

[35] M. Chu and C.C. Wang, J. Appl. hys. 51, 2255 (1980).

[36] S.H. Shin, M. Chu, A.H.B. Vanderwyck, M. Lanir and C.C. Wang, J. Appl. Phys. 51, 3772 (1980).

[37] S.H. Shin, A.H.B. Vanderwyck, J.C. Kim and D.T. Cheung, Appl. Phys. Lett. 37, 402 (1980).

[38] M. Chu, A.H.B. Vanderwyck and D.T. Cheung 37, 486 (1980).

[39] J.G. Pasko, S.H. Shin and D.T. Cheung, Proc. SPIE int. Soc. Opt. Eng. 282, 89 (1981).

[40] M.E. Kim, Y. Taur, S.H. Shin, G. Bostrup, J.C. Kim and D.T. Cheung, Appl. Phys. Lett. 39 336 (1981).

[41] S.H. Shin, J.G. Pasko and D.T. Cheung, IEEE Trans. Electron. Dev. EDL-2, 177 (1981).

[42] J. Bajaj, S.H. Shin, G. Bostrup and D.T. Cheung, J. Vac. Sci. Technol. 21, 244 (1982).

[43] S.H. Shin, J.G. Pasko, H.D. Law and D.T. Cheung, Appl. Phys. Lett 40 965 (1982).

[44] B.J. Feldman, J. Bajaj and S.H. Shin, J. Appl. Phys. 55, 3873 (1984).

[45] D.D. Edwall, E.R. Gertner and W.E. Tennant, J. Appl. Phys.. 55, 1453 (1984).

[46] D.D. Edwall, E.R. Gertner and W.E. Tennant, Proc. IRIS Detector Specialty Meeting, Seattle, WA (1984).

[47] L.O. Bubulac, W.E. Tennant, R.A. Reidel and T.J. Magee, J. Vacuum Sci. Technol. 21, 251 (1982).

[48] E.L. Polisar, N.M Boinykh, G.V. Indenbaum, A.V. Vanyukov and V.P. Schastlivyi, Izv. Vyssh. Ucheb. Zaved. Fiz. 11, 81 (1968).

INTERDIFFUSED MULTILAYER PROCESSING (IMP) IN ALLOY GROWTH

J B MULLIN, J GIESS, S J C IRVINE, J S GOUGH AND A ROYLE
RSRE, St Andrews Road, Malvern, Worcs, UK

ABSTRACT

The Interdiffused Multilayer Processing (IMP) technique has been developed as a way of growing uniform layers of cadmium mercury telluride (CMT). The principle of the technique is discussed in the context of an interdiffusion coefficient $\tilde{D}$ so as to assess the potential applicability of the technique to the growth of other II—VI or III—V alloys. The development of IMP for the growth of CMT layers, together with the properties of the layers and their suitability for use in the fabrication of IR detectors, are reviewed.

INTRODUCTION

The Interdiffused Multilayer Processing (IMP) technique [1,2] was developed as a way of growing uniform epitaxial layers of the alloy cadmium mercury telluride ($Cd_xHg_{1-x}Te$ or CMT). The principle of the technique involves the successive growth of alternate layers, of the binary compounds CdTe and HgTe under conditions which permit their complete interdiffusion in the course of a growth run. In this paper we analyse the reasons for the technique as well as the kinetics of the process in order to assess the potential suitability of IMP for the growth of alloys in general. Those systems that have been studied are reconsidered in the light of this analysis. Finally the results and achievements that have followed the application of the IMP technique to CMT are reviewed.

REACTION MECHANISMS

The original choice [3] of reactants for the MOVPE of CMT were Hg (vapour) Me_2Cd and Et_2Te. The thermal decomposition characteristics of the alkyls [4] and the general features of the growth process have been reviewed [5] previously. Essentially, the formation of HgTe requires the independent decomposition of Et_2Te which achieves maximum efficiency in H_2 around 410°C whereas CdTe forms by what appears to be an adduct—like catalytic reaction between Et_2Te and Me_2Cd, a reaction which occurs with maximum efficiency for the formation of CdTe down to 350°C. The composition of the alloy $Cd_xHg_{1-x}Te$ is found to be critically dependent on gas flow and temperature, making it very difficult to achieve good lateral uniformity of x when depositing over large substrate areas.

Control of composition in the epitaxial CMT layers is constrained by the gas phase composition of the heterogeneously decomposing gaseous mixture at the substrate surface but in the incorporation process during crystal growth the proportion of Cd to Hg in the alloy is ultimately determined by the interface kinetics and the thermochemistry of alloy formation. The incorporation process in a number of III−V alloys grown from the vapour has been shown [6] to occur under a condition close to thermodynamic equilibrium. This characteristic could well be a feature of $Cd_xHg_{1-x}Te$ formation. The weak bond between Hg and Te ($\Delta H^\circ_{298} = -33.9$ kJ mole^{-1} [7] compared with the strong bond between Cd and Te ($\Delta H^\circ_{298} = -102.6$kJ mole^{-1} [8]) means that an excess of [Me_2Cd] over [Et_2Te] results in CdTe formation with little or no HgTe. The composition x is thus primarily determined by the ratio [Me_2Cd]/[Et_2Te]. Further, since Hg pressure has to be controlled, it is impractical for the growth of CMT of specific composition to independently vary the anion [Te] to cation [Hg+Cd] ratio to optimise crystal quality. It is noteworthy that the control of anion to cation ratio is a feature, indeed requirement, for obtaining good quality epitaxial GaAs and $Ga_{1-x}Al_xAs$ layers for example, where the anion to cation ratio needs to be large (5−100).

The two principal requirements for growing uniform alloys by a non−IMP process are thus the use of metalorganics with similar decomposition characteristics and binary compounds having similar heats of formation. Neither of these two conditions are even approximately fulfilled in the $Hg/Me_2Cd/Et_2Te$ system.

THE IMP TECHNIQUE

In the IMP process the individual layers of CdTe and HgTe are grown sequentially under the optimum deposition conditions for each compound. The production of CMT uniform in depth requires the use of thin layers, the combined CdTe and HgTe thicknesses being of the order of $0.2\mu m$ or less for growth at 410°C. In order to ensure complete interdiffusion of the final layers, a short period at the growth temperature (~10 minutes) is included at the end of each growth run. The critical parameters for the interdiffusion process are layer thickness and the magnitude of the interdiffusion coefficient $\widetilde{D}_x$ which is a function of temperature and composition x.

Interdiffusion

Interdiffusion between CdTe and HgTe can be characterised by a concentration− dependent interdiffusion coefficient $\widetilde{D}_x$. The magnitude of $\widetilde{D}_x$ will range between the value for the self diffusion coefficient for Cd in CdTe, D_{Cd}(CdTe) and Hg in HgTe, D_{Hg}(HgTe). The respective values of these self−diffusion coefficients at 410°C are $4\times10^{-16}cm^2s^{-1}$ and $1\times10^{-12}cm^2s^{-1}$ giving an approximate effective mean interdiffusion coefficient of $2\times10^{-14}cm^2s^{-1}$.

In the original IMP paper by Tunnicliffe et al [2] an average value of the interdiffusion coefficient $\tilde{D}$ was derived from analysis of Cd profiles of SIMS data measured on CMT/CdTe interfaces. At 410°C the average value of $\tilde{D}$ was $4\times10^{-14}\text{cm}^2\text{s}^{-1}$.

More recently Fleming and Stevenson [9] have made a specific experimental study of the concentration dependence of $\tilde{D}_x$ using a novel diffusion couple method. They found that $\tilde{D}_x$ could be represented by $\tilde{D}_x = 1380\exp(-7.8x)\exp(-2.0/kT)$. Results of $\tilde{D}_x$ tabulated (see Table I) from this expression for different values of x show very good agreement with equivalent data computed from Zanio's expression for $\tilde{D}_x(\tilde{D}_x = 3.15\times10^2\times10^{-3.53x}\exp(-2.24\times10^{14}/K)$ which has been used in predicting the effect of growth time on superlattice structures [10]. Further, assuming that a mean value of the interdiffusion coefficient for growth of $x = 0.2$ CMT can be represented by $\tilde{D}_{0.6}$ there is very good agreement between all three sets of data (see Table I).

Table I: Interdiffusion Coefficients $\tilde{D}_x(\text{cm}^2\text{s}^{-1})$: CdTe/HgTe

x	0.1	0.2	0.4	0.6	0.65	
410°C	1.1E−12	5.0E−13	1.1E−13	2.2E−14	1.5E−14	ref 9*
410°C	8.0E−13	3.5E−13	7.0E−14	1.4E−14	9.1E−14	ref 10
410°C				4.0E−14		ref 2
350°C	3.4E−14	1.5E−14	3.0E−15	5.8E−16	3.9E−16	ref 10

*extrapolated data

The use of an average value of $\tilde{D}_{0.6}$ has been justified experimentally. A model of interdiffusion using $\tilde{D}_{0.6} = 4\times10^{-14}\text{cm}^2\text{s}^{-1}$ has been used to predict the extent of interdiffusion between layers of different thicknesses, 0.1, 0.3 and $0.5\mu\text{m}$ of CdTe and HgTe (Fig.1). They show that complete interdiffusion for CMT occurs in less than 1 hour if the layer thickness is $0.1\,\mu\text{m}$. Experimental results with somewhat thicker layers show that this period of interdiffusion is quite adequate for achieving uniform layers. The value of $\sqrt{D_{0.6}t}$, the characteristic diffusion distance, can therefore be used as a criterion for establishing complete interdiffusion; in the present example $\sqrt{D_{0.6}t} = 0.12\mu\text{m}$ $(\tilde{D}_{0.6} = 4.\times10^{-14}\text{cm}^2\text{s}^{-1}$, t = 3600s). Using this guideline one can establish a minimum interdiffusion coefficient for a set period (3600s) that will enable an IMP process to succeed for a chosen layer thickness. Assuming that the minimum layer thickness that would characterise an IMP process is the lattice parameter distance say 6Å, then IMP processes are possible for the production of alloy layers if $\sqrt{Dt} > 6\times10^{-8}\text{cm}$ which for 3600s gives a diffusion coefficient of $D > 10^{-18}\text{cm}^2\text{s}^{-1}$. We will use this critical diffusion condition to assess the potential applicability of IMP.

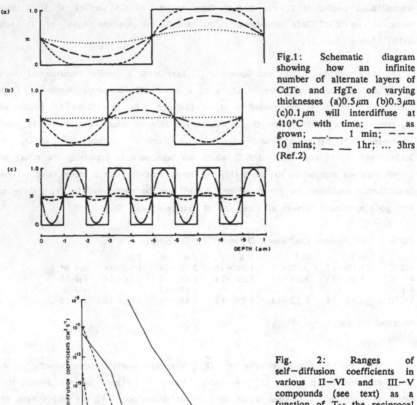

Fig.1: Schematic diagram showing how an infinite number of alternate layers of CdTe and HgTe of varying thicknesses (a)0.5μm (b)0.3μm (c)0.1μm will interdiffuse at 410°C with time; _____ as grown; __.___ 1 min; - - - 10 mins; _ _ 1hr; ... 3hrs (Ref.2)

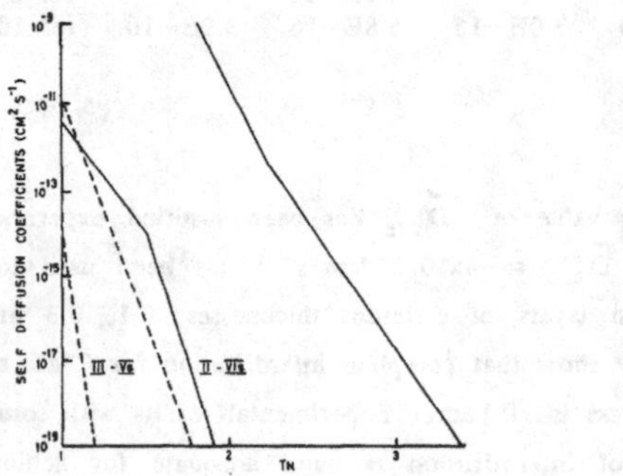

Fig. 2: Ranges of self−diffusion coefficients in various II−VI and III−V compounds (see text) as a function of T_N the reciprocal reduced temperature

The suitability of IMP in other alloy systems

The suitability of IMP for other alloy systems requires primarily that the diffusion criterion expression be satisfied. In considering the application to II−VI and III−V alloy systems generally, it is worth noting the difference in the relative magnitude of the self diffusion coefficients in the two systems. This difference is graphically illustrated in Fig.2. Here the self−diffusion coefficients of the anion and cation components of the binary compounds are plotted as a function of T_N, the reciprocal of the reduced temperature $T_N = T(\text{melting point})/T(\text{actual})$. The range of self diffusion coefficients for the elements of the II−VI [11] compounds, viz CdTe, CdSe, CdS, ZnTe, ZnSe, ZnS are some 5−6 orders of magnitude larger than the self−diffusion coefficients in the III−V compounds [12] InSb, InAs, InP, GaSb, GaAs and GaP. This restricts the scope for IMP application in the III−Vs to the use of extremely thin layers and/or very long annealing times, or alternatively growth at elevated temperatures.

Table II: Diffusion Coefficients (cm²s⁻¹): III-Vs

| T(K) | GaInAs | | | T(K) | InAsP | | |
	Ga(GaAs)	In(InAs)	$\tilde{D}$		As(InAs)	P(InP)	$\tilde{D}$
1023(C)	2E-21	1E-14	4E-18	1000(C)	1E-15	2E-18	4E-17
873(M)	8E-26	7E-18	7E-22	873(M)	9E-19	3E-22	2E-20

Some idea of the significance of reduced diffusivity in the III−Vs compared to the II−VIs can be appreciated from the data listed in Table II, for the growth of two alloys $Ga_{1-x}In_xAs$ and $InAs_{1-x}P_x$ by MOVPE(M) and conventional chloride (C) processes. The estimates of the interdiffusion coefficients in the conventional systems, 4×10^{-18} cm² s⁻¹ for GaInAs at 1023K and 4×10^{-17} cm² s⁻¹ for InAsP at 1000K mean that IMP could be used, although in the case of GaInAs exceptionally thin layers would be needed. Using a limiting thickness of a monolayer IMP growth can be compared with Atomic Layer Epitaxy [13]. A modest increase in growth temperature (~70K) would permit the use of layers 7−10 times thicker. This situation contrasts strongly with the lower temperature (873K) MOVPE growth. Here the significant reduction in temperature reduces $\tilde{D}$ below a practical level, thus precluding the use of IMP. Also the use of higher growth temperatures in MOVPE would probably be countervailing because of the likelihood of premature alkyl decomposition.

An indication of the potential applicability of IMP to the growth of II−VI alloys can be gained from the magnitude of the self diffusion coefficients at 410°C. Thus in the binary compounds containing cations of Zn or Cd and anions of S, Se or Te, values of D $>10^{-18}$ cm²s⁻¹ occur for Cd and Te in CdTe, Cd and Se in CdSe, Zn in ZnTe

Table III: Diffusion Coefficients (cm²s⁻¹): II-VIs

| | CdSeTe | | | | CdZnTe | | |
T(K)	Se(CdSe)	Te(CdTe)	D	T(K)	Zn(ZnTe)	Cd(CdTe)	D
683	9.4E-15	1.1E-14	1.0E-14 *1	683	1.9E-19	5.7E-16	1.04E-17*3
683	9.4E-15	2.8E-17	5.2E-16 *2	683	2.0E-15	5.7E-16	1.06E-15*4

maximum Te pressure *1; high Zn pressure *4,
minimum Te pressure *2; low Zn pressure *3,

Table IV: Diffusion Coefficients (cm^2s^{-1}): II–VIs

	ZnSSe				CdZnS		
T(K)	S(ZnS)	Se(ZnSe)	$\tilde{D}$	T(K)	Cd(CdS)	Zn(ZnS)	$\tilde{D}$
623	1.0E-22	1.6E-22	1.3E-22	573	1.4E-17	1.5E-23	1.5E-20
773	4.8E-19	3.2E-18	1.2E-18	773	4.0E-13	1.6E-17	2.5E-15

	CdZnSe		
T(K)	Cd(CdSe)*	Zn(ZnSe)	$\tilde{D}$
573	2.3E-16	4.5E-26	3.2E-21
773	1.3E-12	2.6E-19	5.7E-16

* Cd saturated

and Cd in CdS.   These binaries therefore represent the most favoured compounds for IMP.   Compounds containing elements having slower diffusivities may still be used for IMP if longer diffusion times are used, or if the complementary diffusing element has a large self diffusion coefficient resulting in a larger mean interdiffusion coefficient.   By increasing the deposition temperature to 500°C, the self diffusion coefficients for Se in ZnSe, S in CdS, Zn in ZnS and S in ZnS, fulfil the critical diffusion condition.

Self diffusion data, together with estimated mean interdiffusion coefficients that are relevant to the alloys CdSeTe, CdZnTe, ZnSSe, CdZnS and CdZnSe, and listed in Tables III and IV.   IMP for the alloy CdSeTe is potentially only slightly less favourable than for CMT, provided higher Te pressures are used since the self–diffusion coefficient of Te in CdTe is very sensitive to Te pressure [14].   For the CdTe/ZnTe system there is an additional pressure dependence associated with the diffusion of Zn in ZnTe.   Thus optimum IMP performance will require high pressures of Zn, together with thinner layers ($< 0.1\,\mu m$) and/or somewhat elevated growth temperatures.

The remaining alloys listed in Table IV, ZnSSe, CdZnS and CdZnSe, have been studied experimentally by Wright et al [15].   For the ZnS/ZnSe system growth at 350°C followed by annealing for 1 hour at 500°C, produced no measurable change in the interdiffusion.   This is predicted from the data in Table IV. In the CdZnS alloy, a similar growth and annealing procedure was adopted.   Here some evidence of annealing was evident at 500°C, as one might expect from the value of the estimated interdiffusion coefficient of $2.5 \times 10^{-15} cm^2 s^{-1}$.   In the case CdSe/ZnSe system some interdiffusion was noticeable again at 500°C.   Here more diffusion occurred than one might have expected from the data in Table IV, but it is possible here that the presence of mismatch dislocations initiated by the ZnSe/GaAs lattice mismatch, the ZnSe was only $\sim 0.75\,\mu m$ thick, could have affected the interdiffusion.

GROWTH OF CMT BY IMP

Almost all the published work to date on IMP growth of CMT [1,2, 15-25] has been carried out in the UK, although a number of the US laboratories are known to be using the technique.

Compositional control and crystal quality

Three potential criticisms of IMP are firstly that the initial growth starts with maximum compositional non-uniformity; secondly, that the lattice mismatch between the CdTe and HgTe is a potential source of dislocations and, thirdly, that the interdiffusion will preclude useful heterojunction formation. There is now much evidence to answer all these criticisms.

The model of depth uniformity has been covered by the section on interdiffusion. Numerous SIMS profiles measured on IMP layers put the depth non-uniformity on a ~0.1 μm scale as < ±1% in x. Larger scale variations when seen can be attributed to flow changes and other causes. The lateral uniformity in layers is a matter of good flow and hydrodynamic control. A good but not untypical result from the work of Whiffin et al [25] is reproduced in Fig.3. Here Δx varies by less than 0.002 over most of the slice area. One of the advantages of MOVPE, the ability to independently control x is illustrated by the heterostructures reproduced in Fig.4.

The lack of sharpness of the composition change at the IMP-grown heterostructure interface appears at first sight incompatible with good heterojunction formation. A typical CdTe/Cd$_{0.2}$Hg$_{0.8}$Te interface can be 0.3 μm wide and CMT(x$_1$)-CMT(x$_2$) interfaces in Fig. 4 are larger (~1 μm). However, for ideally-sharp heterostructures Migliorato and White [26] have shown that the conduction and/or valence band offsets between the semiconductors can result in energy band spikes which may limit effective device operation. Interdiffusion, as is found in IMP, is a potentially important way of overcoming this problem. If sharper junctions are needed then by lowering the growth temperatures using, for example, less stable Te alkyls [27] very significant reduction in interdiffusion could be achieved.

The problem of lattice mismatch between CdTe, CdSe$_x$Te$_{1-x}$ or Zn$_x$Cd$_{1-x}$Te and Cd$_x$Hg$_{1-x}$Te has been the subject of discussion by Booyens and Basson [28, 29] in LPE growth. In the case of the lower temperature MOVPE growth it would appear that the critical thickness for misfit dislocation introduction originally considered by Matthews [30] is about 0.1 μm. No evidence to our knowledge has been reported which has

374

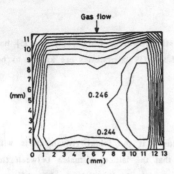

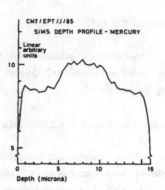

Fig.3 Layer uniformity in x from IR transmission measurements (Edinburgh Instruments). Contour lines are separated by 0.002 in x (acknowledgments to PRL ref 25)

Fig.4 SIMS depth profile demon－strating heterostructures between CMT layers of x values x= 0.35, x= 0.22 and x= 0.35

attributed the initial misfit at the interfaces of CdTe/HgTe IMP layers to resultant dislocation formation. It would appear that the misfit is taken up initially by strain. Even misfit dislocations generated at GaAs/CdTe interfaces [31] where the lattice mismatch is 14.7% are generally confined to a region close to the initial interface (< 1 μm) in IMP layers. Whilst the single crystal nature of IMP grown CMT has been reported detailed studies on their structural quality have not been made. In a recent study by Hails et al [23] of the growth of CdTe and CMT on (111)A CdTe twinned regions were identified from RHEED patterns. The twin domains are rotated 180° within the plane of the substrate. Twinning was shown to be present in layers only 0.5 μm thin, indicating the phenomenon was associated with nucleation. The significance of the twins and their propagation through subsequently deposited MOVPE layers has yet to be established.

Raccah et al [22] have made electrolyte electrode reflectance measurements on IMP layers. They noted compositional uniformity of x with depth of ±0.005 in "standard" IMP layers, and also that the layers were strain－free. The density of "polarisable defects" was negligible but there was a significant amount of alloy clustering. It would be helpful to complement this work with X－ray studies.

Growth on alternative substrates

It is appropriate to mention here that IMP is well suited to the growth of CMT onto sapphire, GaAs, CdSe$_x$Te$_{1-x}$ and Zn$_x$Cd$_{1-x}$Te substrates, where buffer layers of controlled composition and/or the use of lower temperature growth [27] may be needed to prevent diffusion from the substrate [32]. The latter two substrates can be used for lattice matching. The MOVPE growth of II−VIs on alternative substrates is covered in a recent review [33].

Purity and electrical properties

For a particular composition x, the fundamental electrical properties are controlled by the temperature and Hg vapour pressure in the course of the growth process. The generally accepted model of CMT is of a lattice dominated by Frenkel defects where the excitation of Hg from its lattice site gives rise to ionised Hg vacancies (V''_{Hg}). The equilibrium can be represented by equation (1).

$$Cd_xHg_{1-x}Te \rightleftharpoons V''_{Hg} + Hg(v) + 2h \qquad (1)$$

An expression for $[V''_{Hg}]$ in terms of K_g the equilibrium constant for reaction(1), where K_g is given by equation (2) and K_i the electron hole equilibrium constant has been derived by Mullin et al [34] as equation (3) following a previous study by Yang, Yu and Tang [35].

$$[V''_{Hg}] \, [h]^2 \, p_{Hg} = K_g \qquad (2)$$

$$[V''_{Hg}]K_i \, p_{Hg} + [V''_{Hg}]^{3/2}(2K_g^{\frac{1}{2}} \, p_{Hg}^{\frac{1}{2}}) - K_g = 0 \qquad (3)$$

This has enabled one to predict the hole concentration as a function of Hg pressure p_{Hg} for various equilibrium growth/annealing temperatures. These predictions are depicted in Fig. 5. At 410°C pure MOVPE CMT grown under a pressure of p_{Hg} of 5×10^{-2} atm, will be p−type (~10^{17} carriers cm^{-3}) The growth of pure n−type material (n < 10^{15} carriers cm^{-3}) at this Hg pressure would require an n−type dopant/impurity and a growth temperature around 300°C or below. This situation has not yet been proved experimentally.

Early results on the MOVPE of CMT, which were mainly grown by a non−IMP process were significantly n−type rather than as expected p−type. In a few cases preparation−related causes [34] were identified. These included system contaminations, low x regions in the layers and, possibly, substrate contamination. More recently various groups [2,17,21,24,34,36,37] working on IMP growth of CMT have grown p−type layers.

However, classical p−type Hall curves are rarely obtained and the interpretation is difficult. Typical curves are shown in Fig.6. Curve A is almost classical p but the exhaustion region E is not clearly resolved. Curve B has a distinct p−type region but

376

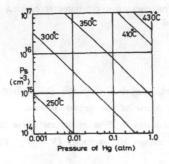

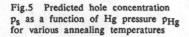
Pressure of Hg (atm)

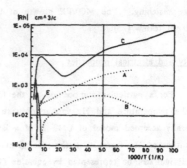
1000/T (1/K)

Fig.5 Predicted hole concentration p_s as a function of Hg pressure p_{Hg} for various annealing temperatures

Fig.6 Hall coefficient R_H cm^{3}c^{-1} versus 1000/T. Curve A good as−grown p−type layer; curve B inversion effects with tendency to second sign in R_H at low temperatures; curve C full inversion, n−type ____ curve throughout tempera−ture range; ... p−type region

R_H decreases in the low temperature region, suggesting that a second sign change is possible. Curve C shows that R_H is negative throughout the temperature range but exhibits a dip. All these curves can be explained on the basis of a two−layer model [38−40] where the structure being measured is regarded as a p−type layer with an n−type layer in parallel. Curve A has virtually no 'n−type skin', whereas B has a little, and curve C is dominated by n−type conduction. The cause of the inversion has been attributed by Mullin and Royle [38] to oxidation and/or surface damage, giving rise to a positive surface charge.

The identification of the exhaustion level (region p curve A) in a curve of type C is impossible and in B speculative. Thus only an estimate of the approximate value of the p−type carrier levels can be made. The general apparent carrier level of p−type IMP layers grown in a number of laboratories [34,36,37] ranges from $2x10^{16}-2x10^{17}$ carriers cm^{-3} for CMT layers (x from 0.2−0.25). The apparent maximum mobilities range from 350 to 900 cm^{2}v^{-1}s^{-1} but these values must be treated with caution because they can be so readily influenced by unidentified inversion effects. The p−type layers have been shown by the PRL [36] and GEC [21] groups to anneal in Hg to yield n−type layers with carrier levels in the range $(1-10)x10^{15}$cm^{-3} more often at the lower level. SIMS and laser mass spectrometry results reported by Capper et al [36] and confirmed by Blackmore [41], show levels of Li, Na, Al, Si, Ga, Ag, In, Cl, Br and Cr at levels less than $2x10^{15}$cm^{-3} although no correlation with the electrical properties was made. Vincent et al [37] did, however, suspect p−type contamination of layers as a result of annealing on CdZnTe substrates.

There is now clear evidence that the as—grown bulk properties of MOVPE material is p—type as expected, that the residual n—type impurity background is generally low $(1-5x10^{15}cm^{-3})$ and that the dominant problems are inversion effects and the difficulty of interpreting conventional Hall measurements.

DEVICE APPLICATIONS OF IMP

The MOVPE growth of CMT is potentially well suited to the preparation of device structures. Its primary role is aimed at IR detectors working in the $3-5\mu m$ and $8-14\mu m$ atmospheric windows. A notable result previously reported [33] is of the fabrication of a 32x32 starring array fabricated from IMP CMT by Mullard and demonstrated and reported by RSRE as the first focal plane array in MOVPE material. More recently Specht et al [42] have reported results of near background limited detector performance (80% and 60% of BLIP) for photoconductive long wavelength arrays fabrication from MOVPE CMT grown by what appears to be a non—IMP procedure. Both GEC and Mullard have made diodes having cut—off wavelengths at $10\mu m$ and R_0A values greater than 6 ohm cm^2. MOVPE material is now capable of producing diodes in the $3-5\mu m$ region having R_0As as good as the best LPE or bulk CMT. An exciting new application for MOVPE CMT is the development of avalanche photodiodes by the SAT group in France using GEC MOVPE—grown layers. Diodes working in the $1.3-1.55\mu m$ region are now showing performance figures [43] comparable with the best InGaAs diodes.

CONCLUSIONS

The IMP technique has proved to be highly successful for the growth of uniform CMT which is of practical and, potentially, very important value in the preparation of IR devices. Application of the technique need not, however, be restricted to the growth of CMT. Its full potential will only be realised by further development and use in other alloy systems.

Copyright C Controller HMSO, London, 1986

REFERENCES

1. S.J.C.Irvine, J.Tunnicliffe and J.B.Mullin, Mat.Lett. 2, 305 (1984)
2. J.Tunnicliffe, S.J.C.Irvine, O.D.Dosser and J.B.Mullin, J.Cryst.Growth 68, 245 (1984)
3. S.J.C.Irvine and J.B.Mullin, J.Cryst.Growth, 55, 107 (1981)
4. J.B.Mullin, S.J.C.Irvine and A.Royle, J.Cryst.Growth 57, 15 (1982)
5. S.J.C.Irvine, J.Tunnicliffe and J.B.Mullin, 65, 479 (1983)
6. J.B.Mullin and D.T.J.Hurle, J.Luminescence, 7, 176 (1973)
7. E.Ratajczak and J.Terpilowski, Rocziki Chem. 43, 1609 (1969)
8. O.Kubaschewski, E.Ll.Evans and C.B.Alcock, Metallurgical Thermo-chemistry (Pergamon, Oxford, 1967),p314.

378

9. J.G.Fleming and D.A.Stevenson - in course of publication
10. K.Zanio, J.Vac.Sci.Technol. A4, 2106 (1986)
11. D.A.Stevenson "Diffusion in the chalcogenides of ZnCd and Pb" in
    'Atomic Diffusion in Semiconductors' ed. D.Shaw, Plenum Press, London
    1973, p431.
12. H.C.Casey,Jr. "Diffusion in the III-V Compound Semiconductors" in
    'Atomic Diffusion in Semiconductors' ed.D.Shaw, Plenum Press, London 1
    973, p351.
13. M.A.Tischler and S.M.Bedair, J.Cryst.Growth, 77, 89 (1986)
14. M.Brown and A.F.W.Willoughby, J.Cryst.Growth 59, 27 (1982)
15. P.J.Wright, B.Cockayne, A.J.Williams and G.W.Blackmore, J.Cryst.Growth
    79, 357 1986
16. M.J.Bevan and K.T.Woodhouse, J.Cryst.Growth 68, 254 (1984)
17. J.Giess, J.S.Gough, S.J.C.Irvine, G.W.Blackmore, J.B.Mullin and
    A.Royle, J.Cryst.Growth 72, 120 (1985)
18. M.J.Hyliands, J.Thompson, V.Vincent, K.T.Woodhouse and M.J.Bevan, J.
    Vac.Sci.Technol. A4:2217 (1985)
19. J.Thompson, K.T.Woodhouse and C.Dineen, J.Cryst.Growth 77, 452 (1986)
20. S.J.C.Irvine, J.Giess, J.S.Gough, G.W.Blackmore, A.Royle, J.B.Mullin,
    N.G.Chew and A.G.Cullis, J.Cryst.Growth 77, 437 (1986)
21. M.J.Hyliands, J.Thompson, M.J.Bevan, K.T.Woodhouse and V.Vincent, J.
    Vac.Sci.Technol. A4, 2217 (1986)
22. P.M.Raccah, Z.Zhang, J.W.Garland and Army H.M.Chu, J.Vac.Sci.Technol.
    A4, 2226 (1986)
23. J.E.Hails, G.J.Russell, A.W.Brinkman and J.Woods, J.Appl.Phys. 60, 262
    4(1986)
24. G.T.Jenkin, J.Thompson, M.J.Hyliands, K.T. Woodhouse and V.Vincent,
    2nd. Int.Symp.on Optical and Electro-optical Applied Science and
    Engineering. SPIE, Cannes, France (1986)
25. P.A.C.Whiffin, B.C.Easton, P.Capper and C.D.Maxey, J.Cryst.Growth 79,
    935 (1986)
26. P.Migliorato and A.M.White, Solid State Electronics 26, 65 (1983)
27. W.E. Hoke, This Proceedings p.
28. J.H.Basson and H.Booyens, Phys.Status Solidi (a) 80, 663 (1983)
29. H.Booyens and J.H.Basson, Phys.Status Solidi (a) 85, 449 (1984)
30. J.W.Matthews 'Coherent interfaces and misfit dislocations' in
    "Epitaxial Growth Pt.B" ed. J.W.Matthews, Academic Press, 1975, p560
31. A.G.Cullis, N.G.Chew, J.L.Hutchison, S.J.C.Irvine and J.Giess, Inst.
    Phys.Conf.Series No.76, Section 1, 29 (1985)
32. J.Giess, J.S.Gough, S.J.C.Irvine, J.B.Mullin and G.W.Blackmore, this
    Proceedings
33. J.B.Mullin, S.J.C.Irvine, J.Giess and A.Royle, J.Cryst.Growth 72, 1
    (1985)
34. J.B.Mullin, A.Royle, J.Giess, J.S.Gough and S.J.C.Irvine, J.Cryst.
    Growth 77, 460 (1986)
35. J.Yang, Z.Yu and D.Tang, J.Cryst.Growth 72, 275 (1985)
36. P.Capper, B.C.Easton, P.A.C.Whiffin and C.D.Maxey, J.Cryst.Growth 79,
    508 (1986)
37. V.Vincent, C.Wilson and J.M.Lansdowne, 3rd Int.Symp. on Optoelectronic
    Applied Science and Engineering. SPIE, Innsbruck, 1986
38. J.B.Mullin and A.Royle, J.Phys.D. 17, L69 (1984)
39. L.F.Lou and W.H.Frye, J.Appl.Phys. 56, 2253 (1984)
40. R.L.Petritz, Phys.Rev. 110, 1258 (1958)
41. G.W.Blackmore - unpublished data
42. L.T.Specht, W.E.Hoke, S.Oguz, P.J.Lemonias, V.G.Kreismanis and R.
    Korenstein, Appl.Phys.Lett. 48, 417 (1986)
43. P.Gori, T.N.Guyen Duy, J.Thompson, P.Mackett and G.Jenkins - to be
    published

A QUALITATIVE MODEL FOR PREDICTING ALKYL STABILITY AND ITS RELEVANCE TO ORGANOMETALLIC GROWTH OF HgCdTe AND GaAlAs

WILLIAM E. HOKE
Raytheon Co., 131 Spring St., Lexington, MA 02173

ABSTRACT

A qualitative stability model established for hydrocarbon molecules and extended to organometallic compounds is applied to alkyl organometallic compounds used in thin film growth. For hydrocarbon molecules the strength of a carbon-hydrogen bond is reduced by neighboring alkyl groups. The mechanism for this effect is delocalization of the unpaired electronic charge in the resultant free radical by neighboring alkyl groups. The delocalization effect is present in organometallic compounds and is illustrated here for several alkyl organometallic systems. Two applications of the delocalization effect in thin film growth are discussed. First the thermal stability of organotellurium compounds is substantially reduced in branched compounds which permits significantly lower HgCdTe ·metalorganic growth temperatures. Also carbon incorporation in GaAs and GaAlAs is reduced with ethyl instead of methyl organometallic compounds. An important factor contributing to this result is the weaker carbon-metal bonds in the ethyl compounds.

INTRODUCTION

The MOCVD (metalorganic chemical vapor deposition) technique has been used to grow epitaxial thin films of a wide range of compounds. Considerable interest in recent years has been on the MOCVD growth of III-V compounds such as GaAs and GaAlAs as well as II-VI compounds such as HgCdTe. In the MOCVD process the starting organometallic sources are pyrolyzed in the substrate region. Typically the weakest bonds in these compounds are the carbon-metal bonds. Consequently the strengths of these bonds can directly affect the pyrolysis temperature and film growth process. Since the strength of a carbon-metal bond is affected by neighboring organic groups, the choice of organometallic sources is an important consideration. As discussed below the stability properties of the organometallic compounds determine the low temperature limit on MOCVD HgCdTe growth temperature as well as affect carbon incorporation in GaAs and GaAlAs films.

Epitaxial films of HgCdTe can be degraded by relatively high growth temperatures, which are typically 400-440°C for MOCVD. At these temperatures diffusional processes broaden heterojunction interfaces and make the growth of novel quantum well structures infeasible. The growth of HgCdTe films on foreign substrates can be contaminated from substrate autodoping. Furthermore the equilibrium density of mercury vacancies increases sharply with temperature. Consequently the use of alternative organometallic sources which permits lower HgCdTe growth temperatures is attractive [1].

Carbon is an acceptor impurity in epitaxial films of GaAs and GaAlAs. Incorporation of carbon from the organometallic sources has been observed in MOCVD growth of these materials [2]. Organometallic sources are also used in an ultra-high vacuum thin film growth technique termed gas-source molecular beam epitaxy (GS-MBE). Significant carbon incorporation in GS-MBE films from the organometallic sources has also been reported [3] and the use of organometallic sources which minimizes carbon incorporation is necessary.

STABILITY MODEL

Less stable organometallic sources are needed to reduce the HgCdTe growth temperature as well as reduce the carbon incorporation in the GaAlAs system. The minimum usable MOCVD HgCdTe growth temperature is limited not by surface kinetics but by the thermal stability of the most stable compound which is the tellurium source [4],[5]. Since the carbon-tellurium bonds are the weakest bonds in the organotellurium molecule, reducing these bond strengths will reduce the thermal stability of the molecule and permit lower HgCdTe growth temperatures. Similarly it is reasonable to expect that carbon incorporation in MOCVD grown GaAs and GaAlAs films is affected by the strengths of the carbon-gallium and carbon-aluminum bonds. Molecules containing weaker carbon-metal bonds should provide for reduced carbon incorporation.

One approach to weakening carbon-metal bonds is to use organometallic compounds containing appropriate organic groups which destabilize the molecule. The pyrolysis of organometallic compounds occurs through the breaking of the carbon-metal bonds with the formation of free radicals. The rate determining step is the breaking of the first carbon-metal bond with the formation of two free radicals. Consequently the activation energy for this step will determine the stability of the molecule. The activation energy is the difference in potential energy content of the parent compound and the free radicals. Neighboring organic groups which reduce the activation energy for free radical formation will destabilize the molecule.

An important mechanism for reducing the activation energy is delocalization of the free radical unpaired electronic charge [6]. With delocalization the unpaired electronic charge does not reside primarily at the atomic site of the broken bond. Instead the electronic charge is spread over neighboring nuclei resulting in stronger bonds and a more stable free radical. Since this stabilization mechanism is not possible for the parent compound, the activation energy is reduced.

The organometallic compounds considered here are alkyl molecules containing only single bonds and no heteroatoms other than the central metal atom. The compounds are analogous to alkyl hydrocarbon molecules. Let C_{RAD} represent the carbon atoms at which the molecular bonds are broken. These atoms are classified by the number of carbon-carbon bonds they possess and consequently the number of alkyl groups bonded to them. A C_{RAD} atom which possesses no carbon-carbon bonds is termed zero order or $0°$, one carbon-carbon bond is primary or $1°$, two carbon-carbon bonds are secondary or $2°$, and three carbon-carbon bonds are tertiary or $3°$. For alkyl hydrocarbon free radicals, delocalization of the electronic charge occurs through overlap of the P orbital of the unpaired electron on the C_{RAD} atom with the sigma orbitals of the alkyl groups bonded to the C_{RAD} atom. Consequently the delocalization increases with the number of alkyl groups bonded to C_{RAD} atom which increasingly destabilizes the parent molecule.

Considerable experimental evidence supports the predictions of the delocalization model that the stabilities of free radicals are in the order $0° > 1° > 2° > 3°$. Table I lists several hydrocarbon radicals and the bond strengths of the parent compounds [7]. The bond strength is the activation energy for formation of the free radical, E_{ACT}. The last column in Table I gives the reduction in the activation energy by delocalization. This quantity was determined by subtracting the activation energy for a given radical from the value for the methyl radical which has negligible delocalization. As expected the bond strengths and consequently stabilities of the parent molecules are in the order $0° > 1° > 2° > 3°$. Furthermore the activation energies are equal for the ethyl and n-propyl radicals which are both primary radicals. Similarly the two secondary radicals, isopropyl and sec-butyl, have the same activation energies.

The E_{ACT} values given in Table I pertain to breaking carbon-hydrogen bonds. The last column lists the reduction in bond strength from delocalization. Similar values pertain to organometallic compounds [8]. These compounds have considerably weaker carbon-metal bonds than carbon-hydrogen bonds. For example the carbon-tellurium bond strength in organotellurium molecules is approximately 30 kcal/mole [9]. Therefore the reduction in organometallic bond strength by delocalization should be significant and result in reduced thermal stability.

TABLE I. Activation energies for formation of hydrocarbon free radicals.

Radical Order	Radical Name	Chemical Equation	E_{ACT} (Kcal/mole)	Reduction in E_{ACT} By Delocalization (Kcal/mole)
0^0	Methyl	$CH_4 \longrightarrow$ $CH_3 \bullet + H \bullet$	105	0
1^0	Ethyl	$CH_3CH_3 \longrightarrow$ $CH_3CH_2 \bullet + H \bullet$	98	7
1^0	N-Propyl	$CH_3CH_2CH_3 \longrightarrow$ $CH_3CH_2CH_2 \bullet + H \bullet$	98	7
2^0	Isopropyl	$(CH_3)_2CH_2 \longrightarrow$ $(CH_3)_2CH \bullet + H \bullet$	95	10
2^0	Sec-Butyl	$CH_3CH_2CH_2CH_3 \rightarrow$ $CH_3CH_2CHCH_3 + H \bullet$	95	10
3^0	Tertiary-Butyl	$(CH_3)_3CH \longrightarrow$ $(CH_3)_3C \bullet + H \bullet$	92	13

MOCVD GROWTH OF HgCdTe

The growth temperatures for HgTe and HgCdTe are determined by the pyrolysis characteristics of the organotellurium source. In the initial work of Mullin et al. [4] the pyrolysis efficiencies of dimethyltelluride and diethyltelluride were examined. Diethyltelluride was found to pyrolize at a temperature approximately 70°C lower than dimethyltelluride. This result is consistent with the expected thermal stability order since pyrolysis produces primary radicals for diethyltelluride and zero order radicals for dimethyltelluride. Because of its capability for lower growth temperatures, MOCVD growth of HgCdTe using diethyltelluride was pursued [4,5]. However, a typical growth temperature with diethyltelluride is 400°C which is relatively high and can adversely affect material properties. Interdiffusion in the HgCdTe system is rapid at 400°C which causes broadened heterojunctions and homojunctions and makes the growth of quantum-well structures impossible. HgCdTe films grown on foreign substrates can be contaminated by substrate outdiffusion. Also the density of mercury vacancies is significant at 400°C.

Lower HgCdTe growth temperatures have been achieved using higher order alkyl organotellurium compounds [1]. In Figure 1 are plotted HgTe growth rates as a function of growth temperature for diethyltelluride, di-n-propyltelluride, diisopropyltelluride, and ditertiarybutyltelluride. The films were grown with an excess mercury pressure so that the growth rates are indicative of the cracking efficiencies of the organotellurium sources. The growth rate for the plateau region, which represents efficient pyrolysis, is higher for diisopropyltelluride than ditertiarybutyltelluride since a higher concentration of diisopropyltelluride was used.

The results in Figure 1 are consistent with the expected stability order and parallel the hydrocarbon results in Table I. No significant difference is observed in the thermal stabilities of diethyltelluride and di-n-propyltelluride which both pyrolyze to produce primary radicals. The growth temperature can be reduced by approximately 50°C by using diisopropyltelluride, which has secondary order alkyl groups, instead of diethyltelluride. Finally the growth temperature can be reduced approximately 70°C with ditertiarybutyltelluride, which has tertiary order alkyl groups, instead of diisopropyltelluride. Therefore, the experimental evidence indicates that the stability order for alkyl organotellurium compounds is $0° > 1° > 2° > 3°$. Additional evidence supporting the stability model is provided in Figure 2 in which the HgTe growth rate is plotted versus inverse temperature for ditertiarybutyltelluride. An activation energy for this process of 20 Kcal/mole is obtained which is lower than the reported value of 30 Kcal/mole for HgTe growth using diethyltelluride [9].

The effect of free radical delocalization has significant implications for MOCVD growth of HgCdTe. Using dimethyltelluride a growth temperature of approximately 450°C would be necessary. Consequently, the deposited films would have broadened heterojunction interfaces and a high density of mercury vacancies. Film growth on foreign substrates would be hampered by substrate autodoping. With ditertiarybutyltelluride a growth temperature of 250°C or less is possible which significantly reduces these problems. Even though this MOCVD growth temperature is quite low, the reported electrical properties for HgTe films grown with ditertiarybutyltelluride are quite good indicating no significant degradation of transport properties from kinetic limitations[1].

MOCVD growth of HgCdTe with compounds such as ditertiarybutyltelluride may make feasible HgCdTe growth with organomercury compounds since the required mercury partial pressure is significantly reduced at 250°C. Only organometallic sources would then be used which would permit a cold wall reactor system. One requirement of the chosen organomercury source would be that it pyrolyzes efficiently at low temperature. Therefore it is of interest to examine the thermal stability of organomercury compounds.

Pyrolysis studies of several organomercury compounds have been performed and results are given in Table II [10]. The thermal decomposition rate constant is given by the Arrhenius relation $C = A \exp(-E/RT)$, where A is the frequency factor, E is the activation energy, T is the absolute temperature, and R is the gas constant. Using this formula the rate constant at 600°K, $C_{600°K}$, is determined and tabulated. $D_1 + D_2$ is the energy required to break both carbon-mercury

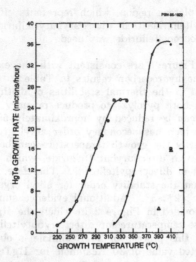

Figure 1. HgTe growth rate as a function of growth temperature for
diethyltelluride (△), di-n-propyltelluride (□), diisopropyltelluride (•), and
ditertiarybutyltelluride (○). For the first three compounds the
organotellurium concentration, mercury concentration, and total flow were
respectively 3.2×10^{-3} atm, 2.6×10^{-2} atm, and 1.8 slm. For
ditertiarybutyltelluride the organotellurium concentration, mercury concentration, and total flow were 2×10^{-3} atm, 5×10^{-3} atm, and 1.2 slm.

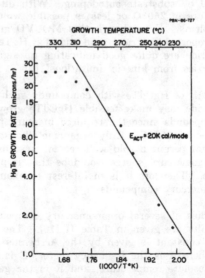

Figure 2. Arrhenius plot of HgTe growth rate as a function of inverse
temperature for ditertiarybutyltelluride. The growth conditions are given
in Figure 1.

bonds. The values for $D_1 + D_2$ are determined from E and thermochemical data [10].

The results in Table II are in very good agreement with expectations. The bond strengths given by $D_1 + D_2$ are largest for dimethylmercury which has zero order alkyl groups. The bond strengths in diethylmercury and di-n-propylmercury, which both have primary alkyl groups, are nearly the same. Finally the bond strengths in diisopropylmercury, which has secondary alkyl groups, are the weakest. The calculated thermal decomposition rate constants for the five organomercury compounds are in the order $0° < 1° < 2°$. The rate constants for the three primary alkyl organomercury compounds are within a factor of 4 of each other at $600°K$. The rate constant for dimethylmercury is approximately four orders of magnitude smaller than the rate constants for the primary alkyl compounds. Diisopropyltelluride has the largest rate constant being approximately two orders of magnitude greater than for the primary alkyl organomercury compounds. The values for the activation energy E are in the order $0° > 1° > 2°$. Diethylmercury has a lower value for the activation energy than the other two primary alkyl organomercury compounds. It has been suggested [10] that this difference may be due to a different pyrolytic mechanism for diethylmercury in which one carbon-mercury bond is initially broken whereas the other primary alkyl compounds decompose by simultaneous rupture of both bonds. The lower value of the activation energy for diethylmercury is offset by a lower frequency factor resulting in a similar rate constant as the other two primary alkyl compounds.

TABLE II. Arrhenius parameters and bond energies for several organomercury compounds.

Alkyl Name and Formula	Radical Order	A (sec^{-1})	E Kcal/mole	C$_{600°K}$ (sec^{-1})	$D_1 + D_2$ Kcal/mole
Dimethylmercury Hg(CH$_3$)$_2$	0°	5.2×10^{15}	57.7	5.0×10^{-6}	60.2
Diethylmercury Hg(CH$_2$CH$_3$)$_2$	1°	2.0×10^{14}	42.2	8.5×10^{-2}	48.5
Di-n-propylmercury Hg(CH$_2$CH$_2$CH$_3$)$_2$	1°	2.6×10^{15}	46.5	3.0×10^{-2}	49.4
Di-n-butylmercury Hg(CH$_2$CH$_2$CH$_2$CH$_3$)$_2$	1°	6.3×10^{15}	47.8	2.4×10^{-2}	————
Diisopropylmercury Hg(CHCH$_3$CH$_3$)$_2$	2°	2.5×10^{16}	40.7	3.7	40.4

Organocadmium compounds are the third source in MOCVD growth of HgCdTe. Pyrolytic studies have been performed on dimethylcadmium [10], although the higher order compounds have not been as extensively examined. As expected the mean bond energy of the cadmium-carbon bonds in dimethylcadmium (35.5 Kcal/mole) is greater than in diethylcadmium (26.5 Kcal/mole) [8].

GROWTH OF GaAs AND GaAlAs USING ORGANOMETALLIC COMPOUNDS

Similar relative stability trends with alkyl order are observed in III-V organometallics as in II-VI compounds. In a mass spectrometer study [11], triethylgallium was observed to thermally decompose at a temperature approximately 150°C lower than trimethylgallium which is consistent with the average strength of the carbon-gallium bond in triethylgallium (53.8 Kcal/mole) being weaker than in trimethylgallium (61.3 Kcal/mole) [8]. Trimethylaluminum is reported to be more thermally stable than triethylaluminum or higher alkyl organoaluminum compounds [12]. Increased antimony incorporation in GaAsSb films was obtained using triethylantimony instead of trimethylantimony [13]. This result was explained by the higher pyrolytic efficiency of triethylantimony due to its weaker Sb-C bond strength (43 Kcal/mole) compared to trimethylantimony (54 Kcal/mole). Also the pyrolysis temperature of tertiarybutylphosphine was found to be approximately 50°C lower than for isobutylphosphine [14]. Isobutylphosphine pyrolyzes to form primary radicals compared to tertiary radicals for tertiarybutylphosphine. There are several other examples of the destabilizing effect of neighboring alkyl groups on carbon-metal bonds in III-V organometallic compounds [8].

An important issue related to thermal stability is carbon incorporation. Since carbon is an acceptor impurity in the GaAlAs system, carbon incorporation must be minimized in the growth of high quality GaAs and GaAlAs films. This is particularly challenging in GaAlAs growth since carbon readily incorporates in this material. The starting organometallic compounds have been shown to be important sources of carbon incorporation in films grown by MOCVD [2] as well as by gas-source MBE [3]. It is reasonable to expect that carbon incorporation should be reduced by using higher order alkyl organometallic compounds which have weaker carbon-metal bonds.

The available experimental evidence supports this expectation. In an early study Seki et al. [15] examined GaAs MOCVD growth using triethylgallium instead of trimethylgallium. The motivation was to reduce the carbon incorporation from trimethylgallium. GaAs films with low carrier concentrations and high mobilities were obtained. The quality of MOCVD-grown GaAs and GaAlAs films using trimethylgallium and triethylgallium has been directly compared for carbon incorporation [2].

Using photoluminescence and SIMS measurements the carbon incorporation was found to be significantly less in the films grown with the triethyl alkyls. The difference in carbon incorporation was particularly significant in the AlGaAs films. With triethylalkyls higher 2-dimensional electron gas mobilities in GaAs/GaAlAs selectively-doped heterostructures were obtained [16]. The improvement in mobility was attributed to reduced carbon incorporation in the AlGaAs layer with the triethylalkyls.

Carbon incorporation is even more pronounced in gas-source MBE. Using trimethylgallium Putz et al. [3] obtained GaAs films which were heavily doped P-type at 10^{20} cm^{-3} due to carbon incorporation. However when triethylgallium was used the background doping in the GaAs films dropped to 10^{15} cm^{-3} because of significantly less carbon incorporation. Intentional P-type doping with trimethylgallium was also reported [17]. Similar results for carbon incorporation in GaAs films grown by gas-source MBE using trimethylgallium and triethylgallium were reported by Kondo et al. [18].

The difference in carbon incorporation from methyl and ethyl organometallic sources has been proposed to be caused by different pyrolytic mechanisms for these compounds [2,3,13]. The ethyl alkyl compounds can undergo a β-elimination process with the production of ethylene which is not possible with the methyl alkyls. As discussed here an important factor is that three carbon-metal bonds are weaker in the ethyl compounds than the methyl compounds. The general property of weaker bond strengths assists pyrolytic processes such as β-elimination.

SUMMARY

A qualitative model developed for hydrocarbon and organometallic molecules has been applied to alkyl organometallic compounds used in thin film growth. According to the model the strengths of carbon-metal bonds are weakened by the number of neighboring alkyl groups due to charge delocalization by the alkyl groups. The thermal stabilities of organotellurium and organomercury compounds have been examined in detail and found to be in agreement with the model. Also published results for relative thermal stabilities of some organometallic compounds containing cadmium, gallium, aluminum, antimony, and phosphorous are also consistent with the model. Significantly lower MOCVD HgCdTe growth temperatures have been achieved using higher order alkyl organometallic compounds which have lower thermal stability. Also the significantly lower carbon incorporation in GaAs and GaAlAs films using triethyl compounds instead of trimethyl compounds is qualitatively correlated with the bond strengths in these molecules.

REFERENCES

1. W. E. Hoke and P. J. Lemonias, Appl. Phys. Lett. 46, 398 (1985); 48, 1669 (1986).
2. N. Kobayashi and T. Makimoto, Jap. J. Appl. Phys. 24, L824(1985).
3. N. Putz, H. Heinecke, M. Heyen, P. Balk, M. Weyers, and H. Luth, J. Cryst. Growth 74, 292(1986).
4. J. B. Mullin, S. J. Irvine, and D. J. Ashen, J. Cryst. Growth 55, 92 (1981).
5. W. E. Hoke and R. Traczewski, J. Appl. Phys. 54, 5087 (1983).
6. R. T. Morrison and R. N. Boyd, Organic Chemistry, 2nd ed. (Allyn and Bacon, Boston, 1969), p. 118-130, 393-394.
7. Handbook of Chemistry and Physics, edited by R. C. Weast (CRC Press, Boca Raton, Florida, 1986), p. F178.
8. The Chemistry of the Metal-Carbon Bond, Vo. 1, edited by F. R. Hartley and S. Patai (John Wiley and Sons, New York, 1982), p. 53-71.
9. I. Bhat and S. K. Ghandi, J. Electrochem. Soc. 131, 1923 (1984).
10. Comprehensive Chemical Kinetics, Vol. 4, edited by C. H. Bamford and C. F. Tipper (Elsevier Scientific Publishers, New York, 1972), p. 215-233.
11. M. Yoshida, H. Watanabe, and F. Vesugi, J. Electrochem. Soc. 132, 677 (1985).
12. G. E. Coats, Organo-Metalic Compounds, 2nd ed. (Butler and Tanner Ltd., London, 1960), p. 70, 133.
13. L. M. Fraas, P. S. McLeod, J. A. Cape, and L. D. Partain, J. Cryst. Growth 68, 490 (1984).
14. C. H. Chen, C. A. Larsen, G. B. Stringfellow, D. W.Brown, and A. J. Robertson, J. Cryst. Growth 77, 11 (1986).
15. Y. Seki, K. Tannno, K. Iida, and E. Ichiki, J. Electrochem. Soc 122, 1108 (1975).
16. N. Kobayashi and T. Fukui, Electron. Lett. 20, 887 (1984).
17. M. Weyers, N. Putz, H. Heinecke, M. Heyen, H. Luth, and P. Balk, J. Electron. Mat. 15, 57 (1986).
18. K. Kondo, H. Ishikawa, S. Sasa, Y. Sugiyama, and S.Hiyamizu, Jap. J. Appl. Phys. 25, L52 (1986).

ORIENTATION EFFECTS ON THE HETEROEPITAXIAL GROWTH OF
$Cd_xHg_{1-x}Te$ ON TO CdTe AND GaAs

J GIESS, J S GOUGH, S J C IRVINE, J B MULLIN AND G W BLACKMORE
Royal Signals and Radar Establishment, St Andrews Road, Great Malvern,
Worcs WR14 3PS, UK

ABSTRACT

For MOVPE growth of CdTe on to (100) GaAs the orientation of the epitaxial layer is dependent on a combination of the growth conditions and
substrate treatment prior to growth, (100) being the preferred orientation.
(111) oriented CdTe layers can be grown on to (111) GaAs substrates. The
effect of orientation on the choice of buffer composition, electrical properties and the extent of Ga diffusion from the substrate is discussed.
Epitaxial CMT layers grown on to suitably buffered GaAs substrates are
compared to similar layers grown on to CdTe substrates.

INTRODUCTION

The versatility of metal organic vapour phase epitaxy (MOVPE) has
enabled the growth of high quality cadmium mercury telluride (CMT) on to CdTe
substrates that are oriented either (111) or (100) 2° towards (110) [1-5].
The epitaxial growth of single crystal CdTe and CMT has also been demonstrated over much larger areas than normally available with CdTe substrates by
using alternative substrate materials such as GaAs [6-8] or sapphire [9-11].
GaAs has an additional advantage as an alternative substrate as it offers a
choice of orientation of the epitaxial layer: only (111) oriented CdTe/CMT
layers can be grown on to sapphire substrates. It has recently been shown
that both (100) and (111) oriented CdTe films can be grown by MOVPE on to
(100) GaAs, the orientation being dependent on the substrate's treatment
prior to growth [12-14].
In this paper results will be presented to show that for growth on to
(100) GaAs, the orientation of the epilayer depends on a combination of the
growth conditions and the substrate's treatment prior to growth. (111)
oriented CdTe layers have also been grown on to (111) GaAs substrates.
Selection of the layer orientation limits the choice of suitable buffer
structure for separating the active CMT layer from the GaAs substrate. The
surface morphology, structure and electrical properties of epitaxial layer
and buffer are also dependent on orientation. Suitably buffered CMT layers
grown on to both (100) GaAs and (111) GaAs will be compared with similar
layers grown on to (100) and (111) CdTe substrates.

EXPERIMENTAL

These growth studies have been carried out in a fully automated horizontal atmospheric pressure MOVPE equipment using the Interdiffused Multilayer
Process (IMP) which has been described in a previous publication [3]. The
metalorganic sources used were dimethylcadmium and diethyltelluride at
concentrations of approximately 1 torr.
The GaAs substrates oriented (100) 2° towards (110) and (111)B were
supplied as polished by Mining and Chemical Products. After degreasing these
substrates were etched in 5:1:1 H_2SO_4, water, H_2O_2 before loading into the
reactor. The CdTe substrates oriented (100) 2° towards (110) and (111)B
(using the notation of Fewster and Whiffin [15]) were prepared by pad
polishing using a 2% bromine in methanol solution.
A Philips single crystal x-ray diffractometer was used to determine the
peak heights for the (400) and (111) reflections from the epitaxial layer.

In calculating the proportion of each orientation present, allowance was made for the relative intensities for the (400) and (111) reflections expected, assuming equal quantities of material had been present [16].

The Ga concentration in the buffer and CMT layers was determined using a Cameca 3F secondary ion mass spectrometer (SIMS). The electrical properties of the buffer were determined by making capacitance voltage measurements on a bevelled edge using evaporated gold contacts.

THE ORIENTATION OF CdTe EPILAYERS GROWN ON TO (100) GaAs SUBSTRATES

Growth of CdTe on to (100) GaAs substrates can result in the nucleation of either (111) or (100) CdTe depending on the heat treatment of the GaAs substrate prior to growth [12-14]. We have recently shown [16] that the orientation is in fact dependent on the selected combination of growth temperature and heat treatment of the GaAs. The resultant orientations for various growth runs on to (100) GaAs are summarised in Table I.

If CdTe is nucleated on to GaAs which has been heated directly to the growth temperature the epilayer always grows with a (100) orientation. When the GaAs has prior to growth been heat cleaned at 580°C for 10 minutes in the presence of residual CdTe deposits, then the orientation of the CdTe is determined by the growth temperature; for growth temperatures above 400°C the epilayer grows with a (111) orientation whereas for growth temperatures below 400°C the epilayer grows oriented (100). A prolonged or higher temperature heat clean or heat cleaning the GaAs in a prebaked reactor all tend to produce (100) oriented CdTe layers.

It has been shown [12,16] that when GaAs is heat cleaned in the presence of CdTe, Te is adsorbed on to the GaAs surface. Feldman et al [12] have proposed that the formation of a Ga-As-Te interfacial layer controls the orientation of CdTe epitaxy layer. If this interfacial layer were completely formed during the heat clean, the orientation would not be expected to depend on the growth temperature. The transition at a growth temperature of 400°C corresponds to the temperature above which the diethyltelluride pyrolysis becomes efficient [2] thus enabling Te rich conditions; below 400°C diethyltelluride can only pyrolyse significantly with dimethylcadmium. These

Table I: A Summary of the Orientation Obtained at Various Growth Temperatures Following Different Treatments of the (100) GaAs Substrate Prior to Growth, Expressed as a Percentage of Layer Oriented (100)

Heat clean prior to growth	Growth temperature (°C)					
	350	370	390	410	465	500
None prebaked reactor	100	-	-	-	-	-
None CdTe in reactor	100	-	-	100	100	>99.9
10 mins at 580°C prebaked reactor	-	-	99.9	-	29	-
10 mins at 580°C CdTe in reactor	100	100	100	0	0.2	0
30 mins at 580°C CdTe in reactor	-	>99.9	-	99	-	-
10 mins at 600°C CdTe in reactor	100	>99.9	-	99.8	-	-

results are consistent with Feldman's theory of formation of a Ga-As-Te interfacial layer determining the orientation of the epilayer but suggest that for MOVPE growth the interfacial layer is only partly formed at the heat clean temperature. In order to obtain (111) oriented epilayers it is also essential to select growth conditions where Te rich growth is possible otherwise the Ga-As-Te layer will not be formed.

A COMPARISON OF CMT GROWTH ON TO (100) AND (111) BUFFER LAYERS

Substrate Orientation

GaAs substrates oriented (100) 2° towards (110) have proved to be a very suitable alternative to CdTe for the growth of high quality (100) CMT epilayers [6].

For the growth of (111) CMT epilayers there is a choice of substrate orientation: (111) GaAs or heat cleaned (100) GaAs substrates. The former nucleates (111) epilayers that grow reproducibly at a growth temperature of 410°C without the necessity of any heat cleaning process. Although the growth of (111) epilayers has been demonstrated on (100) GaAs substrates it requires a well controlled heat clean at elevated temperature followed by nucleation of CdTe under well defined growth conditions as shown in Table I. Any contamination such as carbon or residual oxide on the substrate surface will nucleate (100) oriented growth as its presence will prevent the essential reaction between the GaAs and Te necessary for the formation of the Ga-As-Te interfacial layer. Growth of (111) epilayers on to (111)B GaAs substrates have therefore been chosen as the more reproducible and practical route.

Buffer Composition

Growth of CMT epilayers on to any alternative substrate requires the growth of a buffer layer. The buffer layer which is usually a few microns thick is required to isolate the active CMT layer from any impurity or matrix elements in the substrate that may diffuse and dope the active layer. It also separates the active CMT layer from the complex dislocation structure that can be generated at the substrate/buffer interface as a consequence of the large lattice mismatch. An essential requirement for a buffer layer for MOVPE growth is a smooth surface morphology as any topographic irregularities will propagate and become more pronounced as subsequent epitaxial layers are grown.

For growth on the (111) orientation a CdTe buffer layer satisfies these requirements. A typical growth rate of 45 μm/hour at 410°C for alkyl concentrations just over 1 torr allows the growth of a fairly thick CdTe buffer with an excellent surface morphology.

For growth on the (100) orientation CdTe is not suitable as a buffer layer as its tendency to produce (111) facets results in a surface morphology consisting of pyramids which completely cover the surface. Growing slightly off orientation has been found to elongate the facet features with some suppression of the surface roughening [1]. A much smoother surface morphology can be obtained by growing a CMT buffer. Thus the buffer layer chosen for growth on to (100) GaAs has usually been high x $Cd_xHg_{1-x}Te$, the band gap of the buffer layer being larger than that of the active layer and therefore transparent to the long wavelength infrared radiation.

Ga Diffusion Through the Buffer Layer

It is essential that the buffer layer is sufficiently thick to prevent the diffusion of any electrically active elements from the substrate into the epilayer. For growth on to (100) GaAs we have shown [6] that although the As level falls rapidly at the substrate/buffer interface the Ga diffuses much further into the buffer layer, its concentration still falling 7.5 μm away from the interface. By growing a sufficiently thick buffer (~8 μm), however, it has been possible to obtain Ga levels in the active CMT layer that are measured to be less than 0.01 ppma.

Figure 1 shows that for growth on to (111) GaAs, the Ga level falls much more rapidly at the substrate/buffer interface to a measured concentration of 0.01 ppma only 1 μm away from the interface. This very low Ga level is maintained when an active CMT layer is grown subsequent to the buffer.

A thinner buffer layer is clearly sufficient to prevent significant Ga diffusion from the substrate when a CdTe layer is used to buffer (111) GaAs substrates compared to the relatively thick high x $Cd_xHg_{1-x}Te$ buffer that is required to buffer (100) GaAs substrates.

Electrical Properties of the Buffer Layer

The electrical properties of the buffer layer determine its suitability for Hall measurements of the active CMT layer and predict the behaviour of a detector structure grown in this way. Capacitance voltage measurements were made on a bevelled edge through the CdTe buffer used for (111) GaAs and through the high x $Cd_xHg_{1-x}Te$ buffer on (100) GaAs. The CdTe buffer was shown to be of high resistivity similar to MOVPE CdTe layers grown on to (111) CdTe substrates; the carrier concentration is estimated to be below 10^{12} carriers/cc. The high x $Cd_xHg_{1-x}Te$ buffer layer is however measured to be conducting n type, the carrier concentration in the vicinity of the GaAs/buffer interface being approximately 10^{18} carriers/cc. This carrier concentration is of a similar magnitude to the Ga level as detected by SIMS. The electrical properties of the high x buffer appear to be dominated by the high Ga concentration and would not be suitable for Hall measurements of the

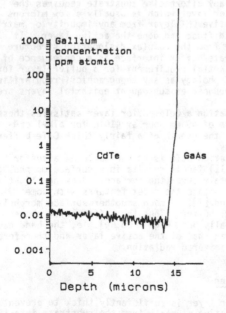

Fig 1. SIMS depth profile of ^{69}Ga for a 15 μm CdTe buffer grown on to a (111)B GaAs substrate.

CMT layer due to the conducting path through the buffer layer. The conducting buffer layer would also constrain possible detector structures fabricated from CMT prepared in this way.

Surface Morphology of the Active CMT Epilayer

The surface morphology of the CMT epitaxial layer is strongly dependent on orientation. Figure 2(a) is an optical micrograph of a CMT layer grown on to a buffered (100) GaAs substrate. The surface morphology is very similar to that of IMP layers grown on to CdTe substrates (fig 2(b)). The morphology consists of a low density of pyramid features ($\sim 10^2$ cm^{-2}) with the epitaxial layer between these pyramids being slightly rippled. These pyramid features nucleate at the substrate/buffer interface at certain substrate defects. As their size is proportional to the total thickness of the epilayer they are potentially more troublesome for buffered layers grown on to GaAs than for unbuffered layers grown on to CdTe substrates.

Figure 2(c) is an optical micrograph of a CMT layer grown on to a buffered (111)B GaAs substrate. The surface morphology is comparable to that of a similar CMT layer grown on to a (111)B CdTe substrate (fig 2(d)). The morphology is markedly different to that for (100) oriented layers with an

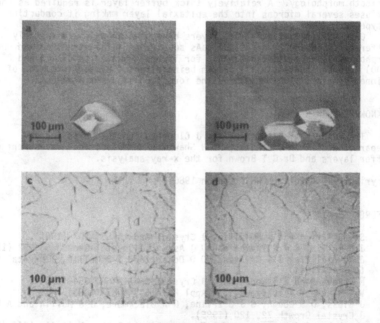

Fig 2. Optical micrographs with Nomarski contrast for 10-15 μm thick Cd$_x$Hg$_{1-x}$Te epilayers (x ~ .21) grown on to a) buffered 2° off (100) GaAs, b) 2° off (100) CdTe, c) buffered (111)B GaAs and d) (111)B CdTe.

absence of large pyramid features. The epitaxial layer consists of large domains; the epitaxial layers have been shown to be twinned with the domains rotated 180° with respect to each other [17]. The very smooth surface morphology obtainable for growth on to (111) substrates would facilitate device fabrication. It is however not yet known whether the twin boundaries are electrically active or whether their presence has a detrimental effect on device performance.

CONCLUSIONS

The orientation of CdTe layers grown by MOVPE on to (100) GaAs substrates is dependent on both the growth conditions and substrate treatment prior to growth. (111) oriented layers will only be nucleated if conditions are chosen that allow the formation of a Ga-As-Te interfacial layer; Te must be adsorbed by the GaAs substrate during a heat clean prior to growth and the epitaxial layer must be nucleated under Te rich growth conditions. Growth of (111) layers on to (111)B GaAs substrates, however, appears to be a more reproducible and practical route.

For growth on to GaAs substrates the selection of substrate orientation limits a) the choice of buffer layer composition, b) the conductivity and c) the extent of Ga diffusion. For (111) growth CdTe is suitable for the buffer; it has a very smooth surface morphology, grows with a high electrical resistivity and confines the Ga within 1 μm of the GaAs/buffer interface. For (100) growth a high x $Cd_xHg_{1-x}Te$ buffer must be grown in order to obtain a smooth morphology. A relatively thick buffer layer is required as the Ga diffuses several microns into the epitaxial layer making it conducting n type.

Good quality epitaxial CMT layers have been grown on to suitably buffered (111)B and 2° off (100) GaAs substrates. The structure and surface morphology is very similar to that for layers grown on to (111)B and 2° off (100) CdTe substrates; (111) growth being effected by the formation of twinned domains, (100) growth by the formation of pyramid defects.

ACKNOWLEDGEMENTS

The authors wish to thank Mrs J Clements for the polishing and preparation of the substrates, Mr N Shaw for the electrical assessment of the buffer layers and Dr G T Brown for the x-ray analysis.

REFERENCES

1   S J C Irvine and J B Mullin, J Crystal Growth 55, 107 (1981).
2   J B Mullin, S J C Irvine and D J Ashen, J Crystal Growth 55, 92 (1981).
3   J Tunnicliffe, S J C Irvine, O D Dosser and J B Mullin, J Crystal Growth 68, 245 (1984).
4   M J Bevan and K T Woodhouse, J Crystal Growth 68, 254 (1984).
5   W E Hoke and R Traczewski, J Appl Phys 54, 5087 (1983).
6   J Giess, J S Gough, S J C Irvine, G W Blackmore, J B Mullin and A Royle, J Crystal Growth 72, 120 (1985).
7   W E Hoke, P J Lemonias and R Traczewski, Appl Phys Lett 44, 1046 (1984).
8   S K Ghandhi, I B Bhat and N R Tasker, J Appl Phys 59, 2253 (1986).
9   H S Cole, H H Woodbury and J F Schetzina, J Appl Phys 55, 3166 (1984).
10 W E Hoke, R Traczewski, V G Kreismanis, R Korenstein and P J Lemonias, Appl Phys Lett 47, 276 (1985).
11 J Thompson, K T Woodhouse and C Dineen, J Crystal Growth, in press.

12   R D Feldman, R F Austin, D W Kisker, K S Jeffers and P M Bridenbaugh,
     Appl Phys Lett 48, 248 (1986).
13   P Y Lu, L M Williams, S N G Chu, J Vac Sci Technol A4, 2137 (1986).
14   P L Anderson, J Vac Sci Technol A4, 2162 (1986).
15   P F Fewster and P A C Whiffin, J Appl Phys 54, 4668 (1983).
16   J Giess, S J Barnett, G T Brown, D C Rodway and S J C Irvine, to be
     published.
17   J E Hails, G J Russell, A W Brinkman and J Woods, J Crystal Growth,
     in press.

Hg$_{1-x-y}$Cd$_x$Zn$_y$Te: GROWTH, PROPERTIES AND POTENTIAL FOR INFRARED DETECTOR APPLICATIONS

DEBRA L. KAISER[*] AND PIOTR BECLA[**]

[*] IBM Thomas J. Watson Research Center, Yorktown Heights, NY 10598
[**] MIT Francis Bitter National Magnet Laboratory, Cambridge, MA 02139

ABSTRACT

Close-spaced isothermal vapor phase epitaxy (VPE) was used to grow quaternary Hg$_{1-x-y}$Cd$_x$Zn$_y$Te epilayers on Cd$_{1-z}$Zn$_z$Te substrates. Composition, resistivity, and carrier concentration depth profiles were determined in the epilayers. p-n junctions were produced from material with appropriate properties using the Hg diffusion method. The junctions showed excellent I-V characteristics and high spectral detectivities.

INTRODUCTION

The quaternary alloys Hg$_{1-x-y}$Cd$_x$Zn$_y$Te have considerable potential for infrared applications. Hg$_{1-x}$Cd$_x$Te is currently used for infrared detectors, however, Cd destabilizes weak Hg-Te bonds [1,2]. This leads to low mechanical strength [3] and long-term changes in transport properties [4] of these alloys. Zinc is reported to strengthen the CdTe [5,6] and HgTe [2] lattices. Thus, Zn additions to Hg$_{1-x}$Cd$_x$Te may stabilize the Cd-Te and Hg-Te bonds, resulting in improved structural and mechanical properties. Furthermore, the semiconducting properties of Hg$_{1-x-y}$Cd$_x$Zn$_y$Te are expected to be similar to those of Hg$_{1-x}$Cd$_x$Te.

In this communication, we report on the growth conditions and electronic properties of quaternary Hg$_{1-x-y}$Cd$_x$Zn$_y$Te epilayers and their potential for device applications. p-n junctions were fabricated from this material and associated electrical and photoelectrical properties were determined.

EXPERIMENTAL DETAILS

The Hg$_{1-x-y}$Cd$_x$Zn$_y$Te epitaxial layers were grown on Cd$_{1-z}$Zn$_z$Te substrates using a two-zone, close-spaced isothermal vapor phase epitaxy method described in detail elsewhere [7]. The sample and HgTe source material were held in a graphite crucible at one end of a sealed quartz ampule and pure mercury was contained in the other end of the ampule. Isothermal experiments were carried out at growth temperatures, T_o, ranging from 600 to 640°C for times of 10 to 100 hours. The Cd$_{1-z}$Zn$_z$Te substrate compositions ranged from z = 0.05 to 0.52. For each T_o and substrate composition, a series of runs were performed at various mercury temperatures, T_{Hg}, providing mercury pressures in the range of 1.0 to 5.0 atm. Epilayers with preselected compositions and electronic properties were used to fabricate p-n junctions. Junctions were made by annealing as-grown p-type samples in Hg-saturated atmospheres [8]. The diffusion process was carried out at 305°C for 0.5 to 1.0 hr.

RESULTS AND DISCUSSION

Characterization of the epilayer material is described first, including composition profiles and electronic properties. Electrical and photoelectrical characteristics of p-n junctions made from the epilayers are then discussed in detail.

Epilayers

The photomicrograph in Fig. 1 shows a typical cross-section of a $Hg_{1-x-y}Cd_xZn_yTe$ epilayer grown on a $Cd_{1-z}Zn_zTe$ substrate. The distinct boundary between the epilayer and substrate allows an accurate measurement of the thickness of the epilayer. Composition profiles for Hg, Cd, Zn and Te shown in Fig. 2 were obtained using wavelength dispersive X-ray analysis on an electron microprobe. The epilayers had composition gradients due to interdiffusion of Hg from the deposited material and Cd and Zn from the ternary substrate. The thickness of the epilayer as determined from the compositional analysis (Fig. 2) corresponded precisely with the thickness measured from the photomicrograph (Fig. 1). Consequently, the epilayer region identified in the micrograph defines the combined thickness of the deposited material and the depth of interdiffusion in the CdZnTe substrate.

The sharpness at the epilayer/substrate interface and the shapes of the composition profiles indicate that diffusion occurred more rapidly in the quaternary phase than in the ternary substrate. The high diffusivity in the epilayer is probably due to a high concentration of vacancies on the metallic sublattice which is typical of $Hg_{1-x}Cd_xTe$ grown by VPE [9]. Although it is difficult to calculate the interdiffusion coefficient in complex systems, the profiles in Fig. 2 indicate that the diffusivity of Cd is higher than that of Zn in this quaternary system.

The energy band gap in the $Hg_{1-x-y}Cd_xZn_yTe$ epilayers varied along the growth direction. For example, for the epilayer in Fig. 2, the energy band gap at room temperature varied from approximately 50 meV at the surface to 1.8 eV at the interface with the substrate. Our range of interest for device applications was from 120 to 700 meV. Step-etching [7] was used to evaluate the energy gap, conductivity type and carrier concentration

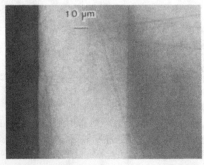

epilayer   substrate

Fig. 1.   Polished, unetched cross-section of a $Hg_{1-x-y}Cd_xZn_yTe$ epilayer on a $Cd_{1-z}Zn_zTe$ substrate.

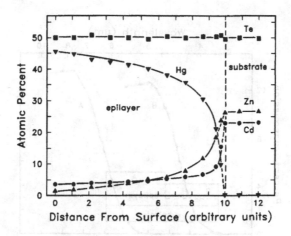

Fig. 2. Composition profiles in a $Hg_{1-x-y}Cd_xZn_yTe$
epilayer grown on a $Cd_{0.48}Zn_{0.52}Te$ substrate
at 600°C under a mercury pressure of 1.5 atm.
1 distance unit = 5.5 μm.

profiles in the epilayers. In each step, the energy band gap was determined
from the optical transmission cutoff wavelength and the conductivity type,
resistivity and carrier concentration were determined from Hall effect and
resistivity measurements.

Typical optical transmission data for two slices cut from the central
portion of an epilayer are shown in Fig. 3. The cutoff wavelengths for
slices A and B at room temperature were 7.1 and 4.8 μm, respectively. These
cutoffs were very sharp, indicating good radial compositional homogeneity.
The thickness of sample A was approximately 50 μm, so the gradient of the
energy gap in this region was 1.7 meV/μm.

For slices A and B, temperature-dependent Hall coefficient and
resistivity results are shown in Fig. 4. In sample A, a transition from p-
to n-type conductivity occurred at 260 K. In sample B, which is a wider gap
material, p-type conductivity was observed up to 300 K. The acceptor
concentration and hole mobility at 77 K calculated from the Hall coeffi-
cient and resistivity data were 4.3×10^{17} cm^{-3} and 86 $cm^2/V \cdot s$ for sample A
and 1.5×10^{17} cm^{-3} and 114 $cm^2/V \cdot s$ for sample B.

Samples A and B were annealed at 305°C under a mercury pressure of
0.2 atm for one hour. These conditions should produce a near-surface n-type
layer with a thickness of about 10 μm. Typical Hall coefficient and resistivity results for this near-surface region are shown in Fig. 5. As
expected, both samples showed n-type conductivity from 77 to 300 K.
Annealing also caused a decrease in resistivity of one to two orders of
magnitude. The calculated carrier concentration and mobility at 77 K
were 2.4×10^{16} cm^{-3} and 1.2×10^5 $cm^2/V \cdot s$ for sample A and 2.5×10^{16} cm^{-3} and
2.6×10^4 $cm^2/V \cdot s$ for sample B.

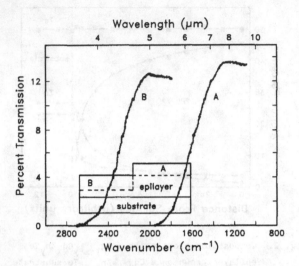

Fig. 3. Optical transmission data at 300 K for two slices A and B from the central part of a thick epilayer.

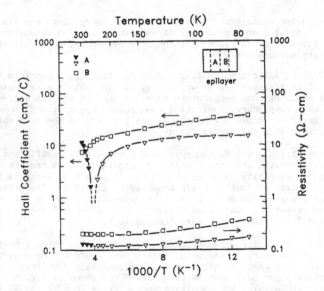

Fig. 4. Temperature-dependent Hall coefficient and resistivity results for the two epilayer slices A and B (see Fig. 3). Closed and open symbols are for n- and p-type conductivity, respectively.

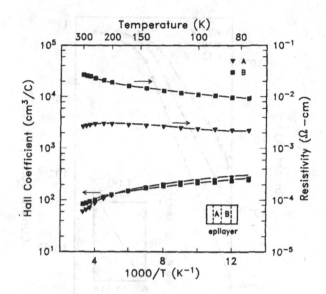

Fig. 5. Hall coefficient and resistivity as a function of temperature for the near-surface n-type region in annealed epilayer samples A and B.

p-n Junctions

Typical current-voltage (I-V) characteristics as a function of temperature for a mesa-type junction made from $Hg_{1-x-y}Cd_xZn_yTe$ epilayers are shown in Fig. 6. The cutoff wavelength of the junction (PV-2) was 2.2 µm at 77 K. The forward I-V characteristics fit the standard diode equation

$$I = I_{sat}[\exp(qV/\beta kT) - 1] \tag{1}$$

where the coefficient β varied from 1.6 to 1.9 in the temperature range 63 to 131 K. Since $\exp(qV/\beta kT) \gg 1$, the saturated current, I_{sat}, could be determined by extrapolating the linear portion of the forward I-V curves for various temperatures to zero voltage. These values of I_{sat} were used in the equation [10]

$$R_o = \left(\frac{\partial V}{\partial I}\right)_{V=0} = \frac{\beta kT}{qI_{sat}} \tag{2}$$

to calculate zero-bias resistivity, R_o, versus temperature. For the junction area, A, of 3.5×10^{-3} cm², the temperature-dependent R_oA values are shown in the inset in Fig. 6. These R_oA products are extremely large.

The spectral sensitivity S_λ and the detectivity D_λ^* were determined from the standard photoelectrical equations

$$S_\lambda = \left(\frac{V_D}{V_T}\right)\left(\frac{A_T}{A_D}\right)S_T \tag{3}$$

402

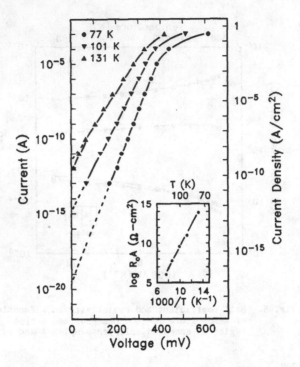

Fig. 6. Current-voltage curves as a function of
temperature for the p-n junction (PV-2)
with a cutoff wavelength of 2.2 μm at
77 K. The inset shows the temperature
dependence of the R_oA product.

and

$$D_\lambda^* = S_\lambda (A_D \, \Delta f)^{\frac{1}{2}} / V_N \qquad (4)$$

where (V_D/V_T) is the ratio of the detector-to-thermocouple voltage signals,
(A_T/A_D) is the ratio of surface areas of the thermocouple and detector, S_T
is the absolute sensitivity of the Zeiss VTH-1 vacuum thermocouple at 12 Hz,
and V_N is the noise voltage in the bandwidth Δf. It was assumed that under
zero applied voltage the noise V_N of the junction was thermally limited;

$$V_N = (4kT \, R_o \, \Delta f)^{\frac{1}{2}} \qquad (5)$$

Typical spectral D_λ^* characteristics of p-n junctions made in $Hg_{1-x-y}Cd_xZn_yTe$
are presented in Fig. 7. The field of view of the diodes was 60°, the
chopping frequency was 12 Hz and the background temperature was 300 K.
Under these conditions, the detectivities of junctions PV-2, PV-5 and PV-8
at 77 K were 7.6×10^{11}, 2.7×10^{11} and 1.2×10^{10} cm $Hz^{\frac{1}{2}}W^{-1}$, respectively. These
detectivities were comparable to those of high-quality $Hg_{1-x}Cd_xTe$ junctions
[11].

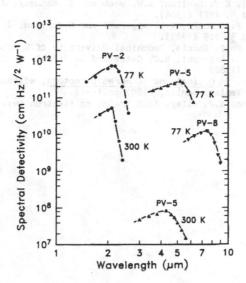

Fig. 7. Spectral detectivities of three p-n
junctions at 300 and 77 K.

CONCLUSIONS

Quaternary $Hg_{1-x-y}Cd_xZn_yTe$ epilayers were grown on $Cd_{1-z}Zn_zTe$
substrates by the close-spaced isothermal VPE method. The optical and
electrical properties showed that the material was of high quality. p-n
junctions were constructed from epilayers with suitable compositions and
electrical properties. These junctions had excellent I-V characteristics
and high spectral detectivities, demonstrating that $Hg_{1-x-y}Cd_xZn_yTe$ is a
potential material for infrared device applications.

ACKNOWLEDGEMENTS

The authors acknowledge the support of DARPA through Contract No.
N00014-83-K-0454.

REFERENCES

1.   A. Wall, C. Caprile, A. Franciosi, R. Reifenberger and U. Debska, J.
      Vac. Sci. Technol. A 4, 818 (1986).
2.   A. Sher, A.B. Chen, W.E. Spicer and C.K. Shih, J. Vac. Sci. Technol.
      A 3, 105 (1985).
3.   W.E. Spicer, J.A. Silberman, I. Lindau, A.B. Chen, A. Sher and J.A.
      Wilson, J. Vac. Sci. Technol. A 1, 1735 (1983).
4.   G. Nimtz, B. Schlicht and R. Dornhaus, Appl. Phys. Lett. 34, 490
      (1979).
5.   S.B. Qadri, E.F. Skelton, A.W. Webb and J. Kennedy, Appl. Phys. Lett.
      46, 257 (1985).

404

6. S.B. Qadri, E.F. Skelton, A.W. Webb and J. Kennedy, J. Vac. Sci. Technol. A $\underline{4}$, 1971 (1986).
7. P. Becla, P.A. Wolff, R.L. Aggarwal and S.Y. Yuen, J. Vac. Sci. Technol. A $\underline{3}$, 119 (1985).
8. P. Becla, Ph.D. Thesis, Technical University of Wroclaw, 1976.
9. P. Becla, J. Lagowski, H.C. Gatos and L. Jedral, J. Electrochem. Soc. $\underline{129}$, 2855 (1982).
10. P.W. Kruse, in Optical and Infrared Detectors, edited by R.J. Keyes, (Springer-Verlag, Berlin, 1980) pp.14-17.
11. M. Lanir and K.J. Riley, IEEE Trans. on Electron Devices $\underline{ED-29}$, 274 (1982).

HgTe-CdTe SUPERLATTICES GROWN BY PHOTO-MOCVD

William L. Ahlgren,* J.B. James,* R.P. Ruth,* E.A. Patten,* and
J.-L. Staudenmann**
* Santa Barbara Research Center, 75 Coromar Drive, Goleta, CA 93117
**Ames Laboratory and Department of Physics, Iowa State University, Ames,
IA 50011

ABSTRACT

HgTe-CdTe superlattices have been grown, for the first time, by photo-assisted MOCVD. The substrate temperature was 182°C. Superlattices were obtained despite low growth rates requiring long growth times (~10 hours). Interdiffusion during growth may be slowed down by growing under saturated Hg vapor to minimize cation-vacancy formation. The nominal superlattice structures were 70Å HgTe-30Å CdTe, 40Å HgTe-40Å CdTe, and similar. Actual superlattice structures were verified by cross-sectional TEM and diffractometer x-ray diffraction patterns. The x-ray diffraction patterns showed satellite peaks up to third order. The actual structures had HgTe layers ~20% thicker than the nominal (target) values. A grid-like array of dislocations at the substrate-epilayer interface, suggesting operation of a dislocation-blocking mechanism, was observed. Deficiencies in the superlattice growths include a low growth rate, nonuniform layers, high dislocation density (~10^8 cm^{-2} in best layers), and high n-type carrier concentration (~10^{18} cm^{-3} with mobilities up to 3.5×10^4 cm^2 V^{-1} s^{-1} in the best layer) which may reflect the presence of donor impurities in the material.

APPARATUS AND GROWTH CONDITIONS

The reactor used is shown schematically in Figure 1. It is a quartz reactor with horizontal flow, containing a graphite susceptor in which a thermocouple is embedded for temperature control. The susceptor is heated from below by a quartz-halogen lamp, as described by Beneking, et al. [1]. The substrate, mounted on the susceptor, can be illuminated from above by a 1000W HgXe short-arc lamp. The UV illumination passes through a suprasil window which can be kept free of deposit by a flow of pure carrier gas. The spectrum of the HgXe lamp is modified by reflecting it from a dichroic mirror, with the result that most of the radiation reaching the substrate is in the 200 to 250 nm range.

The reactants used were elemental mercury, dimethyl cadmium (DMC), and either diethyl tellurium (DET) or di-isopropyl tellurium (DIPT). Palladium-diffused hydrogen was used as the carrier gas. Elemental mercury was introduced into the reactor just upstream from the substrate and was kept at the same temperature as the substrate, so that the gas flowing over the substrate was saturated with mercury vapor. All superlattice growths were carried out at a substrate temperature of 182°C.

Growth of HgCdTe under Hg-saturated conditions and low temperatures is desired in order to minimize the formation of cation vacancies in the lattice, which act as acceptor defects resulting in p-type conductivity [2], and whose presence also enhances Hg self-diffusion in the lattice [3]. Thus, it is desirable to suppress the formation of cation vacancies so that the conductivity type can be controlled by intentionally added dopants rather than native point defects, and also to reduce interdiffusion rates in multilayer structures.

The reactor design used is intended to provide a Hg-saturated vapor phase adjacent to the substrate while minimizing Hg consumption during a growth run. The Hg vapor cannot be switched on and off, but is continually

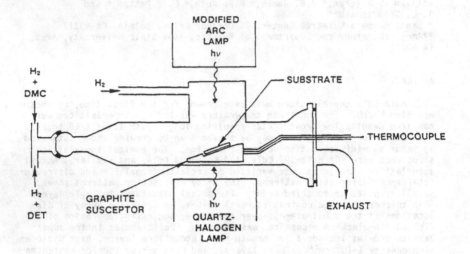

Figure 1.   Reactor Configuration for Photo-Assisted MOCVD

present at a uniform level. The flow of the tellurium-bearing reactant (DET or DIPT) is also constant. The flow of the cadmium-bearing reactant (DMC) is, however, interrupted by switching between vent and run. As Mullin and Irvine [4] have reported, CdTe is deposited rather than HgCdTe when DMC is present in the reactor in significant concentrations, even though Hg vapor may be present in a far greater concentration. This is consistent with the fact that the free energy of formation of CdTe is much more negative than HgTe, providing a stronger driving force for crystallization [4]. Thus, a superlattice structure can be grown by switching only the DMC flow between vent and run.

A principal problem experienced with the apparatus and growth conditions described above is low growth rate. The growth rate at low temperature (182°C) is believed to be limited by very inefficient decomposition of the tellurium-bearing reactant [5,6]. Using DET, growth rates were <0.1 μm h^{-1}, so that growth runs of 6 to 12 hours duration were needed at 182°C to produce layers thick enough for characterization. Using DIPT, growth rates were somewhat higher, consistent with the report of Hoke and Lemonias [7], but were still <0.25 μm h^{-1} so that growth runs of at least 6 to 8 hours duration at 182°C were still required. Despite the low growth rate and consequent long growth times, superlattice structure has been observed in the layers grown.

SURFACE MORPHOLOGY AND STRUCTURE OF SUPERLATTICE LAYERS

The surface morphology of one of the best superlattice layers grown is shown in Figure 2. This layer is macroscopically specular, but has a faceted surface when viewed in the SEM. The substrate was a CdTe wafer oriented to a {100} direction. Superior surface morphology, which would be flat and featureless at 10,000x (as in Figure 2) has not yet been achieved.

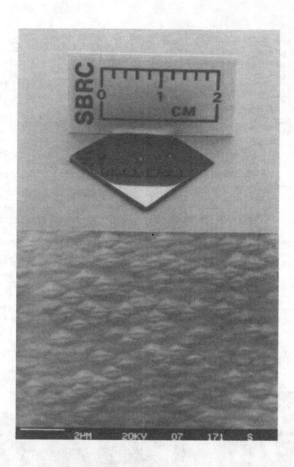

Figure 2. Surface Morphology of HgTe-CdTe Superlattice Run No. 333, Grown by Photo-MOCVD at 182°C on a CdTe {100} Substrate. The nominal structure was 70A HgTe-30A CdTe. The actual structure is shown in Figure 3.

A TEM cross section of the same layer is shown in Figure 3. The CdTe layers appear bright in this figure, and the HgTe layers are dark. The layer thicknesses can be measured; they are 30Å CdTe and 90Å HgTe. The layer thicknesses which had been aimed for during growth were 30Å CdTe and 70Å HgTe; thus, the HgTe growth rate was higher than expected. Although not clearly visible in Figure 3, at lower magnification layer waviness associated with the faceted surface morphology is evident.

A TEM cross section of a different layer, grown on a $Cd_{0.96}Zn_{0.04}Te$ {111}B substrate, is shown in Figure 4. Again the structure was nominally 70Å HgTe-30Å CdTe and this structure can be measured in part of the figure, but the layers are non-uniform in thickness. In other portions of this layer even greater non-uniformity was observed. Poorer surface morphology and higher dislocation densities were obtained with {111}-oriented substrates than with {100}.

The superlattice structure is confirmed by diffractometer x-ray diffraction patterns. An example is shown in Figure 5, which was made from the same superlattice shown in Figures 2 and 3. This superlattice consists of 58 periods grown on a CdTe {100} substrate in 11.2 hours at 182°C. A high-resolution single-crystal x-ray diffractometer, using a monochromator described in reference [8] together with a rotating-anode x-ray source and a Huber four-circle diffractometer, was used to produce these diffraction patterns. The incident wavelength was set to 1.47639Å (W-Lαl) and extended θ - 2θ scans were performed on the (0,0,2), (0,0,4), and (0,0,6) reflections. The (0,0,2) scan shown in the figure has three orders of satellite peaks. The data analysis was done as in Knox, et al. [9], yielding a period of 119.4Å, in good agreement with the TEM cross-section in Figure 3. A quantitative estimate of the degree of interdiffusion is still lacking, but the superlattice structure is clearly evident.

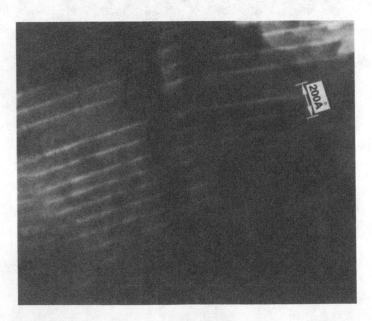

Figure 3. TEM Cross Section of HgTe-CdTe Superlattice Shown in Figure 2, Run No. 333

Figure 4. TEM Cross Section of HgTe-CdTe Superlattice Run No. 328, Nominally 70A HgTe-30A CdTe

NOMINAL SUPERLATTICE STRUCTURE 70A HgTe — 30A CdTe
STRUCTURE DETERMINED BY TEM: 90A HgTe — 30A CdTe
SUPERLATTICE PERIOD FROM POSITION OF SATELLITE PEAKS (BELOW): 119.4A

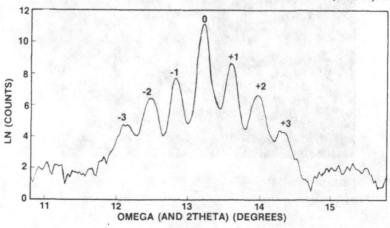

Figure 5. Diffractometer X-ray Diffraction Pattern of HgTe-CdTe Superlattice Run No. 333

410

The TEM examination of the superlattices reveals an interesting feature which is illustrated in Figure 6. This shows the same superlattice layers as in Figures 2, 3, and 5, consisting of 58 periods of 120A each (90A HgTe-30A CdTe) grown on a CdTe {100} substrate. A grid-like array of dislocations has formed, which lies in a plane near the substrate-epilayer interface. This is reminiscent of the pattern expected when dislocation blocking occurs. There is a critical thickness which occurs during epitaxial layer growth on a lattice-mismatched substrate, whereby threading dislocations from the substrate are bent into the plane of the interface, thus helping to accommodate the misfit and simultaneously being removed from the epitaxial layer. This phenomenon is known as dislocation blocking. It is well known that such a mechanism can in principle be used to produce dislocation-free epitaxial layers [10,11]. To our knowledge this is the first observation of the phenomenon in a HgTe-CdTe superlattice structure (see James and Stoller [12] for previous observations in a HgCdTe epitaxial layer grown by LPE). The operation of the blocking mechanism is shown in Figure 7. This figure is one member of a stereo pair from which the three-dimensional dislocation structure has been determined. A dislocation is seen emerging from the substrate which has bent over into the plane of the substrate-epilayer interface (feature C in the figure). The dissociation of a dislocation into two partial dislocations, which define the edges of a stacking fault is seen in the figure (feature D). Also visible are two dislocation half-loops which nucleated at the top of the epilayer and moved down into the epitaxial layer (features A and B). Possibly because of such incipient misfit dislocations, the dislocation density in the epitaxial layer ($\sim 10^8$ cm^{-2}) is higher, not lower, than the expected dislocation density in the substrate ($\sim 10^6$ cm^{-2}). By the use of a properly designed graded buffer layer, however, it may be possible to reduce the dislocation density in the epitaxial superlattice layer below that in the substrate [13].

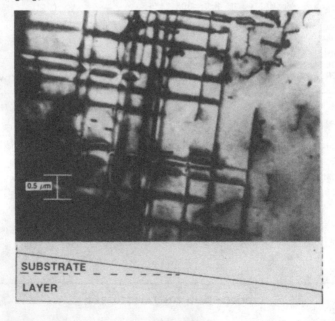

Figure 6. Grid-like Dislocation Network at Interface Between
Substrate and Layer in Run No. 333

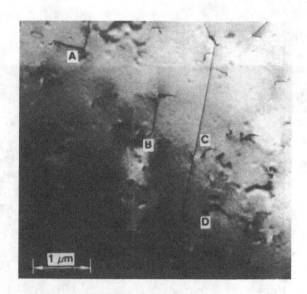

Figure 7.   Dislocation Structures in Run No. 333. A) Dislocation generated at layer surface propagating part way to substrate/layer
interface; B) Dislocation generated at layer surface which has
propagated to the substrate/layer interface; C) Threading dislocation bent over into substrate/layer interface; D) Single
dislocation generated at substrate/layer interface split into
two partial dislocations with stacking fault.

ELECTRO-OPTICAL PROPERTIES

The electrical properties of the superlattices were measured by determining the Hall coefficient and resistivity in a standard van der Pauw configuration. No special care was taken in forming the contacts to the superlattice, which were assumed to have connected all the superlattice layers in parallel. At a temperature of 77K most of the superlattices measured
were n-type with a carrier concentration of about 1.0×10^{18} cm^{-3}. The mobilities varied markedly, from 5×10^2 to 4×10^4 cm^2 V^{-1} s^{-1}. The variation in electron mobility could correlate with dislocation density, as a
similar variation in defect structure of different layers was observed. An
example is presented in Figure 8, which shows TEM micrographs of two superlattices. Dislocation density in the superlattice on the left is 10^8 cm^{-2},
and the electron mobility is 1.4×10^4 cm^2 V^{-1} s^{-1}. Dislocation density in
the superlattice on the right is 10^{10} cm^{-2}, and the electron mobility is
4.7×10^2 cm^2 V^{-1} s^{-1}. This observation merely confirms that high dislocation density adversely affects electron mobility, as expected.

The observed electron concentrations at 77K ($\sim 10^{18}$ cm^{-3}) are higher
than expected for intrinsic HgTe at this temperature ($\sim 7 \times 10^{16}$ cm^{-3}
[14]). The 300K resistivity, 3.5×10^{-4} Ω-cm, is about 50 times lower than
the value reported in [15] for bulk HgTe annealed in saturated Hg vapor at
300°C. This might be due to donor impurities in the HgTe layers, or to
transfer of electrons from the relatively high-energy conduction band of
the CdTe layers down into the "wells" formed by adjacent HgTe layers, as
occurs in modulation-doped superlattices [16]. The latter assumes there

412

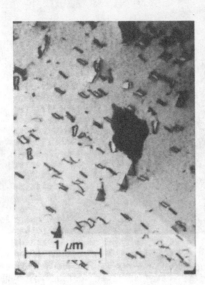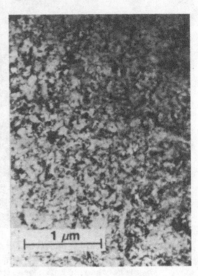

Figure 8. Superlattices with Widely Different Dislocation Densities as
Viewed in TEM. A) Run No. 325, dislocation density ~10^8 cm^{-2};
B) Run No. 327, dislocation density ~10^{10} cm^{-2}.

are donor impurities in the CdTe layers. The estimates of carrier concentration and mobility are based on simple analysis of the Hall effect data
($n = 1/(qR_H)$; $\mu = R_H/\rho$). Complementary measurements of electrical properties are needed, as well as analysis of layer purity.

Because growth takes place at low temperature under Hg-saturated
vapor, the formation of cation-vacancies (which would tend to make the layers p-type) is minimized. If the layers really do grow in equilibrium with
saturated Hg vapor, then a subsequent anneal at the growth temperature in
an evacuated and sealed quartz ampoule to which an excess of Hg has been
added should not alter the electrical properties. Figure 9 shows the measured carrier concentrations and mobilities of four superlattice layers in
the as-grown condition and after annealing for 2, 4, and 8 hours at the
growth temperature (182°C) in saturated Hg vapor. The electrical properties do not change significantly, consistent with (but not proving) the hy-
pothesis that growth occurs in equilibrium with saturated Hg vapor. It is
assumed that the interdiffusion which occurs during these anneals is not
overwhelming, but this has not yet been confirmed.

The infrared transmission and reflection of one of the superlattices
is shown in Figure 10. This particular superlattice consists of 129 periods, each nominally 39A HgTe-39A CdTe on a CdTe {111} substrate. TEM
cross-section measurements show that the actual layer thicknesses are about
80A HgTe-30A CdTe. The overall epilayer thickness is then 1.4 μm. The interference pattern in the reflectance curve confirms that the layer thickness is about 1.4 μm, assuming an index of refraction n ≈ 4, which is the
value for HgTe. The positions of the superlattice bandgap calculated for
the nominal superlattice structure [17], and of the homogeneous alloy [18]
that would result from complete interdiffusion, are shown for comparison.

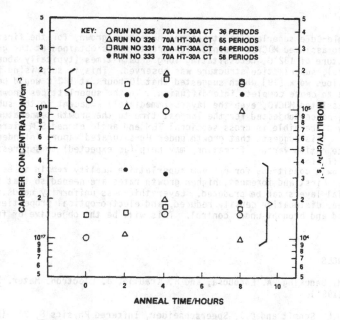

Figure 9. Carrier Concentration and Mobility as a Function of Anneal Time for Four Different Superlattices, All with the Same Nominal Structure

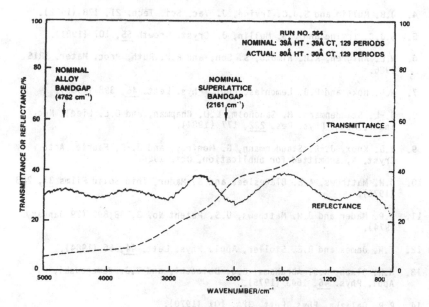

Figure 10. Infrared Transmittance and Reflectance of HgTe-CdTe Superlattice, Run No. 364

414

CONCLUSION

HgTe-CdTe superlattice structures have been grown, for the first time, by photo-assisted MOCVD. Despite low growth rates obtained at the growth temperature of 182°C, necessitating long growth times (typically about 10 hours), superlattice structure was observed. This is surprising in view of previous work [19] which suggested that 10 hours at 182°C would be sufficient to cause complete interdiffusion. In the superlattices grown by photo-assisted MOCVD, even the layers immediately adjacent to the substrate, hence subjected for the longest time to the growth temperature, are clearly discernible in cross sectional TEM and angle etched SEM micrographs. This suggests that growth under Hg-saturated vapor, intended to minimize cation-vacancy formation, may help (as expected) to suppress layer interdiffusion.

Growth conditions for optimum superlattice quality remain to be established. First and foremost, higher growth rates are needed so that thicker epitaxial layers can be produced. Layer thickness uniformity needs to be improved, dislocation density reduced, and electro-optical properties understood and brought under control. This will be the objective of future work.

REFERENCES

1. H. Beneking, A. Escobosa, and H. Krautle, J. Electron. Mater. 10, 473 (1981).

2. J.L. Schmit and C.J. Speerschneider, Infrared Physics 8, 247 (1968).

3. J.-S. Chen, Defect Studies on Crystalline Solids, PhD thesis, University of Southern California, 1985.

4. J.B. Mullin and S.J.C. Irvine, J. Vac. Sci. Tech. 21, 178 (1982).

5. S.J.C. Irvine and J.B. Mullin, J. Cryst. Growth 55, 107 (1981).

6. W.L. Ahlgren, R.H. Himoto, S. Sen, and R.P. Ruth, Proc. Mater. IRIS, 1986.

7. W.E. Hoke and P.J. Lemonias, Appl. Phys. Lett. 46, 398 (1985).

8. J.-L. Staudenmann, M. Sandholm, L.D. Chapman, and G.L. Liedl, Nuc. Instr. Meth. Phys. Res. 222, 177 (1984).

9. R.D. Knox, J.-L. Staudenmann, G. Monfroy, and J.-P. Faurie, Acta Cryst. A, submitted for publication, Oct. 1986.

10. J.W. Matthews, A.E. Blakeslee, and S. Mader, Thin Solid Films 33, 253 (1976).

11. S.R. Mader and J.W. Matthews, U.S. Patent No. 3,788,890 (29 January 1974).

12. T.W. James and R.E. Stoller, Appl. Phys. Lett. 44, 56 (1984).

13. G.H. Olsen, M.S. Abrahams, C.J. Buiocchi, and T.J. Zamerowski, J. Appl. Phys. 46, 1643 (1975).

14. R.R. Galazka, Phys. Lett. 32A, 101 (1970).

15. T. Okazaki and K. Shogenji, J. Phys. Chem. Solids 36, 439 (1975).

16. R. Dingle, H.L. Störmer, A.C. Gossard, and W. Wiegmann, Appl. Phys. Lett. <u>33</u>, 665 (1978).

17. Calculated using code provided by J.N. Schulman, Hughes Research Laboratories.

18. G.L. Hansen, J.L. Schmit, and T.N. Casselman, J. Appl. Phys. <u>53</u>, 7099 (1982).

19. J.-L. Staudenmann, R.D. Horning, R.D. Knox, J.-P. Faurie, J. Reno, I.K. Sou, and D.K. Arch, Trans. Met. AIME, 1986. In press.

Epitaxial Growth II

MBE GROWTH OF MERCURY CADMIUM TELLURIDE: ISSUES AND PRACTICAL SOLUTIONS

J. W. COOK, JR., K. A. HARRIS, and J. F. SCHETZINA
North Carolina State University, Department of Physics, Raleigh, NC
27695-8202

ABSTRACT

The growth of thin films of mercury-based materials by molecular beam epitaxy (MBE) presents significant experimental problems which must be overcome in order to successfully grow infrared detector materials such as mercury cadmium telluride (MCT). Many of the problems associated with the use of Hg in MBE arise from its high room temperature vapor pressure (2 mTorr) and its low sticking coefficient. The MBE system must be designed for Hg usage by considering such things as the ultra high vacuum pumping system, the Hg source, Hg containment, and Hg removal. In addition, Hg is a toxic heavy metal and must be handled appropriately. Other problems involved with the growth of MCT are associated with the design of the MBE furnaces which are used to evaporate cadmium telluride and tellurium.

A system designed specifically for the growth of Hg-based materials has been designed and constructed at North Carolina State University. The features of this system which were included specifically to deal with the problems stated above will be discussed. In addition, information on the growth and characterization of MCT films deposited in the system will be presented.

INTRODUCTION

The first MBE growth of MCT by Faurie and Million[1] in 1981 demonstrated the feasibility of using this elegant thin film growth technique for depositing epitaxial layers of this important infrared detector material. However, there has been very little discussion in the literature of the significant practical problems which must be overcome in order to successfully deposit MCT films by MBE. In this paper the problems are identified and solutions are presented. The solutions to the problems are discussed from the experience obtained at North Carolina State University in designing and constructing a MBE system specifically for the growth of MCT and other Hg-based materials.

Many of the problems in the growth of MCT by MBE arise from the high vapor pressure of Hg at room temperature. Since MBE growth usually takes place at pressures in the 10^{-10} Torr range, the use of Hg necessitates the use of extensive liquid nitrogen shrouding for Hg containment and special sources which allow the Hg be confined when film growth is not in progress. In addition, care must be taken in choosing a suitable ultra high vacuum pumping system for the growth chamber. It should be noted that the ion pumping systems commonly used in the MBE growth of III-V materials can be severely damaged if Hg inadvertently enters the pumps.

Other problems arise because of the low sticking coefficient of Hg.

While Faurie and Million first overcame the high vapor pressure problem of Hg by using HgTe and a source of Hg, they found that it was necessary to use elemental Hg in order to obtain a large enough flux to compensate for the very low sticking coefficient of Hg[2]. Thus, an additional requirement on the design of the Hg source is that it be capable of delivering a large flux of Hg to the substrate. The Hg flux often needs to be 1000 times that of the Te or Cd sources.

An additional problem with using Hg in MBE is that a suitable way must be employed to remove the Hg from the growth chamber before the system is vented to the atmosphere in order to clean out the system and refill the evaporation sources. Two common solutions to this problem are the use of an external cold trap into which the system is baked for Hg removal or the use of a Hg diffusion pump to remove the Hg.

Not only are there problems with dealing with Hg in the growth of MCT, but there are also problems with the evaporation of Te and Cd (or CdTe). These problems arise from the high vapor pressures that these materials have at low temperatures. Because of this the furnaces used for evaporation must have excellent temperature control in order to assure a stable flux and no significant temperature gradient near the top of the furnace in order to prevent recondensation and clogging of the furnace.

In the next section, the Hg compatible MBE system that has been designed and constructed at North Carolina State University is described along with why certain features were incorporated in order to overcome the problems mentioned above. In the final section growth parameters and characterization data on MCT films deposited in the system are presented.

DESCRIPTION OF MBE SYSTEM

A block diagram showing the main components of the MBE system is shown in Figure 1. The MBE system was designed in a modular form to allow for easy modification of the components. The system consists of three vacuum chambers: the main growth chamber, a preparation chamber, and a load lock. Only aspects of the main chamber design were significantly modified for MBE of Hg-based materials. A schematic of the main chamber is shown in Figure 2. The source flange has positions for up to seven furnaces. None of the furnace ports is angled more than 12.5 degrees from the center line of the chamber. The source to substrate distance is 25 cm. Each furnace position has a computer operated shutter for control of the flux of the evaporated material. The sample mount and an ionization gauge are mounted on opposite sides of a rotatable plate so that either the sample mount or the ionization gauge can be moved into the

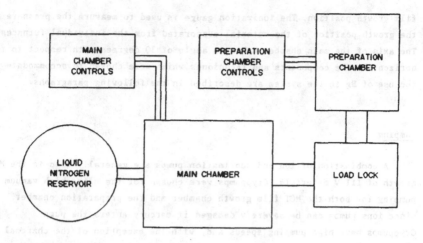

Figure 1. Block diagram of MBE system.

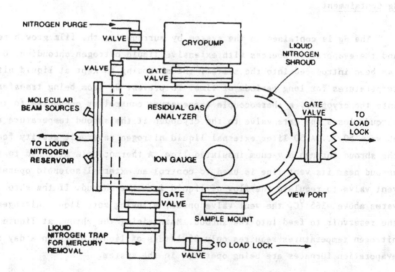

Figure 2. Schematic diagram of main chamber of MBE system.

film growth position. The ionization gauge is used to measure the pressure at the growth position of the material evaporated from the individual furnaces. The axis of the main chamber makes an angle of 30 degrees with respect to the horizontal. The components and techniques which were chosen to accommodate the use of Hg in the system are described in the following paragraphs.

Pumping

A combination of ion and sublimation pumps are generally used in the MBE growth of III-V materials. Cryopumps were chosen for the ultra high vacuum pumping for both the MCT film growth chamber and the preparation chamber since ions pumps can be severely damaged if mercury enters the pump. Cryopumps have high pumping speeds and, with the exception of the charcoal adsorber used for pumping hydrogen and helium, no parts in the vacuum system which are incompatible with Hg. This adsorber is replaceable.

Hg Containment

The Hg is contained in the system by surrounding the film growth region and the evaporation sources with extensive liquid nitrogen shrouding. Once Hg has been introduced into the system, the shrouding is kept at liquid nitrogen temperatures for long periods of time. To prevent Hg from being transferred into the cryopump, a thermocouple temperature controller connected to the shroud closes the gate valve to the cryopump if the shroud temperature rises above -150 $^{\circ}$C. A 50 liter external liquid nitrogen reservoir gravity feeds the shroud through a vacuum insulated line. A thermocouple attached to the shroud near its vent line is used to control an external solenoid operated vent valve to regulate the flow of liquid into the shroud. If the shroud warms above -165 $^{\circ}$C, the vent valve opens, allowing more liquid nitrogen from the reservoir to feed into the shroud. Maintaining the shroud at liquid nitrogen temperatures requires only 25 liters of liquid nitrogen a day if no evaporation furnaces are being operated in the system.

Hg Removal

The Hg is removed from the system by closing the gate valve to the cryopump, disconnecting the external liquid nitrogen reservoir from the main chamber, and baking the main chamber at 150 °C into an external liquid nitrogen cooled trap. A turbomolecular pump removes any gases which do not freeze on the liquid nitrogen cooled surfaces of the cold trap. The chamber is baked for two to three days to remove the Hg since the rate of removal of Hg is limited by the fine powder of Te and Cd which floats on the surface of the liquid Hg. At a bake-out temperature of 150 °C one would expect to be able to remove a few kilograms of Hg in several hours. Baking the chamber at a higher temperature may not be desirable since it can cause a significant redeposition of the Te and Cd. This may cause problems for bearing surfaces in the vacuum system (such as those in rotary feedthroughs). When the pressure in the main chamber falls to near 10^{-4} Torr, the cold trap is isolated by closing the connecting valves. The trap is then vented to the atmosphere, removed from the system, covered, and allowed to warm to room temperature in a fume hood. A mercury detector is used to confirm that all the Hg has been removed from the main chamber before the source flange is disassembled and cleaned.

Hg Source

The main features of the source include a shut-off valve to isolate the Hg reservoir from the film growth chamber, temperature stability of better than 0.1 °C, Hg fluxes giving beam equivalent pressures at the film growth position greater than 1 mTorr, refilling with Hg without venting the main growth chamber, and Hg contact only with stainless steel. A brief description of the source and its usage follows. Details of the source construction have been published elsewhere[3].

A schematic of the Hg source is shown in Figure 3. The source is mounted on a 4.5" Conflat flange. It consists of two Hg reservoirs and a Hg vapor transport tube separated by all stainless steel valves. The volume of the internal Hg reservoir is 68 cm^3 while the volume of the external reservoir is 92 cm^3. The cap on the fill port of the external reservoir is sealed with an o-ring. The internal reservoir is surrounded by a heated aluminum block, which is covered by a layer of fiberglass for thermal insulation. The transport tube is surrounded by a heated aluminum tube. The internal diameter of the Hg transport tube is 6.4 mm, and it extends 38 cm into the vacuum chamber from the 4.5" flange. Platinum resistance thermometers (PRT) are used to control the temperatures of the Hg reservoir and the Hg transport tube. Platinum resistance thermometers rather than thermocouples are used as temperature sensors because of their greater sensistivity in the 100 to 200 °C range.

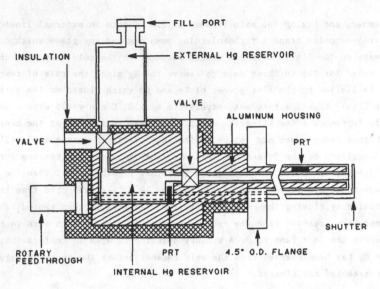

Figure 3. Schematic diagram of the mercury evaporation source for MBE.

The following describes the operation of the mercury source. With the
valve between the internal Hg reservoir and the transport tube open and the
valve to the external reservoir closed, the main vacuum chamber is evacuated
to UHV pressures. Next, the internal reservoir and transport tube are baked
at 200 °C for 24 hours to minimize contamination of the Hg vapor by any
volatile materials that may remain in the source from cleaning procedures.
The temperature controllers are set to 20 °C and the temperature of the Hg
reservoir is allowed to fall to 40 °C or less. Then the valve between the
internal reservoir and the transport tube is closed, and the external
reservoir is filled with Hg. Next, the cap is secured on the fill port and
the valve between the internal and external reservoirs is opened to fill the
internal reservoir. Since the internal reservoir has a smaller volume than
the external reservoir, no air is trapped in the internal reservoir when the
valve between the two reservoirs is closed. Furthermore, since the mercury
flows into the internal reservoir from the bottom of the external reservoir,
any mercury oxide which floats on the surface stays in the external
reservoir. The internal reservoir can be refilled at any time by closing both
valves and filling the external reservoir and then opening the valves between
the reservoirs. With the valve between the reservoirs closed, the temperature
of the internal reservoir is set to the value necessary to obtain the desired
Hg flux. The valve to the main chamber must be open while the temperature of

the internal reservoir is being increased since the greater thermal expansion of Hg relative to stainless steel might cause the reservoir to rupture if this valve is closed. The beam equivalent pressures at the film growth position range from 3×10^{-5} Torr at an internal reservoir temperature of 150 °C to 1×10^{-3} Torr at 193 °C. The distance from the end of the Hg transport tube to the film growth position is 16 cm. The temperature of the transport tube is maintained at 200 °C during the use of the source.

Te, Cd, CdTe Sources

Furnaces which have a large temperature gradient near their open end may clog when source materials such as Te, Cd, and CdTe are used. This clogging results in a time dependent decrease in the flux of the material being evaporated. This problem arises because of the high vapor pressures that these materials have in the solid state. The problem can be eliminated either by adding additional windings at the open end of the furnace or by adding an additional heating element. The latter solution was chosen at NCSU since it is only possible to add the proper number of additional windings to eliminate this temperature gradient for one operating temperature of the furnace. Using an additional heating element and temperature controller, the temperature gradient can be eliminated for any operating temperature. Figure 4 shows a

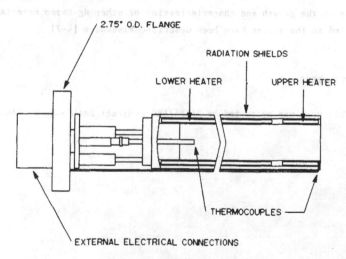

Figure 4. Schematic diagram of MBE furnace for the evaporation of materials such as Cd, Te, and CdTe.

schematic of the NCSU furnace used for the Te, Cd and CdTe sources. The furnace has two separately heated regions with an inconel sheathed chromel-alumel thermocouple sensor for each region. The small independently heated region at the open end of the furnace elminates the condensation of the evaporated material at the top of the furnace. The heating elements are 0.020" tantalum wire supported by 3/32" O.D. alumina tubes. Two independent Eurotherm Model 980 Temperature Controllers and Eurotherm Model 931 SCR Assemblies power the two heaters. A pyrolytic boron nitride crucible designed for the furnace has a useful volume of 40 cm^3.

MCT FILM GROWTH

Both (111) and (100) substrates of GaAs, CdTe, and CdZnTe have been used in the MCT film growth. X-ray rocking curves are measured for the CdTe and CdZnTe substrates before use to insure that the epitaxy is being attempted on a single crystal surface. Typical beam equivalent source pressures of MCT growth on (100) CdTe at 175 $^\circ$C having a mole fraction x=0.20 of Cd are as follows: Hg, 2 x 10^{-4} Torr; Te, 2 x 10^{-6} Torr; and CdTe, 5 x 10^{-7} Torr. The x=0.2 MCT films are characterized by sharp absorption edges and specular surfaces. The temperature dependences of the Hall mobility and carrier concentration for two MCT films with x values near 0.2 are shown in Figures 5 and 6. Note that the mobility for the film shown in Figure 6 rises to 3.4 x 10^5 cm^2/V-s near 50 K and remains essentially constant at lower temperatures. Details of the growth and characterization of other Hg-based materials deposited in the system have been described elsewhere [4-7].

ACKNOWLEGMENTS

This work was supported by DARPA/ARO Contract DAAL03-86-K-0146.

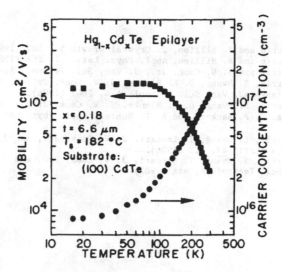

Figure 5. Temperature dependences of Hall mobility and carrier concentration
MCT with x=0.18.

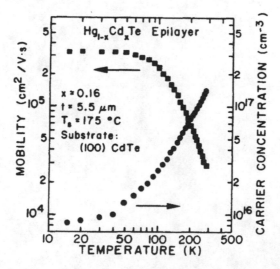

Figure 6. Temperature dependences of Hall mobility and carrier concentration
for MCT with x=0.16.

428

REFERENCES

1. J. P. Faurie and A. Million, J. Crystal Growth $\underline{54}$, 582 (1981).
2. J. P. Faurie and A. Million, Appl. Phys. Lett. $\underline{41}$, 264 (1982).
3. K. A. Harris and J. W. Cook, Jr., J. Vac. Sci. Technol. A, xxx (1987).
4. K. A. Harris, S. Hwang, D. K. Blanks, J. W. Cook, Jr., J. F. Schetzina, and N. Otsuka, J. Vac. Sci. Technol. A $\underline{4}$, 2061 (1986).
5. K. A. Harris, S. Hwang, D. K. Blanks, J. W. Cook, Jr., J. F. Schetzina, N. Otsuka, J. P. Baukus, and A. T. Hunter, Appl. Phys. Lett. $\underline{48}$, 396 (1986).
6. K. A. Harris, S. Hwang, Y. Lansari, J. W. Cook, Jr., and J. F. Schetzina, Appl. Phys. Lett. $\underline{49}$, 713 (1986).
7. K. A. Harris, S. Hwang, Y. Lansari, J. W. Cook, Jr., and J. F. Schetzina, J. Vac. Sci. Technol. A, xxx (1987).

InSb: A KEY MATERIAL FOR IR DETECTOR APPLICATIONS

S.R. JOST*, V.F. MEIKLEHAM* AND T.H. MYERS*
*Electronics Laboratory, General Electric Company, Syracuse, NY 13221

ABSTRACT

InSb has served as an important mid-wave IR (λ=3-5µm) detector material for several decades. In this presentation, we will briefly review General Electric's InSb Charge Injection Device technology. Emphasis will be placed on device performance as a function of material parameters. A new InSb materials technology utilizing liquid phase epitaxy will be described. This epitaxial growth technology improves InSb material parameters and increases minority carrier lifetimes by more than two orders of magnitude to near the Auger limit. Comparisons will be made between available bulk material parameters and that of the epitaxial material.

INTRODUCTION

Indium Antimonide photodiodes have found numerous applications in the defense industry for more than 20 years. Perhaps the best known (and most successful) of these systems has been the Sidewinder air to air antiaircraft missile. The sensor in this system typically employs one single element InSb photodiode that is cooled by an argon gas Joule-Thompson cooler. More recently, single element diodes have also been employed in commercially available scanned thermal imaging systems. However, progress in penetrating the commercial market has been slow as a result of high detector costs and the necessity of cryogenic cooling for higher performance.

To increase the sensitivity and resolution of an infrared imaging system, more high performance detectors are required on the focal plane. This may be achieved by adding additional detectors with a signal wire out of the dewar for each element in the focal plane. At some point, the number of dewar penetrations will become impractical from a thermal loading and packaging standpoint. This has led to the development of advanced focal planes where the signals from many detectors are multiplexed in the dewar to reduce the complexity of the focal plane packaging. At General Electric (GE), the development of focal plane detector arrays consisting of InSb Charge Injection Devices (CID) was initiated in the early 1970's to address these issues[1]. In addition, silicon integrated circuits have been developed for operation at cryogenic temperatures to perform the signal preamplification and multiplexing functions. One of these advanced focal plane detectors is shown in figure 1.

InSb CHARGE INJECTION DEVICES

The InSb CID is a simple Metal Insulator Semiconductor (MIS) capacitor that integrates photogenerated charge in a potential well formed beneath a semi-transparent metal gate. This is shown schematically in figure 2. The potential well is formed by pulsing the gate into deep depletion; at this time charge integration begins. Both photogenerated minority carriers and minority carriers due to dark current are collected beneath the gate during integration. These carriers change the surface potential of the gate electrode. The well is then collapsed by returning the gate voltage to the inversion level causing the integrated charge to recombine (charge injection). The total integrated charge is then determined by measuring the potential difference between the well containing photogenerated and dark carriers and the empty well after the injection pulse.

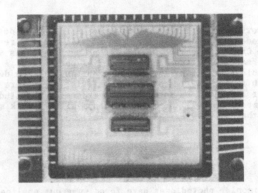

Figure 1. Dual 64 element line array with 2 silicon MUX/Preamp chips.

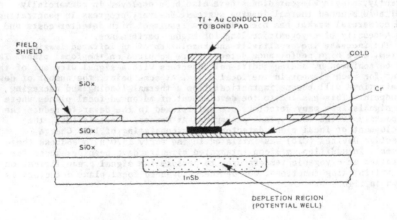

Figure 2. The InSb CID structure.

For area CID detector arrays, two coupled MIS capacitors are used for each element: one capacitor is connected to a column line and the other to a row line. By selective row and column well collapsing, it is possible to read out a single detector element of choice. This permits the readout of a 16384 element detector array with only 256 connections. Such an array is shown in figure 3. The performance of these advanced focal planes is limited by the

flux of background radiation, the noise in the readout electronics, and the
dark current noise in the detector itself. This paper is primarily concerned
with detector noise which, for a given device structure and process, is
directly related to the quality of the semiconductor material.

Figure 3. An advanced focal plane containing a 128x128 InSb CID area array
and four silicon MUX/preamp chips.

The dark current in the InSb CID is proportional to the sum of the
depletion generation of carriers and minority carrier diffusion from the bulk
semiconductor into the depletion layer. For InSb with a generation lifetime
τ_g, diffusion length L, carrier concentration N_D, and depletion width W, the
dark current density may be written as[2]

$$J_{Dark} = \frac{qn_iW}{2\tau_g} + \frac{qn_i^2L}{N_D\tau_g} \quad \text{amps cm}^{-2} \tag{1}$$

where q is the electronic charge and n_i is the intrinsic carrier
concentration. For $N_D = 3 \times 10^{14} \text{cm}^{-3}$, L = 25µm and an operating temperature of
80K, the depletion generation current is nearly three orders of magnitude
higher than the diffusion current. The material parameter that has the most
significant effect on the dark current is τ_g, the generation lifetime. This
parameter is related to the minority carrier lifetime which is very sensitive
to impurities and crystalline defects. Impurities and defects increase the
probability of carrier generation in the depletion layer and hence lead to a
shorter generation lifetime.

InSb, with a cutoff wavelength of 5.3µm, has normally been restricted to
operating temperatures ranging from 78K to 90K. This limitation is mainly due
to the small generation lifetimes that are a result of impurities and defects
in the crystal lattice of available bulk material. At GE, we have found that
these impurities and defects can be greatly reduced by growing epitaxial
layers of InSb on InSb wafers by Liquid Phase Epitaxy (LPE).

Our LPE technology employs the vertical infinite melt LPE technique that
was developed at GE more than 20 years ago for the epitaxial growth of GaAs.
This system has been employed because of the production process requirement
for large wafers and the run to run reproducibility of the infinite melt
technique. In this technique, the substrate is lowered into a supercooled
indium based melt that is 10 to 30% antimony (see figure 4) for epitaxial

growth. Prior to growth, the substrate is chemically etched, mounted on a quartz substrate holder, and placed in the LPE reactor loadlock. The substrate is then baked in a reducing atmosphere to remove residual surface oxide that can impede uniform nucleation of the epitaxial layer. When the reactor reaches the appropriate growth temperature, the substrate is dipped into the melt for growth. Layer growth rates of up to 10μm per minute are common, and layer thicknesses range from 5μm to more than 150μm. Available bulk InSb is an excellent substrate providing that the surface preparation and crystal orientation are maintained within certain limits.

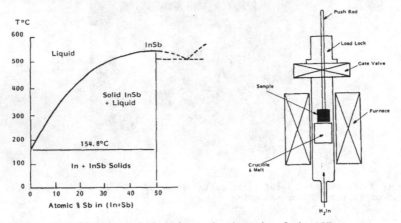

Figure 4. InSb liquidus/solidus plot and schematic of the LPE reactor.

MIS generation lifetimes, which can be directly related to a convolution of the material minority carrier lifetime and surface state generation-recombination, have been measured on both bulk and epitaxial InSb. For bulk InSb, typical lifetimes are less than 20ns, whereas lifetimes in the LPE layers range from 200ns to more than 2μs. This represents a two orders of magnitude increase. Another indication of minority carrier lifetimes can be determined by measuring the decay of a photoconductance signal generated by a light pulse. In the absence of trapping, the decay time of the photogenerated carriers is a function of minority carrier lifetime and surface recombination-generation. Figure 5 is a plot of the temperature dependence of photoconductive lifetime of CVD SiO_2 passivated samples of non-contact polished bulk material and LPE InSb. Here, the bulk lifetime is measured to be 60ns as compared to a value of 400ns for the LPE epilayer.

Identical detector arrays have been fabricated on bulk and LPE InSb material for evaluation. The longer generation lifetimes observed in the epitaxial material translate directly into reduced dark current density in the detectors for a given operating temperature. Figure 7 is a log plot of measured dark current in InSb CID detectors fabricated on bulk and epitaxial material. For any operating temperature, the epi-detector demonstrates more than a two order of magnitude reduction in the dark current density. This translates to an order of magnitude improvement in detector dark current limited detectivity. In addition, for an equivalent level of dark current, the bulk detector must be operated more than 25K colder than the LPE detector. Epitaxial materials technology permits system designers to consider InSb as a candidate for MWIR applications where elevated temperature (>100K) is an operational requirement.

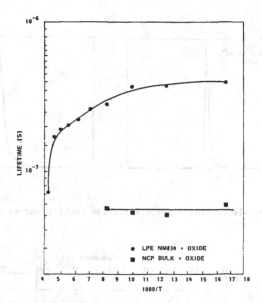

Figure 5. Temperature dependance of photoconductance lifetime for non-contact polished bulk InSb and an LPE layer. Both samples were passivated with CVD SiO_2.

Capicitance-voltage and x-ray diffraction studies have shown that the epitaxial material has fewer impurities in addition to a more perfect crystallinity. This directly leads to the increased carrier lifetimes observed. The apparent reduction in the doping level is a result of the favorable segregation of impurity atoms at the liquid/solid interface during growth rather than compensation. Double crystal x-ray diffraction studies indicate that there is a factor of 64 fewer dislocations in the LPE than in commercially available bulk material. Double crystal x-ray rocking curves are shown for typical bulk and LPE material in figure 6. These diffraction peaks are (333) reflections measured using a BEDE double crystal diffractometer utilizing an InSb first crystal and Cu $K\alpha$ radiation with a beam size of 1x6 mm^2. A summary of typical materials characteristics is shown in Table I.

Table I. Typical Materials Characteristics.

PROPERTY	BULK InSb	LPE InSb
Wafer Diameter	2 inches	2 inches
Minimum Doping Density	$2 \times 10^{14} cm^{-3}$	$3 \times 10^{13} cm^{-3}$
Generation Lifetime	20ns	200-2000ns
Photoconductive Lifetime	60ns	400ns
X-ray Rocking FWHM	25 arc sec	11 arc sec

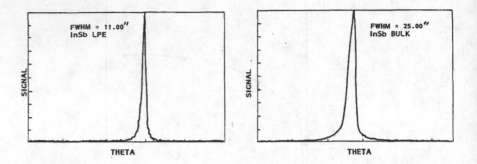

Figure 6. Double crystal x-ray diffraction rocking curves for bulk and LPE InSb.

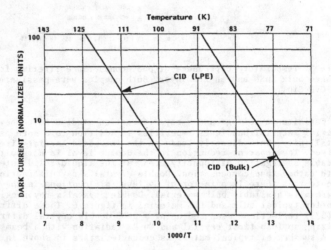

Figure 7. Dark current as a function of temperature for CID detectors fabricated on LPE and typical bulk InSb.

SUMMARY

Large area, epitaxial InSb layers grown by LPE offer significant
detector performance improvements over devices fabricated on commercially
available bulk material. Focal planes incorporating these LPE detector arrays
will permit InSb based systems to operate at temperatures greater than 100K
with performance levels comparable to bulk InSb detectors at lower
temperatures.

REFERENCES

1. J.C. Kim, IEEE J. Solid State Circuits, SC-13, 158 (1978).

2. S.R. Jost, Proceedings of SPIE, 409, "Technical Issues in Infrared
Detectors and Arrays", Esther Krikorian, Editor, 62 (1983).

METALLURGICAL PROFILE MODELING OF Hg CORNER LPE HgCdTe

BURT W. LUDINGTON
Santa Barbara Research Center, 75 Coromar Dr., Santa Barbara, CA 93117

INTRODUCTION

Mercury cadmium telluride ($Hg_{1-x}Cd_xTe$) is an important material for infrared detector applications due to the variable bandgap obtained through simply varying the mercury/cadmium ratio. Thus IR sensing systems with various wavelength requirements, typically from 2 to 12 μm, can potentially all be supplied by one materials technology. Bulk crystal growth technology was originally pursued for these applications. However, large, high quality crystals are difficult to grow because of the high melting point of HgCdTe and the resulting high Hg vapor pressure. Liquid phase epitaxial (LPE) growth techniques can lower the growth temperature through the use of a solvent, and allow the deposition of thin films on large foreign substrates that are grown without any Hg.

The HgCdTe alloy can be conceptualized as a binary compound of CdTe and HgTe. During the LPE growth process, changes in temperature cause the composition of the solid to change, because the segregation coefficients of the two alloys are different. Thus precipitate growth techniques that use cooling to drive growth, exhibit compositional variations in the grown layer.

In addition to variations in x value due to the phase diagram, the HgCdTe system possess a large interdiffusion coefficient for Hg and Cd. This results in significant interdiffusion of Hg from the growing layer with Cd from the substrate at the growth times and temperatures commonly employed for LPE film growth. Interdiffusion will occur for any compositional gradient, and is important for heterojunction and superlattice based devices. The interdiffusion process cannot be modeled by simply error function solutions because the diffusion coefficient is a strong function of x value.

Successful modeling of the growth process is important both as a test of scientific understanding, and as an engineering tool for material fabrication. Simulations of HgCdTe growth have been reported by both Shaw [1] and Zanio [2]. Shaw modeled an unstirred, Te rich melt with an approach that did not incorporate any interdiffusion. The model simulated the diffusion of solute to the growing layer, and could account for supercooled solutions. Zanio reported a model incorporating interdiffusion, that deposited material by the addition of sequential layers of pure CdTe and HgTe. The proper material composition was achieved by the addition of layers in the proper ratio, and by allowing interdiffusion to average out the composition variations. The model described here incorporates interdiffusion, and grows layers with the appropriate compositional value directly.

SIMULATION

Approach

Due to the difficulty in solving the diffusion equations analytically when the diffusion coefficient is a function of concentration, a numerical approach was taken. The numerical approach forces the diffusion and growth processes to be treated as separate, sequential steps. The HgCdTe film is

divided into a series of incrementally added lamina as shown in Figure 1.
After addition of the Nth lamina, the set of lamina 1 through N are allowed
to interdiffuse at the current growth temperature for the time required to
deposit the Nth lamina. This process is continued until the deposition
process is complete.

The growth process has been further subdivided into two parts; the
determination of the x value of the material being deposited in the next
lamina, the composition submodel, and the amount of time required to grow
the next lamina, the deposition submodel. Lamina thickness is selected to
be sufficiently small to assume constant composition, with the composition
given by the phase diagram. The numerical approach requires constant
thickness lamina. Thus a deposition rate is first calculated, and this
value used to compute the time required for lamina growth using the prede-
termined thickness step.

Interdiffusion

To determine the extent of interdiffusion occurring during the growth
process, it is necessary to solve the diffusion equation,

$$d/dz[D(x,T) * dx/dz] = dx/dt, \tag{1}$$

where z = layer thickness
x = composition
T = temperature
D = diffusion coefficient
t = time.

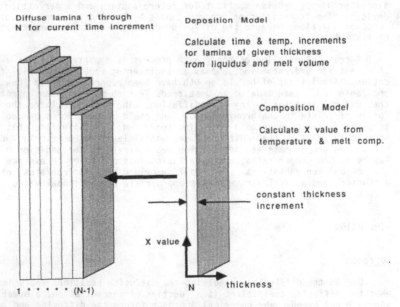

Figure 1. Schematic of growth model submodels and sequence

Since the diffusion coefficient, D, is a strong function of composition, the complementary error function solutions are not valid, and a numerical approach was used. The Crank-Nicholson finite difference technique was chosen because of its computational stability. One equation is generated for each lamina, and at each time step, the equations are solved iteratively for convergence. The set of equations grows as the number of lamina increases. This set of equations is tridiagonal and is solved in a matrix format using the Thomas algorithm.

At the start of the growth process, an initial composition is provided corresponding to the substrate material upon which the HgCdTe film will be grown, i.e. constant composition of x equal 1. The differential equations require boundary values at the two surfaces; the growing HgCdTe surface, and the substrate surface. Since the substrate is much thicker than the interdiffusion length into the substrate, the substrate surface boundary condition sets the composition constant with space and time. As there can be no gradient at the HgCdTe growth surface, the boundary condition at that surface sets the first derivative of the composition with distance to zero.

The interdiffusion coefficient data D, was obtained from Tang[3], and is shown in Figure 2. The functional form is,

$$D(x,T) = \exp [\quad a + (b * x) \quad / (T*k)], \tag{2}$$

where D = interdiffusion coefficient (cm^2/sec)
    a = -1.54
    b = -0.455
    x = composition
    T = temperature (K)
    k = Boltzmann's constant (8.62E-5 eV/K).

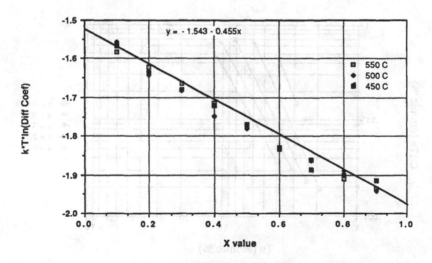

Figure 2. Dependence of diffusion coefficient activation energy on composition

This equation for the interdiffusion coefficient is valid over the temperature range of 400°C to approximately 550°C as shown in Figure 2. The interdiffusion data used from Tang is for Te rich HgCdTe, as LPE material grown at high temperatures is known to be Hg vacancy rich.

COMPOSITION

The composition of the material being deposited at the surface of the epitaxial layer is assumed to be in equilibrium with the entire volume of the melt. This is justified based on the active stirring of the melt during growth, and the lack of melt supersaturation. Homogeneous nucleation of material in the melt at the crystal point has been experimentally observed to limit Hg corner supersaturation. The composition of deposited material is then given by the solidus of the phase diagram. The phase diagram has been empirically determined by Herning [4], empirically modeled by Ludington [5], and analytically modeled by Tung [6]. The empirical results are used here, but they have been found to agree with Tung to within approximately 10K. Figure 3 illustrates the solidus slope for the Hg corner obtained from this data, showing the dependence on temperature and melt composition.

Deposition Rate

The deposition rate is the amount of material deposited on the substrate per degree temperature drop in the melt. The rate can be developed either analytically from solute diffusion and boundary layer calculations, or from an empirical approach using measured deposition rates in addition

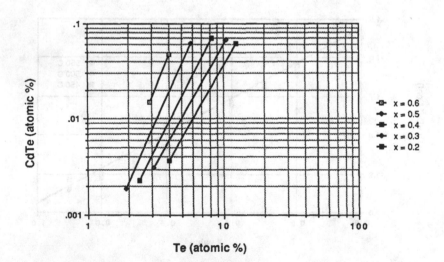

Figure 3. Empirically determined solidus slopes for x values range 0.2 to 0.6 grown from the Hg corner of the phase diagram

to the liquidus slope. The empirical approach is used here due to the difficulty in modeling the stirred melt used in the VLPE technique.

The deposition rate is calculated from,

$$dZ/dT = Ls * V_m * \varepsilon \qquad (3)$$

where T = temperature (°C)
    Z = layer thickness (cm)
    Ls = liquidus slope (cm^3/mole C)
    V_m = melt volume (moles)
    ε = deposition efficiency (cm^{-2})

The liquidus slope is determined from empirically fitting the phase diagram, and predicts the total amount of material crystallizing from the melt per degree drop in temperature. The deposition rate can then be modeled as the product of the liquidus slope, the melt volume, and a deposition efficiency term that indicates the percentage of precipitating material deposited on the substrate. The deposition efficiency has been found to depend upon several variables; start of growth relative to the crystal point, stirring rate and melt thermal profile [5]. Figure 4 shows a typical deposition efficiency curve.

MODEL VERIFICATION

Layer Growth and Characterization

Several HgCdTe films have been grown using the VLPE process and characterized for compositional and thickness. The layers were all grown near

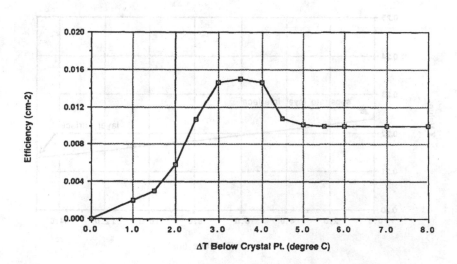

Figure 4. The deposition efficiency is a function of the difference between the current growth temperature and the melt crystal point

515°C, using 7 Kg, 10% Te composition melts, and varied in thickness from 15 to 19 μm. Infrared spectral transmission measurements were used to map the initial lateral thickness and compositional variations. The substrate material was CdZnTe. The layer growth times were typically 30 minutes, with the layers being cooled to room temperature immediately after completion of growth.

Modeling

The simulation model was run with the growth and diffusion parameters described above. The lamina thickness was 0.1 μm in all cases. In general, the model can be run either to mimic an existing layer, or to explore the effects of varying a specific parameter. If simulation of a given layer is desired, the thickness and composition submodels each have one fitting parameter that can be adjusted to obtain the required thickness and composition.

A series of five long wavelength growth runs were simulated to verify the thickness and composition submodels. At this point it has not been possible to verify the interdiffusion model as the program is underflowing due to the small numerical quantities involved. Hence the rest of the discussion will focus on the composition and deposition models.

The x value and composition of the five layers were determined by averaging together the results of approximately 30 spectral transmission points per layer. Layer thickness was obtained from the interference modulation in the IR transmission spectra. Table I lists the actual temperature delta, ending growth temperature, x value, and modeled ending temperature and crystal point. The actual temperature change and x values are were used in the simulation. Figure 5 illustrates the modeled output of composition versus thickness for Layer A.

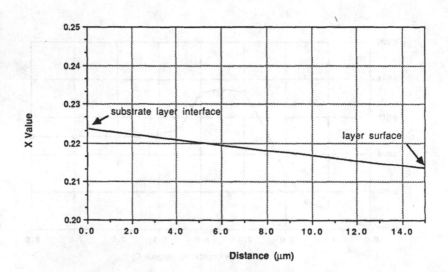

Figure 5. Simulation using Layer A input data showing thickness and composition without interdiffusion

Table I. Measured and simulated growth parameters for
a series of layer growths.

Layer	Actual T	Actual End Temp.	Actual x value	Modeled Ending Temp.	Modeled Crystal Pt.
A	1.4	516.4	.214	513.6	519.1
B	1.8	516.2	.209	513.0	518.8
C	1.5	516.3	.211	513.3	519.0
D	1.6	516.2	.210	513.1	518.5
E	1.7	516.6	.216	513.9	519.5

Thickness

The model was quite successful in predicting layer thickness using
only fixed input parameters such as T, melt volume and melt composition as
shown in Figure 6. Actual layer thickness is plotted against modeled
thickness with a constant crystal point used in the model. Slight changes
in crystal point can account for the rest of the layer thickness variations. This can be seen in Figure 7 which shows the crystal point variation necessary in the five growth runs to remove the residual thickness
discrepancies. Small variations in crystal point are observed in practice
as a melt is depleated of HgCdTe through layer growth and then resaturated.

Composition

The model was also successful in predicting the layer composition
using the actual melt composition and growth temperature. To match the

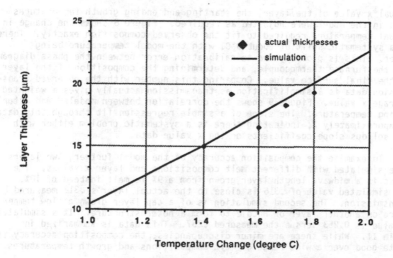

Figure 6. Plot of actual and model layer thicknesses versus
growth temperature with fixed crystal point.
Temperature is more important than crystal point
in determining layer thickness.

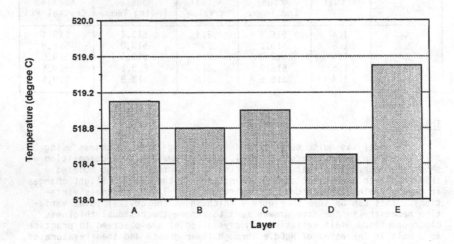

Figure 7. Crystal point variations necessary to obtain observed
thicknesses for the five growth runs modeled

actual x value of the layer, the starting and ending growth temperatures
used in the model were adjusted as required. Figure 8 shows the change in
actual temperature required to fit the observed composition exactly. There
is a systematic shift of about 3°C, with the model temperature being
lower. This is probably due to calibration error between the phase diagram
and the furnace thermocouple, and determining the composition of the layer
as the final surface value. Comparing this number with the observed transmission data is a simplification as transmission actually gives a weighted
average x value. Figure 9 shows the correlation between modeled and actual
ending temperature. The slope of a simple regression fit through this data
is approximately 2, indicating there is a systematic problem either with
the solidus slope coefficients or the x value data.

To examine the composition accuracy of the model further, two layers
were simulated with different melt compositions and layer x values. The
first is a midwave length layer grown from a 9% Te melt instead of 10%.
The simulated value of .316 is close to the actual layer's .312 measured by
transmission. The second simulation is of a cap layer grown at low temperature on another layer of HgCdTe to form a heterostructure. It's simulated
x value is 0.253 versus the measured .267. This data is summarized in
Table II. While there are minor discrepancies, the composition accuracy is
quite good over a wide range of melt compositions and growth temperatures.

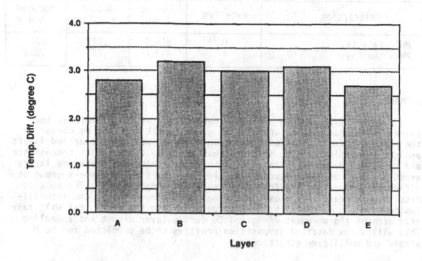

Figure 8. Change in model ending temperature from that actually used
during growth necessary to match observed composition

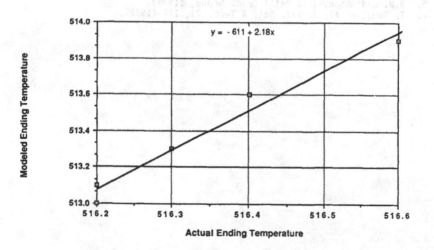

Figure 9. Correlation between actual and modeled ending temperature
with a slope of 2 suggesting systematic error in solidus
slope for x value data

Table II. Measured and simulated composition parameters for MWIR layers.

Description	Melt Te%	Ending Temp.	Actual x value	Modeled x value
MWIR Base Layer	9	514.2	.312	.316
MWIR Cap Layer	4	419.0	.267	.253

SUMMARY

A HgCdTe layer growth model has been presented that simulates the thickness and composition of material grown by LPE from the Hg corner of the phase diagram. Detailed comparisons between well characterized layers and the simulation results indicate that the model can predict composition and thickness accurately, and that with minor tuning can reproduce layers exactly. The composition model was also used to simulate layers grown at different temperature and from different melt compositions. The modeled results were reasonably close to the measured compositions. An interdiffusion model is currently being incorporated into the model that will take into account the movement of Hg and Cd during layer growth and annealing. This will allow detailed composition profiles to be predicted for both single and multilayer structures.

REFERENCES

1. D.W. Shaw, J. Cry. Growth 62, 247 (1983).
2. K. Zanio, J. Vac. Sci. & Tech. A 4, 2106 (1986).
3. M. Tang, 1986 Workshop on the Phy. and Chem. of HgCdTe.
4. P.E. Herning, J. Elec. Mat. 13, 1 (1984).
5. B.W. Ludington, IRIS Mat'l Spec. Group, (1986).
6. T. Tung, et al, J. Vac. Sci. & Tech. 21, 117 (1982).

InAsSbBi and InSbBi: POTENTIAL MATERIAL
SYSTEMS FOR INFRARED DETECTION

S. M. BEDAIR, T.P. HUMPHREYS, P. K. CHAING, and T. KATSUYAMA
Electrical and Computer Engineering Department
North Carolina State University
Raleigh, North Carolina 27695-7911

ABSTRACT

InSb$_{1-x}$ Bi$_x$ (0.01 < x < 0.14) and InAsSbBi quaternary alloys are potentially attractive materials for the development of semiconductor infrared detectors covering the 8-14 µm range [1,2,3].

We report for the first time, MOCVD growth of InSb$_{1-x}$ Bi$_x$ (0.01 < x < 0.14) and InAs$_{1-x-y}$ Sb$_x$ Bi$_y$ with 0.5 < x < 0.7 and 0.01 < y < 0.04 on both GaAs and InSb substrates using AsH$_3$, TMSb, TEI and TMBi. Electrical measurements of the undoped InSb$_{0.99}$ Bi$_{0.01}$ shows a background carrier concentration of approximately 10^{16}/cm^3 and a room temperature mobility of 20,215 cm^2/V.sec. To-date, these are the best reported electrical measurements for this ternary alloy.

The formation of a secondary Bi phase and single crystal growth of metallic bismuth-antimony at the surface of InSb$_{1-x}$ Bi$_x$ which results in deterioration of morphology with increasing values of x is also investigated. A wide range of analytic techniques, including SEM, EDX, electron microprobe and AES have been employed in our surface analysis.

Introduction

This paper reports the first results on the epitaxially growth of thin film InSb$_{1-x}$ Bi$_x$ (0.01 < x < 0.14) and InAs$_{1-x-y}$ Sb$_x$ Bi$_y$ with 0.5 < x < 0.7 and 0.01 < y < 0.04 semiconducting alloys by MOCVD.

Present results include growth parameters, electrical, and crystal characterizations. Surface derived features for InSb$_{1-x}$ Bi$_x$ were also investigated by SEM (scanning electron microscopy) with supplemental element identifications by EDX (energy dispersive X-ray analysis), electron microprobe analysis and AES (Auger electron spectroscopy).

Results and Discussion

Epitaxial layers were grown in a vertical quartz reactor operating at atmospheric pressure. Arsine (AsH_3), Triethylindium (TEI) (Alfa), trimethylantimony (TMSb) (Alfa) and trimethylbismuth (TMBi) (Alfa) were used as arsenic indium, antimony and bismuth sources, respectively. The TMBi bubbler was maintained at approximately $-12°C$, thus keeping the partial pressure of the TMBi vapor in the gas phase low enough to deposit a few percent of Bi in the solid phase. Epilayers were grown on both (100) InSb and Cr doped semi-insulating GaAs at a growth temperature of $445°C$.

Compiled in Table I is a summary of the growth parameters, solid composition, carrier concentration and carrier mobility for several $InSb_{1-x}Bi_x$ and InAsSbBi samples. In the absence of a high resistivity InSb substrate all Hall measurements were conducted on $InSb_{1-x}Bi_x$ and InAsSbBi epilayers deposited on semi-insulating (100) GaAs. All grown $InSb_{1-x}Bi_x$ epilayers were n-type with carrier concentrations in the low $10^{16}/cm^3$ to the $10^{17}/cm^3$ range. The same range of carrier concentration was also observed for the deposition of epilayers of InSb under similar conditions (without Bi doping). In particular, it is clear that the addition of a few percent of Bi to the InSb solid film does not result in a significant change in the carrier concentration. A dramatic improvement however is observed in the electron mobility, even in cases were the solubility limit of Bi is exceeded. For instance, the epilayer $InSb_{0.86}Bi_{0.14}$ has an electron mobility which is approximately twice the mobility of a InSb film ($\mu = 5000$ cm^2/V sec) grown under similar conditions (Table I). Epilayer thickness for both InSb and $InSb_{1-x}Bi_x$ corresponding to sample runs 229, 231 and 233 was approximately 1 μm. When thicker layers of $InSb_{1-x}Bi_x$ were deposited (2 μm, run #253), an electron mobility of 20,215 $cm^2/V.sec$ was recorded at room temperature. The corresponding mobility at 77K was 6800 $cm^2/V.sec$. This decrease in mobility at low temperature is attributed to the predominance of a Dexter-Seitz's dislocation scattering mechanism. Further, it is noted that the mobility of this material is far superior to that grown by existing MBE techniques. For instance, with a comparable Bi concentration in the solid, carrier concentrations and mobilities grown by MBE are $1.6 \times 10^{18}/cm^3$ and 480 $cm^2/V.sec$, respectively [4,5]. This may be compared to $4 \times 10^{16}/cm^3$ and 20,215 $cm^2/V.sec$ in the present study.

Also compiled in Table I are the electrical characteristics of the InAsSbBi alloys. Carrier concentrations of the grown films are in the high $10^{16}/cm^3$ to low $10^{17}/cm^3$ range. The incorporation of Bi in InAsSb, although improving the mobility slightly, resulted in a significant degradation of surface morphology.

TABLE I Growth conditions and electrical Hall measurements for $InSb_{1-x}Bi_x$ and InSbAsBi

RUN	Growth Temperature (°C)	Epilayer Thickness (μm)	Gas Flow Rate (μmole/min)				Solid Composition		Carrier Concentration n (cm^{-3}) [300K]	Mobility μ (cm^2/V.sec) [300K]
			TEI	TNSb	AsH$_3$	TMBi	InSb%	InBi%		
228	445	1	80	8	-	-	-	-	3×10^{16}	5000
229	445	1	80	8	-	1.2	-	12	1×10^{17}	5000
231	445	1	80	8	-	2.4	-	14.5	2×10^{17}	10,940
233	445	1	80	8	-	0.5	-	3	7×10^{16}	4000
253	445	2	80	8	-	0.3	-	<1	4×10^{16}	20,215
235	445	1	130	16	5.5	1.2	52	4.4	6×10^{16}	6000
236	445	1	160	20	5	1.2	52	1.2	2×10^{17}	8000

The surface morphology and microstructure of the $InSb_{1-x}Bi_x$ epilayers corresponding to an InBi mole fraction of 3% (run #233) and 12% (run #229) are shown in Figure 1 (a) and (b), respectively. Three regions are clearly defined on both SEM micrographs. Region A, we identify as polycrystalline coherent precipitates. They have a regular geometry, both square and oblong in appearance, and are randomly distributed over the surface. The corresponding distribution surface density is largest on the $InSb_{1-x}Bi_x$ epilayer for $x = 14.5\%$ (run #239) and is reduced to very low levels with decreasing x. The precipitates are small, ranging from 1 μm to 5 μm. Dimensions perpendicular to the surface are of the order of 1000 $\overset{\circ}{A}$, using the technique of Ar^+ ion milling and alpha step measurements. Orthorhombic single crystal formation with preferred surface orientation is clearly illustrated in region B. Dimensions of these single crystals are large, varying in length from 5 μm to 10 μm. Surface distribution density is largest for the $InSb_{1-x}Bi_x$ ($x = 14\%$) epilayer and as x is reduced below the bulk solubility limit [3,6], the features completely disappear. Region C represents an intermediate area between the polycrystalline phase and single crystal structures that is both featureless and smooth.

Examination of the various surface regions by EDX gives several interesting results (Figure 2). For instance, region A is mainly composed of Bi precipitates segregated at or near the surface. Indium is also incorporated in this polycrystalline precipitate as confirmed by electron microprobe analysis. Furthermore, employing glacing angle x-ray diffraction confirms the presence of polycrystalline phases of InBi and In_2Bi together with metallic Bi [4,5]. Region B has been identified as single crystal (metal-alloy) BiSb, that is formed during growth due to an excess of free antimony at the surface. Auger analysis confirms the presence of an Sb rich surface in all of the $InSb_{1-x}Bi_x$ epitaxial grown films. Coalescence of free Bi and Sb on the surface to form single crystal BiSb on cooling is tentatively proposed. The smooth, mirror-like surface pertaining to region C is depleted of Bi and appears to be composed only of InSb.

In the case of $InSb_{1-x}Bi_x$ samples, where the InBi percentage composition is below the solubility limit of 2.6% in InSb [3,6], the surface morphology was good with only a small surface density of second phase polycrystalline Bi precipitates.

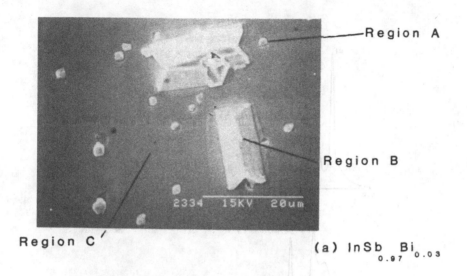

Region A

Region B

Region C

(a) InSb$_{0.97}$Bi$_{0.03}$

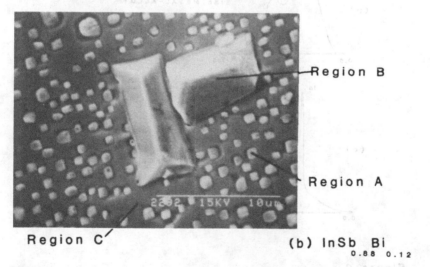

Region B

Region A

Region C

(b) InSb$_{0.88}$Bi$_{0.12}$

Figure 1.
SEM micrographs of InSb$_{1-x}$Bi$_x$ films were (a) x=0.03
(b) x=0.12 of 1μm thickness grown on a (100)
orientated InSb substrate.

452

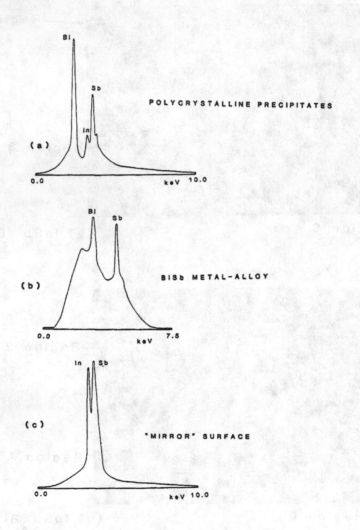

Figure 2
Energy dispersive X-ray (EDX) traces corresponding
to (a) Region A (b) Region B and (c) Region C in
Figure 1.

In conclusion, we have successfully grown InAsSbBi quaternary and $InSb_{1-x}Bi_x$ ternary epilayers over a wide range of x values. High electron mobilities have been recorded for these $InSb_{1-x}Bi_x$ films grown below the InBi% solubility limit. A trend reflecting the deterioration in surface morphology with increasing x has been observed. Investigation of the corresponding surface microstructure reveals the predominance of polycrystalline phases of In_2Bi, InBi and metallic Bi. The presence of the metallic alloy BiSb is also observed on those films where the InBi percentage exceeds the solubility limit.

References

1. J. L. Zilko and J. E. Green, J. Appl. Phys. 51, 1549 (1979).
2. K. Oe, S. Ando and K. Sugiyama, J. Appl. Phys. 20, L303 (1981).
3. A. J. Noreika, W. J. Takei, M. H. Francombe and C. E. C. Wood, J. Appl. Phys. 53, 4932 (1978).
4. A. J. Noreika, J. Greggi, Jr., W. J. Takei and M. H. Francombe, J. Vac. Sci. Technol. A1, 558 (1983).
5. B. Joukoff and A. M. Jean-Louis, J. Cryst. Growth. 12, 169 (1972).
6. J. L. Zilko and J. E. Green, Appl. Phys. Lett. 33, 256 (1978).

MOLECULAR BEAM EPITAXIAL GROWTH OF $Cd_{1-x}Zn_xTe$
MATCHED TO HgCdTe ALLOYS

N. MAGNEA[*], F. DAL'BO[*], J. L. PAUTRAT[*], A. MILLION[+], L. DI CIOCCIO[+]
G. FEUILLET[*]

Centre d'Etudes Nucléaires de Grenoble, 85 X ~ 38041 Grenoble Cédex, France
[*]Département de Recherche Fondamentale
[+]Laboratoire d'Electronique et de Technologie de l'Informatique

ABSTRACT

$CD_{1-x}Zn_xTe$ alloys of various composition have been grown by the Molecular
Beam Epitaxy Technique and characterized by Transmission Electron Microscopy,
C(V) measurements and photoluminescence spectroscopy techniques. The quality
of the thick layers is comparable to that of bulk material. Thin strained layers have
also been grown whose interfaces are structurally good. The recombination within
a CdTe well confined between $Cd_{1-x}Zn_xTe$ barriers is dominated by intrinsic
processes.

I. INTRODUCTION

The fabrication of efficient infrared detectors based on II VI semiconductors
can be achieved with the help of epitaxy techniques if suitable substrates are
available. A good substrate must be of high crystalline quality and lattice matched
to the layer to be grown. If not matched a convenient buffer layer must be grown.
The present work was aimed to the growth by the Molecular Beam Epitaxy technique
of layers of $Cd_xZn_{1-x}Te$ matched to $Cd_{0.8}Hg_{0.2}Te$. The substrate of
$Cd_{0.96}Zn_{0.04}Te$ is also lattice matched.

After a description of the growth and characterization techniques (Part II),
the part III deals with the growth of $Cd_{1-x}Zn_xTe$ buffer layers (0 < x < 10 %) and
the properties of these layers deduced by photoluminescence, C(V) analysis and
Transmission Electron Microscopy.

Some heterostructures have also been fabricated with a CdTe well (50 to 230
nm) sandwiched between barriers of $Cd_{1-x}Zn_xTe$. The optical response of the well
and the quality of the interfaces have been analysed (Part IV).

II. DESCRIPTION OF THE EXPERIMENTS

The (100) $Cd_{0.96}Zn_{0.04}Te$ substrates are carefully polished in a Bromine methanol solution and cleaned before introduction in the MBE system. In situ cleaning is performed by heating to 300° for 10-15 minutes. This allows for the desorption of the excess surface Tellurium and the formation of a good RHEED diagramm as described before [1].

The growth technique uses two effusion cells loaded with CdTe and ZnTe. As these compounds evaporate congruently the ratio of the fluxes coming from both cells defines the composition x of the layer if Zn and Cd have similar sticking coefficients :

$$x = \frac{\Phi_{ZnTe}}{\Phi_{ZnTe} + \Phi_{CdTe}} \tag{1}$$

where Φ_{ZnTe} and Φ_{CdTe} are the equivalent fluxes of undissociated molecules. During the growth, the substrate temperature is held fixed at 300°C and the deposition rate is close to 0.2 nm s^{-1}. The layer thickness reaches 2 to 5 μm.

During the growth on the (100) face the RHEED diagramm displays a well streaked (2x1) pattern and Kikuchi lines and bands. The morphology and cristalline quality are good up to x=0.3 as checked by X-ray diffraction data. The rocking curves half width vary between 3 and 5 minutes angle. We have used several characterization techniques to evaluate the properties of the layers.

For Transmission Electron Microscopy Experiments, cross sections of the samples are prepared by the Ar^{+} ions thinning technique and analysed on a 200 keV microscope (JEM 200 Cs).

The photoluminescence spectra are recorded at 2 K under the cw excitation of a krypton ion laser. The normal excitation power is 0.02 W. The emitted light is focused on the entrance slit of a monochromator and detected with a cooled photomultiplier (GaAs cathode).

For the C(V) measurements, Schottky diodes are obtained by the evaporation of Indium electrodes ($\phi = 0.4$ mm) on the top of the layer. The p-type substrate is contacted with an electroless gold plating and the C-V curves are recorded with an automatic capacitance bridge at 10 KHz.

III. OPTICAL AND ELECTRICAL MEASUREMENTS ON SINGLE THICK LAYERS

As the grown layers may be used as a part of an electronic device, it is important to check the electrical and optical performance of the layer and of the

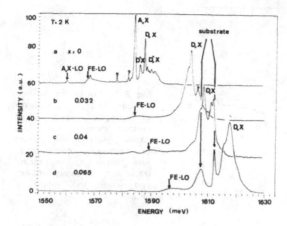

Fig. 1. Photoluminescence spectra of $Cd_{1-x}Zn_xTe$ layers of various composition. The shift of the FE-LO transition is used to determine the bandgap energy.
The other lines are due to excitons bound to neutral donors (D_oX, to ionized donors (D^+X), and to neutral acceptors (A_oX).

various interfaces.

We have used single thick layers of various composition to measure the photoluminescence spectrum and check the composition and overall efficiency. The figure 1 shows the spectra obtained for layers of Zn composition x=0 : 3.2 : 4 : 6.5 % as measured by X ray fluoresence. The main lines of the excitonic spectrum of the layer have been indexed in agreement with earlier work [2.3]. The detection of the LO replica of the free exciton allows to get the bandgap energy and the composition using the following formula

$$Eg(x) = 1.605 + 0.505 \, x + 0.285 \, x^2 \ (eV) \tag{2}$$

The LO energy is taken as 21.2 meV [4] independently of the composition and the free exciton binding energy is 10.4 meV [5].

The relative strength of the lines D_oX and A_oX indicates that the donor content is somewhat higher in the layers than in the substrates. This can be the indication of some contamination but it could also mean that the closeness to stoichiometry of the layer increases the donor solubility [2]. It is worth noting that there is no indication of the broad 1.4 eV band (not shown on the figure) which is often taken as characteristic of defects [6].

The spectrum of fig. 1(a), deals with a pure CdTe layer grown in similar conditions. It can be compared in detail with the high resolution spectra taken on bulk CdTe [2.3]. The attribution of the lines has been done in this way. However the bound exciton lines are slightly shifted with respect to the corresponding ones taken on a bulk CdTe. This is attributed to the built in strains in the layer. Indeed the lattice mismatch between layer (6.482 Å) and substrate (6.464 Å) induces a biaxial compressive stress on the layer [7] which splits and shifts the excitonic lines [8]. There is a critical thickness (300 nm on (100) oriented substrate) above

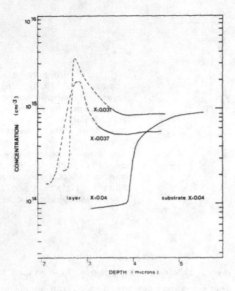

Fig. 2 Profile of the carrier concentration measured on $Cd_{1-x}Zn_xTe$ layers of various composition. The dashed part of the curves is frequency dependent. The layer to substrate transition is easily observed when the misfit is small.

which the stress is mainly released by the generation of dislocations. The residual stress is 8-10 MPa for 5 μm thick layers [7, 8].

The measured linewidth of D_oX (0.4 meV) is comparable to the linewidth measured on some bulk crystals but somewhat larger to what is obtained on the best crystals [2]. The misfit induced dislocations are certainly partly responsible for this broadening.

The electrical properties of these layers have been measured and the resulting profiles are shown on figure 2. If the layer and the substrate have the same composition (x=0.04) the C(V) profile indicates that the doping level of the layer is $\sim 10^{14}$ cm^{-3} much lower than in the substrate (8-9 10^{14} cm^{-3}). The position of the interface deduced of the profile agrees with the grown thickness. If the mismatch $\Delta a/a$ exceeds $2\ 10^{-4}$ there is a pronounced spike close to the interface indicating that the space charge edge is almost pinned there : the mismatch induced dislocations offer many bound states. The position and strength of this apparent peak vary with frequency indicating thus that it is not due to a shallow dopant distribution.

IV. HETEROSTRUCTURES

The above results demonstrate the overall quality of the thick layers although the interface between substrate and layer is dislocated by the misfit induced dislocations when there is a difference in composition between layer and

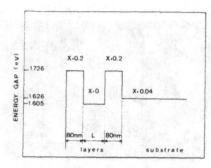

Fig. 3 Heterostructures grown on a (100) $Cd_{0.96}Zn_{0.04}Te$ substrate. The Z_n content x and the thickness of the layers are indicated.

substrate.

Now it is interesting to check the quality of the interfaces between layers of different composition when their thickness is less than the critical thickness where the stresses, begin to be released.

A typical structure is shown on fig. 3. The CdTe well of thickness L is limited by two barriers of $Cd_{1-x}Zn_xTe$ (x=0.12 – 0.2). Due to the lattice parameters difference the barrier are in extension while the well is in compression.

During electron microscopy observation of such structures it is difficult to get a different contrast from the various layers because the mean atomic weight is only slightly modified by the composition. So in most cases the interfaces are perfect and indetectable. In one case shown in fig. 4. one of the interface has been found decorated by precipitates. High resolution imaging of the region shows the continuity of the atomic planes and thus demonstrates the absence of dislocations.

We have measured the optical response of this structure when excited above the $Cd_{1-x}Zn_xTe$ bandgap. The electrons and holes may become trapped inside the CdTe well and most of the recombination occur there.

A typical spectrum recorded at 2K, under standard excitation power (20 w/cm^2) is shown in fig. 5 for several well thicknesses.

In all cases the layer spectrum is strongly modified with respect to the spectrum of thick layer shown on fig. 1a. A new strong line is detected at 1.595 eV. It is well resolved for L=0.23 nm and L=0.1 nm, not resolved for L=0.05 nm. The position of this line is in perfect agreement with the free exciton absorption peak detected on the same layers shown on the figure [6], so we attribute the line at 1.595 eV to the FE line [8].

Usually this transition is hardly distinguished in the photoluminescence of bulk material due to the very strong reabsorption by the crystal itself. Here the reabsorption is very limited since the layers are very thin. In contrast the LO

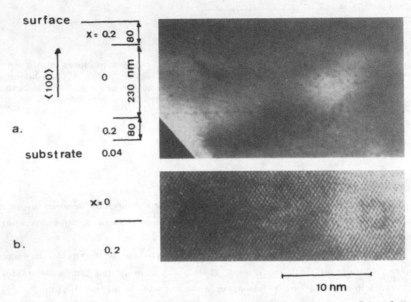

Fig. 4 TEM observation of a cross section of the heterostructure. One of the interfaces is decorated by precipitates of unknown origin. (a) HRTEM imaging of the interface region along a (110) direction (b).

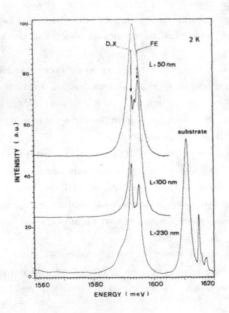

Fig. 5 Luminescence spectra of CdTe wells of various thickness. The line labelled FE is not clearly resolved for the 20 nm well.

replica of the FE are easily observed on bulk materials and not detected here.

There are several reasons for the enhancement of the intrinsic transitions. Firstly, due to the cap layer, the surface recombination is impossible, this increases the lifetime of the carriers [9]. Furthermore, if the well collects efficiently the carriers generated in the barriers, the effective carrier conentration in the well is much higher than in an homogeneously excited bulk sample under comparable laser power. As a consequence the extrinsic transitions become saturated while the intrinsic line increases. This point is confirmed by the effect of a reduction of the excitation power shown in Fig. 6. The FE line disappears when the excitation power is reduced within a factor of 10^2 to 10^3.

The absence of LO replica of the FE line remains an intringuing point. The emission of LO phonons is usually required by the k-vector conservation for the recombination of "hot" excitons, with non zero k-vector [10]. Apparently in these confined layers this condition becomes unnecessary despite the strong excitation. Either the excitons thermalize before recombination or the carriers have been efficiently thermalized before the formation of the exciton. This point has to been explained further.

A similar behavior of the intrinsic/extrinsic lines has been observed in III V compounds. In particular this has been used as an interface quality test for GaAs 10 nm wells sandwiched between $Ga_xAl_{1-x}As$ barriers. For very good interfaces the luminescence is dominated by intrinsic processes while it is reduced and

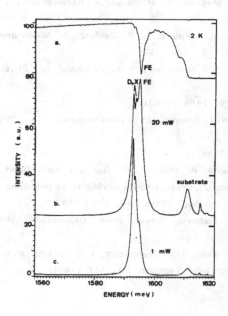

Fig. 6 Absorption (a) and luminescence (b:c) spectra of a CdTe well of 100 nm thickness. For spectrum (c) the excitation power has been reduced (x 1/20). The FE transition is then very small.

dominated by extrinsic lines if one of the interfaces is of poorer quality [11]. Here the well thickness is much higher and the other parameters of the well are very different. From these observations we can conclude that the non radiative interface processes, if any, are much less efficient than the increase of the collection of the carriers, or the surface recombination reduction.

V. CONCLUSION

The growth of CdZnTe layers, which could be used as buffer layers before the growth of $Cd_{1-x}Hg_xTe$ has been described.

Thick layers are almost strain free and reproduce the properties of bulk material. The doping level is close to 10^{14} cm^{-3}. Thin strained layers of very good quality can also be grown. The recombination properties of thin CdTe wells are dominated by intrinsic processes.

REFERENCES

[1] J.P. Faurie, A. Million, R. Bock and J.L. Tissot, J. Vac. Sci. and Technol. A1, 1593 (1983).

[2] J.L. Pautrat, J.M. Francou, N. Magnea, E. Molva and K. Saminadayar, J. Cryst. Growth, 72, 194 (1985).

[3] J.M. Francou, K. Saminadayar, J.L. Pautrat, J.P. Gaillard, A. Million and C. Fontaine, J. Crystal Growth, 72, 220 (1985).

[4] G. Milchberg, K. Saminadayar, E. Molva and H.R. Zeismann, J. Phys. Chem. Solids, 46, 423 (1985).

[5] D.T.F. Marple, Phys. Rev. 129, 2466 (1963).

[6] Z.C. Feng, A. Mascarenhas, W.J. Choyke, J. of Luminescence, 35, 329 (1986).

[7] C. Fontaine, thesis Grenoble (1986).

[8] N. Magnea, F. Dal'bo, C. Fontaine, A. Million, J.P. Gailliard, Le Si Dang, Y. Merle d'Aubigne, S. Tatarenko, J. Cryst. Growth (1986) to be published.

[9] R.J. Nelson and R.G. Sobers, J. Appl. Phys. 49, 6103 (1978)

[10] E. Gross, S. Permogorov, B. Razbirin, Soviet. Phys. Uspekhi 14, 104 (1971)

[11] W.T. Masselink, Y.L. Sun, R. Fisher, T.J. Drummond, Y.C. Chang, M.V. Klein and H.K. Morkoc, J. Vac. Sci. Technol. B2, 117 (1984)

DUAL ION BEAM SPUTTER DEPOSITION OF
CdTe, HgTe AND HgCdTe FILMS

S. V. KRISHNASWAMY, J. H. RIEGER, N. J. DOYLE, AND M. H. FRANCOMBE
Westinghouse R&D Center, 1310 Beulah Road, Pittsburgh, PA 15235

ABSTRACT

Experiments have been performed to assess the feasibility of using ion-beam sputter deposition for the growth of CdTe, HgTe and HgCdTe films. Some simple cryogenically cooled dual-target configurations have been employed in an investigation of epitaxial growth on CdTe substrates. Good-quality epitaxy was achieved for CdTe at temperatures down to 140°C, and for HgTe and HgCdTe at temperatures extending to below 50°C. Based upon compositional and phase analyses, and upon IR absorption measurements, we conclude that, using an excess Hg flux, stoichiometric transfer of the HgCdTe target composition to the substrate is approximately obtained. However, some departure from stoichiometry is produced at higher substrate temperatures (> 150°C) due to thermal re-evaporation of Hg, and under high sputtered Hg fluxes due to selective re-sputtering of HgTe. The good structural quality and excellent compositional uniformity of the films indicate that ion-beam sputter deposition may be suitable for low-temperature processing of IR detector structures.

INTRODUCTION

In recent years considerable efforts have been devoted to the development of $Hg_{1-x}Cd_xTe$ as multi-spectral, high-performance IR detector material. Growth by epitaxy of heterojunction [1] IR detectors embodying precise thickness, composition and doping profiles has required particular attention. There is also strong interest in HgTe/CdTe superlattice [2] as a promising candidate for IR detection at wavelengths beyond 18 μm.

In order to achieve the desired composition and doping profiles in such layered structures, growth temperatures should be below 250°C. In fact, recent results [3] for superlattices suggest that measurable interdiffusion can occur at temperatures even as low as 110°C. The only growth techniques that appear capable of satisfying these conditions are MBE [4], laser-assisted deposition (LADA) [5], energy-assisted MOCVD [6], and sputtering [7]. In general, all these techniques are at the research or exploratory technology stage, and although preliminary detector (and even array) fabrication has been achieved, the full potential and relative merits of these approaches have yet to be demonstrated.

In general for MBE, LADA, and MOCVD growth methods, the deposition rate at the substrate surface, and hence the film composition, is controlled by thermal effects characterized by rather slow time response. On the other hand, sputtering, and in particular ion-beam sputtering, is unique in that thermal influences can be virtually eliminated. The ion-beam current and rate of sputtering can be modulated instantaneously by electronic means, and hence the deposition of sputtering material is amenable to precise control.

In an earlier report [8] we demonstrated that ion-beam sputtering from cryogenically cooled targets of HgTe and HgCdTe results in stoichiometric transfer of these materials onto cooled substrates. If the sputtered material is deposited onto uncooled substrates, partial loss of Hg occurs and the loss increases with increasing substrate temperature.

More recently [9], we presented results on growth of both CdTe and HgCdTe at elevated temperatures. These studies showed that when using targets of CdTe, epitaxial CdTe films of good quality were obtained at temperatures as low as 140°C. By co-sputtering from cryogenically cooled targets of HgCdTe and solid Hg, single-phase epitaxial layers of HgCdTe were grown at temperatures as low as 55°C. In this report we describe in more detail the conditions for growth of CdTe, HgCdTe, and HgTe films and present new information on compositional, structural, and IR absorption properties.

EXPERIMENTAL

The films were deposited in a UHV compatible dual ion-beam system, details of which have been published elsewhere [9]. Provision was made for cooling the targets (sputtered with 3 cm Kaufman-type ion guns) with liquid nitrogen. A vacuum load-lock for introduction of samples was incorporated, and the substrate holder was rotated during deposition.

Several dual-sputtering configurations were considered (Figure 1), only two of which have yet been tested. A third configuration has also been used in the present work, in which the second target was comprised of CdTe to allow pre-deposition of a thin epitaxial CdTe buffer film prior to growth of the main HgCdTe layer. Most of the experiments to date were performed using single-crystal pieces of $Hg_{1-x}Cd_xTe$ of composition x = 0.21 (Cominco, Inc.). Supplementary Hg was provided by floating the pieces in a pool of mercury, which was subsequently frozen, and sputtering from the combined solid target. In selected cases for HgCdTe deposition, and in most of the HgTe film depositions, we used the arrangement shown in Figure 1(a) in which a second solid Hg target was sputtered to obtain supplementary Hg. The CdTe and HgTe target material employed was prepared respectively by hot-pressing (CERAC, Inc., Milwaukee, WI) and by melt-growth (Cominco, Inc.). CdTe (001) pieces (1 cm^2) were used as substrates for film deposition, and additional silicon substrates were used for thickness measurement purposes. In most of our depositions the substrate-to-target distance was ~ 7 cm. Using a beam of 1 kv Ar$^+$ ions at a total ion current of 30 mA, a deposition rate of ~ 1 to 3 μm/hr was obtained, depending on the target configuration.

Phase composition and epitaxial structure of the deposited films were determined from RHEED as well as x-ray oscillation patterns. In addition, double-crystal x-ray rocking curve data were obtained for selected epitaxial films. Film composition and that of the target surface was obtained from EDAX data. In analyzing the composition from EDAX data, the x values of the $Hg_{1-x}Cd_xTe$ targets derived from IR absorption measurements were used for calibration purposes. IR absorption curves for the HgCdTe films were measured with a Nanosec/Infrared microspectrophotometer, and for selected CdTe films photoluminescence spectra at 2 K were obtained.

RESULTS

Table 1 summarizes the results for selected CdTe, HgTe and HgCdTe films on CdTe substrates. Epitaxial CdTe films ~ 0.5 to 1.5 μm thick were grown on CdTe (001) substrates for substrate temperatures as low as 140°C and up to 330°C, with optimum growth occurring at ~ 240°C. For T_s above 300°C the sticking coefficient of Cd and Te atoms decreases, resulting in the film deposition rate dropping significantly. When grown below 140°C loss of epitaxy occurs.

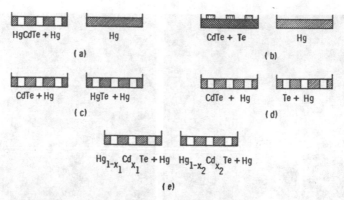

Figure 1. Target arrangements for dual ion beam sputtering.

Table 1. Summary of Selected CdTe, HgTe and HgCdTe Film Depositions

Run #	T_s (°C)	Thickness (μm)	RHEED	Conductivity
CdTe-18	140	0.66	epi	-
CdTe-26	240	1.20	epi	-
CdTe-19	330	0.42	epi	-
HgTe-2	60-74	2.00	epi	n
HgTe-3	85	2.00	epi	p
HgTe-5	110	4.20	poly	p
HgCdTe-1	50	2.50	epi	n
HgCdTe-10	75	0.73	epi/twinned	p
HgCdTe-8	100	0.60	epi	-
HgCdTe-7	125	0.60	poly	-
HgCdTe-5	150	0.70	poly/weak	-
HgCdTe-6	170	0.57	poly	-

A typical RHEED pattern for an ion-beam sputtered CdTe layer is shown in Figure 2a, which reveals the homoepitaxial nature of the film. Figure 2b is an x-ray oscillation pattern for the same film. A double-crystal x-ray rocking curve for a 0.8 μm thick CdTe film grown at 240°C indicated a FWHM for the (400) peak of 46 arc second compared to that of 23 arc second for the substrate. This compares very favorably with our best MBE-grown CdTe on InSb, which yields a FWHM for the same peak of 18.6 arc seconds (cf. the InSb substrate FWHM of 13.2 arc seconds) [10]. No noticeable differences could be seen between films sputtered from a pressed powder target and from a single-crystal target. Photoluminescence spectra were obtained on CdTe films ion-beam sputtered at 185°C which indicated low concentration of radiative defects. The radiative defect density (ratio of the peak of broad band at 1.44 eV to the peak of bound exciton at 1.59 eV), ρ, for these films is ~ .002 to .004. The value of ρ reported in the literature [11] for MBE CdTe films grown at similar temperatures ranges from .006 to .020.

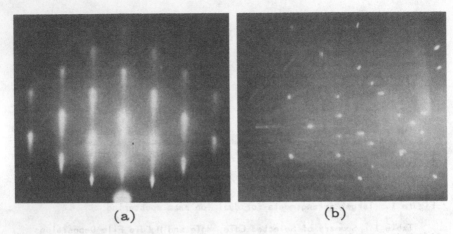

(a)   (b)

Figure 2.   (a) RHEED pattern, [110] azimuth, for ion-beam sputtered CdTe film
(0.5 μm thick) on CdTe (001) substrate grown at T_s = 160°C.
(b) x-ray oscillation pattern of the above sample confirming
homoepitaxial growth.

As seen from the table, epitaxial HgTe films were obtained for T_s as low
as 60°C. Figure 3 shows the RHEED pattern of a HgTe film deposited on a CdTe
(001) substrate at 85°C. From the figure we can see that a (100) 2 x 1
structure is formed in the case of HgTe/CdTe. A similar observation has been
made for MBE-grown ZnTe and CdTe superlattice layers [10]. Films deposited
above 85°C were p-type and those below 85°C were n-type.

HgCdTe films of various thicknesses (0.5 - 14 μm) were deposited on CdTe
(001) substrates with substrate temperatures varying from 50°C to 170°C.
Figure 4 schematically represents the results of the HgCdTe film depositions.
Most of the films were less than 2 μm thick and when deposited at T_s > 100°C,
were polycrystalline, whereas films deposited at T_s ≤ 100°C were epitaxial.
In some cases the substrate was not heated directly, and it was found that
the sample temperature rose during deposition, usually to values of 60-70°C,
due to heat radiated from the ion guns. In other cases the temperature was
held at a constant value over the length of the run. Figure 5 compares a
RHEED pattern and x-ray oscillation pattern for a 2.5 μm thick HgCdTe film
with a 0.5 μm CdTe in-situ buffer layer, both indicating epitaxial growth.
Again for these films we notice a (100) 2 x 1 structure similar to that
observed for HgTe films. However, with no buffer CdTe layer, HgCdTe films
even when epitaxial showed twins and, furthermore, films greater than 2 μm
were polycrystalline, even when they were deposited at temperatures below
100°C.

As can be seen from Table 1 as well as Figure 4, in films deposited at
higher temperatures, loss of Hg occurs as detected by the presence of weak Te
phase. Also, under the deposition condition where excess Hg is provided, re-
sputtering of Hg and Te from the growing film seems to occur. This is seen
from the higher x value, as evidenced by the shift in IR absorption cut-off
as well as EDAX data, for a film deposited with excess Hg, than that for a
film deposited with no excess. The loss appears to increase with
supplementary Hg flux and also with increase in the substrate temperature.

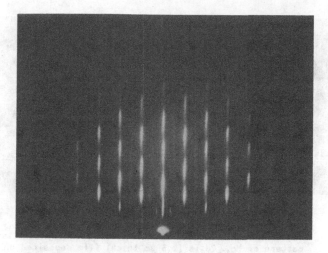

Figure 3. RHEED pattern, [001] azimuth, of HgTe films deposited on
(001) CdTe substrates at 85°C.

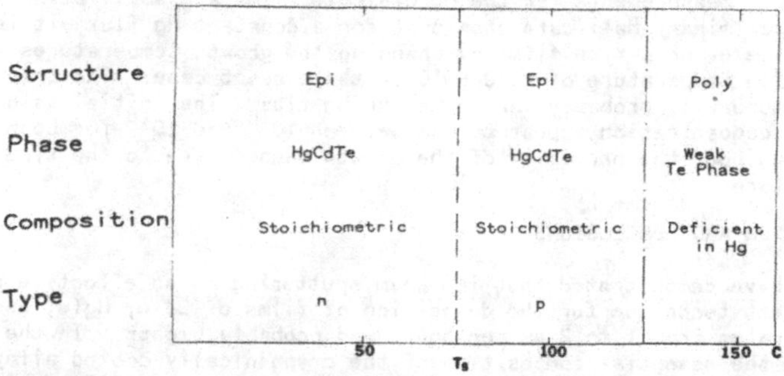

Structure	Epi	Epi	Poly
Phase	HgCdTe	HgCdTe	Weak Te Phase
Composition	Stoichiometric	Stoichiometric	Deficient in Hg
Type	n	p	

                              50   100   150   C
                                        T_s

Figure 4. Schematic description of characteristics of HgCdTe films as a
function of substrate temperature during deposition.

IR absorption spectra obtained for thicker films showed (Figure 6) cutoff wavelengths close to about 5.5 μm at 77 K. This corresponds to an x
value of 0.29, a composition somewhat lower in Hg than the EDAX analyses for
the films indicated, and suggests that the films might contain some HgTe as a
second phase. Also Figure 6, which is a composite of IR data taken from the
sample at four different regions 1 cm apart, indicates that the compositional
uniformity as judged by IR absorption data exists over a 1 cm² area.

468

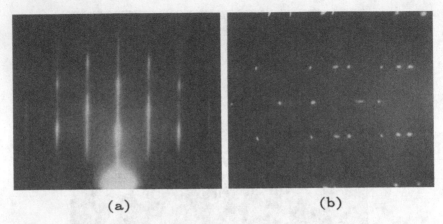

(a)   (b)

Figure 5. (a) RHEED pattern [110] azimuth, and (b) x-ray oscillation
           pattern of $Hg_{1-x}Cd_xTe$ (2.5 μm thick) film deposited on CdTe
           (001) substrate at 50°C. Arrow points to (100) 2 x 1
           structure.

Electrical measurements for the HgCdTe/CdTe films are still being carried
out. Preliminary Hall data show that for a constant Hg flux, it is possible
to produce n- or p-type films by changing the growth temperatures with a
transition temperature of ~ 70-80°C in the present case. The transition
temperature will probably vary with the Hg flux. The initial values of
carrier concentration appear to lie between 10^{16} and 10^{18} for both types,
depending upon the proximity of the growth temperature to the type transition
temperature.

DISCUSSION AND CONCLUSIONS

     We have demonstrated that ion-beam sputtering is an effective and
convenient technique for the deposition of films of CdTe, HgTe, and HgCdTe at
growth rates from 1 to 3 μm per hour, and probably higher. In the case of
HgCdTe, the essential composition of the cryogenically cooled alloy
sputtering target can be preserved intact in the growing film, using excess
sputtered Hg flux ratios of 2 to 3X supplied from a solid source, provided
that the substrate temperature does not exceed 150°C. Higher excess Hg
fluxes can readily be obtained, either by use of a supplementary sputtering
target or by use of a thermal evaporation source; thus, there should be no
problem in achieving growth temperatures as high as those used typically in
MBE, i.e., ≃ 185°C.
     High-quality epitaxial growth of CdTe has been demonstrated for ion-beam
sputtered films deposited at temperatures down to 140°C. Using pre-deposited
CdTe buffer layers, epitaxy of HgCdTe films sputtered from alloy targets was
achieved at surprisingly low temperatures, with twin-free films being
obtained on CdTe substrates reproducibly at temperatures between 30 and
100°C. We suggest that use of a single- composition alloy target facilitates

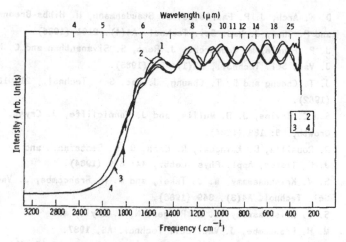

Figure 6. IR transmission spectra of $Hg_{1-x}Cd_xTe$ film (6.2 μm) taken at four different regions of the CdTe substrate.

homogeneous transfer of the HgCdTe composition to the substrate, minimizes the need for thermal diffusion to remove compositional variations, and helps promote epitaxial growth at significantly lower temperatures than for other vapor-phase techniques. This process may be aided by the presence of an excess flux of sputtered Hg atoms possessing high kinetic energy. IR absorption data for thicker films grown at temperatures below 100°C indicate a composition of $x = 0.29$; cf. the chemical composition of $x = 0.21$ for the target. EDAX composition analyses yield a lowest value of x for the films of $x = \simeq 24$, but this value is subject to a rather large uncertainty of at least ± 0.02. From presently available data we conclude that the chemical composition of the films can deviate from that of the target due to two effects, both of which have been observed in the present investigation. First, at higher growth temperatures when an inadequate supplementary flux of Hg is used, Hg is lost due to thermal re-evaporation, leading to the appearance of a second free Te phase. Second, use of higher Hg fluxes appears to result in selective loss of some HgTe (1-3%) from the film due to re-sputtering. This effect is accentuated at higher temperatures.

ACKNOWLEDGEMENTS

We would like to thank C. M. Pedersen for substrate preparation, Dr. W. J. Takei and Dr. A. J. Noreika for helpful discussions, and Dr. W. J. Choyke for performing the photoluminescence measurement.

REFERENCES

1. P. R. Bratt, J. Vac. Sci. Technol., A1: 1687 (1983).

2. T. C. McGill, G. Y. Wu, and S. R. Hetzler, J. Vac. Sci. Technol., A4(4), 2091 (1986).

3. D. K. Arch, J. P. Faurie, J. L. Staudenmann, H. Hibbs-Brenner, and P. Chow, J. Vac. Sci. Technol. A4(4), 2101 (1986).

4. J. P. Faurie, M. Boukerche, J. Reno, S. Sivananthan and C. Hsu, J. Vac. Sci. Technol., A3(1): 55 (1985).

5. J. T. Cheung and D. T. Cheung, J. Vac. Sci. Technol., 21: 182 (1982).

6. S.J.C. Irvine, J. B. Mullin, and J. Tunnicliffe, J. Cryst. Growth, 68: 188 (1984).

7. R. Rousille, D. Amingual, R. Corh, G. L. Destefanis and J. L. Tissot, Appl. Phys. Lett., 44: 679 (1984).

8. S. V. Krishnaswamy, W. J. Takei, and M. H. Francombe, J. Vac. Sci. Technol. A4(3), 849 (1986).

9. S. V. Krishnaswamy, J. H. Rieger, N. J. Doyle, and M. H. Francombe, J. Vac. Sci. Technol. A5, 1987.

10. G. Monfroy, S. Sivanathan, X. Chu, J. P. Faurie, R. D. Knox, and J. L. Staudenmann, Appl. Phys. Letts. 49:152 (1986).

11. Z. C. Feng, A. Mascarenhas, W. J. Choyke, R. F. C. Farrow, F. A. Shirland, and W. J. Takei, Appl. Phys. Lett., 47:25 (1985).

CdTe FILMS GROWN ON InSb SUBSTRATES BY ORGANOMETALLIC EPITAXY

I.B. BHAT, N.R. TASKAR, J. AYERS, K. PATEL, AND S.K. GHANDHI
Rensselaer Polytechnic Institute, Electrical, Computer, and Systems
Engineering Department, Troy, New York 12180

ABSTRACT

Cadmium telluride layers were grown on InSb substrates by organometallic vapor phase epitaxy and examined using secondary ion mass spectrometry
(SIMS), photoluminescence (PL) and double crystal x-ray diffraction (DCD).
The substrate temperature and the nature of the surface prior to growth are
shown to be the most important parameters which influence the quality of
CdTe layers. Growth on diethyltelluride (DETe) stabilized InSb substrates
resulted in CdTe growth with a misorientation of about 4 minutes of arc with
respect to the substrates. On the other hand, the grown layers followed the
orientation of the substrates when a dimethylcadmium (DMCd) stabilized InSb
was used. Growth at 350°C resulted in the smallest x-ray rocking curve
(DCRC) full width at half maximum (FWHM) of about 20 arc seconds.

INTRODUCTION

The growth of CdTe on InSb has attracted much attention [1-4] in recent
years, because of their close lattice match (better than 0.05% at 25°C). In
addition, InSb has higher crystalline perfection than CdTe and is less
expensive. It is also available as a large diameter wafer (40mm). We have
reported earlier that quality of CdTe grown on InSb substrates by organometallic vapor phase epitaxy (OMVPE) depends on the growth temperature
and on the surface quality just prior to growth [5, 6]. In this paper, we
report on additional characterization of CdTe layers, prepared under a
variety of reactor conditions. The layers were characterized by PL at 10 K,
by SIMS, and by double crystal x-ray diffraction. Results of these characterization efforts are used to optimize the growth conditions.

EXPERIMENTAL

The CdTe films were grown in an atmospheric pressure, horizontal reactor, of the type previously described [5], using diethyltelluride (DETe) and
dimethylcadmium (DMCd) as the tellurium and cadmium sources respectively, at
growth temperatures from 350 - 420°C. The substrates were (100) 2°—>(110)
oriented InSb, prepolished on one side. The DMCd and DETe partial pressures
were 0.5×10^{-4} atm. and 2×10^{-4} atm. respectively, and growth times were
adjusted so that the layer thickness was approximately 2 μm in each run.

PL measurements were made using 7 mW He-Ne laser with a 0.75 meter
monochromator and a liquid nitrogen cooled detector with a S-1 photocathode.
The x-ray measurements were made with a computer controlled 6" inch double
crystal diffractometer (Bede Scientific Instruments Ltd., England) using
InSb as the first crystal and Cu K_α radiation. All x-ray curves refer to
(400) reflections. The uniformity of the material was checked by measuring
the double crystal rocking curve (DCRC) at different regions in the sample,
using an x-y translation stage. The beam incident on the sample was 1mm x
2mm in size.

CHARACTERIZATION

Secondary Ion Mass Spectrometry:

Secondary Ion Mass Spectrometry (SIMS) of CdTe layers has been used in order to study the interdiffusion of indium into the layer. Figure 1(a) and Fig. 1(b) show the SIMS data for CdTe layers grown at $350^{\circ}C$ on DMCd stabilized and DETe stabilized InSb substrates respectively. It is seen that the indium level observed in the CdTe layers grown on DMCd stabilized substrates is about 6-7 times higher than that grown on DETe stabilized InSb, under identical growth conditions. We have reported earlier [5] that there is a surface reaction between InSb and DMCd which forms alloys of In-Cd-Sb at the CdTe-InSb interface. Thus, indium is freely available for diffusion, rather than being bonded as InSb. We believe that this is the cause for a higher In level in these layers. This type of visible surface reaction is not observed between DETe and InSb. The interdiffusion of In into the CdTe layer is more pronounced when the growth temperature is increased to $420^{\circ}C$. In addition, autodoping of the CdTe layer occurs because of the evaporation of In from the back surface of the substrate. A significant amount of Indium (10^{15}-$10^{17}/cm^3$) can be incorporated in this manner, depending upon reactor conditions.

Photoluminescence:

PL measurements made on these layers confirm the above SIMS results. The PL spectra of the CdTe grown on DETe stabilized InSb substrates exhibited a very narrow (2.1 meV FWHM) exciton peak near 1.59 eV, indicating a very low doping level. On the other hand, the PL spectra of CdTe grown on DMCd stabilized surface had a relatively wider (5.8 meV FWHM) exciton band, because of the higher doping level (6). Layers grown at $420^{\circ}C$ had much higher doping level (about 2×10^{17} cm^{-3} as determined by capacitance voltage measurements) and larger FWHM, consequently the exciton band in the PL spectra was not resolved (5). Instead, a new level at 1.587 eV was seen, the origin of which is not clear. From the PL and SIMS measurements and also from the study of the layer morphology, it was found that growth at around $350^{\circ}C$ on a DETe stabilized surface results in good quality CdTe layers, with minimum In interdiffusion.

Double Crystal X-ray Diffraction:

PL measurements give only a qualitative measure of both the crystalline perfection as well as its doping level. In order to assess the crystalline quality, x-ray diffraction measurements have been taken. Double crystal x-ray rocking curves (DCRC) give a quantitative measure of epitaxial layer quality. For example, for a perfect InSb crystal the convoluted full width at half maximum of the (400) reflection will be about 14 arc second (7). A higher FWHM of the DCRC indicates poorer crystal quality which in turn affects the epilayer grown on it.

Figure 2(a) shows the DCRC of a 1.7 μm thick CdTe epilayer grown at $350^{\circ}C$ on a DMCd stabilized InSb surface. The (400) reflection of the epilayer exhibit a convoluted FWHM of about 22 arc seconds. The line width shown here is among the lowest value reported for OMCVD grown CdTe and compares very well with the best molecular beam epitaxial CdTe layers (3). The peak separation of 114 arc second is close to the theoretical value

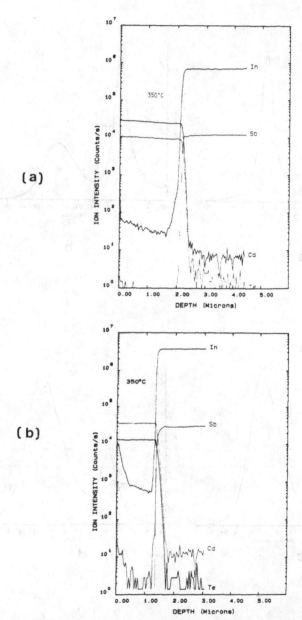

Figure 1. SIMS profile of CdTe grown at 350°C on (a) DETe stabilized InSb substrate, (b) DMCd stabilized InSb substrate.

474

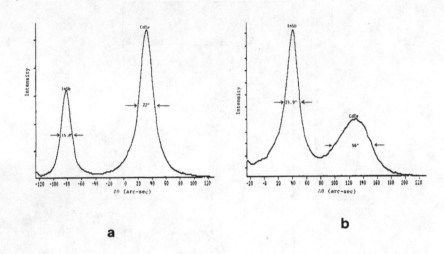

a

b

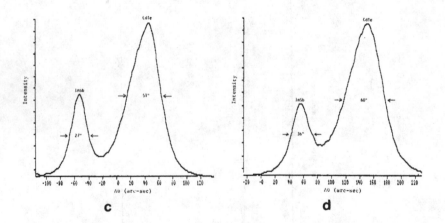

c

d

Figure 2. DCRC of CdTe grown on DMCd stabilized InSb substrate at (a) 350°C,
(b) 380°C, (c) 400°C, (d) 420°C.

expected, assuming a lattice constant of 6.477 Å for InSb and 6.482 A° for CdTe.

In order to study whether the epilayer has grown with the same misorientation as the substrate, the sample was rotated with respect to the surface normal by 180°. The measured DCRC peak spacing was independent of the sample rotation angle, indicating that the epilayer has the same orientation as the substrate.

Figures 2(b), 2(c), and 2(d) show the rocking curves of CdTe layers grown at 380°C, 400°C and 420°C respectively, on DMCd stabilized InSb substrates. The (400) x-ray reflection of both the substrates and the epilayers have a wider FWHM with increasing temperature. All layers exhibit similar x-ray rocking curves, somewhat inferior to those obtained for layers grown at 350°C. The peak separation between the CdTe and InSb x-ray peaks is again close to the theoretical value of 86 arc second. From these results, as well as the PL data, we conclude that optimum growth temperature for the CdTe growth is about 350°C.

Figure 3 shows an x-ray rocking curve of a CdTe layer grown on a DETe stabilized InSb substrates at 350°C. Both the FWHM of the CdTe (400) reflection and the peak separation between the two reflections are different from the previous case. Here, the FWHM of the CdTe layer is 250 arc seconds In addition, the CdTe peak is separated from the InSb peak by 300 arc seconds, which is much higher than the theoretical value of 86 arc secs. The large peak separation shown in this figure will result because of two reasons. First, if the CdTe layer is under compression along the interface because of the lattice mismatch between InSb and CdTe, then the CdTe lattice will expand in the direction perpendicular to the interface. This will result in an increaesd CdTe-InSb peak spacing. Second, if the CdTe layer grows with a lattice misorientation with respect to the substrate, the result will be a larger or smaller peak separation than the theoretical value. To determine the degree of misorientation with respect to the substrate, we have taken the rocking curve with the sample rotated by 180° along the sample normal. The result is shown in Fig. 4. When compared to Fig. 3, it is seen that the CdTe peak now appears before the InSb peak, and the peak separation is 170 arc second. From this data, we conclude that the CdTe layer has grown with a misorientation of about 4 min. of arc with respect to the substrate. It is not clear why the CdTe growth on a DETe-stabilized surface occurs with a small misorientation with respect to the substrate . Other workers have observed much larger misorientation, of

about 48 min. of arc when CdTe layer was grown on (110) InSb substrates [1]. In summary therefore, we find that stress as well as layer misorientation are present in the growth of layers on DETe stabilized substrates. It is interesting to note that the PL spectrum of a CdTe layer grown on a DETe stabilized surface shows a narrower linewidth than the PL spectrum of CdTe grown on a DMCd stabilized surface, whereas the reverse is true in the case of DCR curves. This is not contradictory, since the FWHM of PL spectra is very sensitive to the doping level in the layer, as well as to the crystal quality. This is indeed evident from Fig. 1, which shows the level of In doping for layers grown at 350°C. Hence, PL alone cannot be used to assess the quality of layer. From a device point of view, layers grown with DETe stabilized surface are preferred to layers grown on DMCd stabilized surfaces at 350°C, because they have superior morphology, with no evidence of hillocks [5].

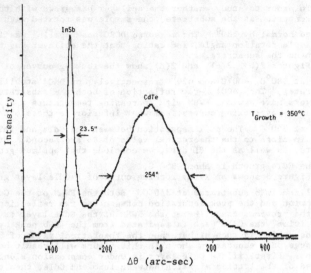

Figure 3. DCRC of CdTe grown on DETe stabilized
InSb Substrate.

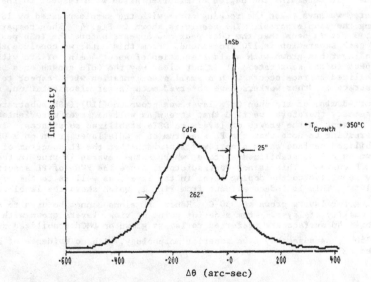

Figure 4. DCRC of CdTe layer and InSb substrate,
when the sample was rotated by 180° with
respect to the condition in Figure 3.

CONCLUSIONS

We have shown the effect of the growth temperature and the nature of the InSb surface prior to growth on the quality of CdTe layers, using PL, SIMS and DCRC. It is found that 350°C is an optimum temperature for the growth of CdTe on InSb. Indium diffusion from the substrate can be reduced by growing on a DETe stabilized surfaces, as seen from SIMS data. However, these layers had wider DCR curves, indicating poorer crystal perfection. The epilayer has an additional misorientation of a few minutes of arc with respect to the substrate. In summary, therefore very high quality CdTe layer can be grown by OMCVD, as evidenced by x-ray rocking curves.

ACKNOWLEDGEMENT

The authors would like to thank J. Barthel for technical assistance on this program, and P. Magilligan for manuscript preparation. This work was sponsored by the Defense Advanced Research Projects Agency (Contract Number N-00014-85-K-0151), administered through the Office of Naval Research, Arlington, Virginia; and by the Solar Energy Research Institute (Grant No. ZL-5-04074-2). Additional funds were provided by Agreement No. 751-RIER-BEA-8 from the New York State Energy Research and Development Authority, in the form of a grant from the International Telephone and Telegraph Company, and from an IBM Fellowship to one of the authors (N.R.T.). This support is greatly appreciated.

REFERENCES

[1] W.E. Hoke, P.J. Lemonias and R. Traczewski, Appl. Phys. Lett., 44, 1046 (1984).
[2] S.K. Ghandhi and I. Bhat, Appl. Phys. Lett., 45, 678 (1984).
[3] A.J. Noreika, R.F.C. Farrow, F.A. Shirland, W.J. Takei, J. Greggi, S. Wood, J. Vac. Sci. Technol., A4, 2081 (1986).
[4] J.M. Ballingall, D.J. Leopold and D.J. Peterman, Appl. Phys. Lett., 47, 262 (1985).
[5] N.R. Taskar, I.B. Bhat and S.K. Ghandhi, J. Crystal Growth, 77, 480 (1986).
[6] S.K. Ghandhi, N.R. Taskar and I.B. Bhat, to be published in Applied Phys. Lett.
[7] J.H. Dinan and S.B. Qadri, J. Vac. Sci. Technol., A4, 2158 (1986).

PLASMA-ASSISTED EPITAXIAL GROWTH OF COMPOUND SEMICONDUCTORS FOR INFRARED APPLICATION

K. MATSUSHITA, T. HARIU, S. F. FANG, K. SHIDA AND Q. Z. GAO
Department of Electronic Engineering, Tohoku University,
Sendai 980, Japan

ABSTRACT

GaSb, InSb and InAs epitaxial layers with mirror surface were grown on GaSb, GaAs, InP, Si and sapphire substrates at relatively low temperatures by plasma-assisted epitaxy (PAE) in hydrogen plasma. Carrier concentrations and Hall mobilities of undoped PAE layers at room temperature are $p=6\times10^{16}cm^{-3}$; $\mu p=750cm^2/Vs$, $n=1\times10^{16}cm^{-3}$; $\mu n=39,000cm^2/Vs$ and $n=7\times10^{17}cm^{-3}$; $\mu n=21,000cm^2/Vs$ for GaSb on GaAs, InSb on GaAs and InAs on InP, respectively. As the first application of PAE layers to optoelectronic devices, p-GaSb/n-GaAs heterojunction photodiodes have been demonstrated to result in remarkable reduction of dark current with photoresponse in the wavelength region between 0.85 and 1.7μm for the light incident from GaAs.

INTRODUCTION

GaSb, InSb, InAs and their alloys are attractive materials for application to optoelectronic devices in the longest wavelength infrared region of all III-V compound semiconductors. These devices can be discrete, as some of them have already been in practical application, but they are desired to be integrated on the same chip of Si IC or higher speed GaAs IC to achieve more intelligent functions with higher sensitivity and higher speed.
It is then required to develop technologies which can grow these materials epitaxially at low temperatures even on lattice mis-matched substrates, as in the case of GaAs on Si which recently has been extensively investigated.

The purpose of this paper is to describe first the growth behaviors and the electronic properties of plasma-assisted epitaxial GaSb, InSb and InAs layers on different substrates, and second the electrical characteristics and photoresponse of p-GaSb/n-GaAs heterojunction photodiodes which are fabricated by PAE GaSb layers directly grown on n-GaAs substrates.

PLASMA-ASSISTED EPITAXY (PAE)

PAE has been developed for the low temperature epitaxial growth of semiconductor crystals by supplying atoms with enhanced internal energy to activate chemical reaction and kinetic energy for migration on the growing surface [1]. Several other advantages of PAE, including the cleaning effect of substrate surface at low temperatures have already been confirmed [2,3].

Similar PAE apparatuses as described elsewhere [4,5] were used for the growth of GaSb, InSb and InAs layers. In order to precisely control the supply rate of constituent atoms independent of plasma conditions, elemental sources (6-nine purity) were used and supplied from resistively heated crucibles, in the same way as MBE, through hydrogen plasma to substrates. The use of hydrogen gas at $10^{-1}-10^{-2}$ Torr was found to give epitaxial layers of much better quality compared with other plasma gases.

Different substrates including GaSb, GaAs, InP, Si and sapphire were used to compare the relaxation of lattice mis-match and the electronic properties of grown layers. They are usually <100> oriented, but sometimes substrates with orientation 3 and 8 degrees off <100> toward (110) were also used. After chemical etching to obtain mirror surfaces, the substrates are treated in hydrogen plasma at a temperature of epitaxial growth or below, to remove native oxide layers before the growth of PAE layers.

Table I summarizes physical constants of grown layers and substrates at 300K which are used in this experiment.

Table I Physical constants of grown layers and substrates at 300 K.

Layers or Substrates	Lattice Constant (A)	Thermal Expansion Coefficient (x10^{-6}/K)
InSb	6.478	5.04
GaSb	6.094	6.7
InAs	6.058	5.19
InP	5.8694	4.5
GaAs	5.653	6.0
Si	5.4309	2.4

RESULTS

It was first confirmed that mirror surfaces of grown layers can be obtained in PAE with less supply ratio of anions relative to cations than in MBE, most likely due to the supply of excited atoms or molecules in PAE and epitaxial layers were obtained in quite a wide range of substrate temperature: 340-440°C for GaSb, 300-550°C for InAs, 250-440°C for InSb. Hydrogen gas was used as a discharging gas and the applied plasma power is 20W, constant throughout this experiment.

Properties of PAE-GaSb on GaSb and GaAs

Use of appropriate gases and optimization of applied plasma power is essential to obtain high quality epitaxial layers in PAE. Fig.1 (a) and (b) respectively show the photo-luminescence spectra of GaSb layers grown by PAE in hydrogen plasma and the maximum luminescence intensity in PAE-GaSb layers grown on (100)GaSb and (100)GaAs substrates as a function of applied plasma power. Photo-luminescence measurement indicates that the applied plasma power should be optimized to remarkably enhance the luminescence intensity near band edge irrespective of substrates.

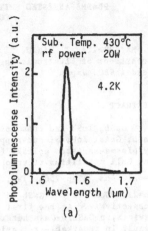

(a)

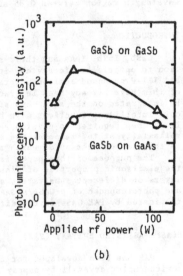

(b)

Fig.1 (a) Photoluminescence spectrum of a GaSb film deposited by PAE
(b) Maximum luminescence intensity in GaSb films deposited on GaSb and GaAs as a function of plasma power.

Fig.2 shows the Hall
mobilities of undoped PAE-GaSb
layers grown on GaAs in
comparison with those by other
methods as a function of
substrate temperature. Undoped
GaSb layers showed p-type
conduction similar to MBE layers
and crystals grown from a melt
[7]. Hole concentrations and
Hall mobilities of undoped GaSb
layers grown at a substrate
temperature of 410°C are
$6 \times 10^{16} \text{cm}^{-3}$ and $750 \text{cm}^2/\text{Vs}$,
respectively, which are
comparable to those obtained by
other methods like MBE, MOCVD, in
spite of a lower substrate
temperature in PAE [4].

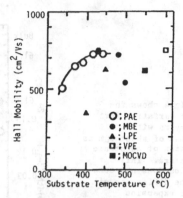

Fig.2 Hall mobilities of undoped p-GaSb
films deposited by PAE in comparison
with those by other methods as a function
of substrate temperatures.

Properties of PAE-InSb and PAE-InAs on different substrates

Fig.3 (a) and (b) show the variation of lattice constants and half-
widths of (400) X-ray diffraction lines of InSb and InAs, respectively,
grown on different substrates, as a function of the thickness of grown
layers. In both cases, the thickness above 1μm is required for the growth
restoring the lattice constants of bulk, except the case of InSb on GaSb.
In the latter case, the difference of thermal expansion coefficients [8] is
considered to be responsible for the strain even in thick layers. However,
the deviation of lattice constants of grown layers cannot be systematically
explained only in terms of the difference of thermal expansion and/or of
lattice constants. It can be suggested that the combination of different
anions between substrate and a grown layer leads to easier relaxation. The
smaller half-widths of InSb on GaSb than on GaAs, and of InAs on InP than on
GaAs can be attributed to the smaller lattice mis-match during the growth.

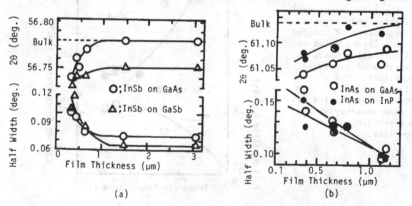

(a)   (b)

Fig.3. The variation of lattice constants (shown by X-ray diffraction
angles) and half-widths of diffraction lines of (a) InSb and (b) InAs on
different substrates as a function of thickness of layers grown at 390°C
and 450°C, respectively.

Fig.4 shows the similar variation for InAs layers with thickness of about 2μm as a function of substrate temperature. This variation does not come from the difference of thermal expansion coefficients, but from the difference of growth behaviors at different temperatures.

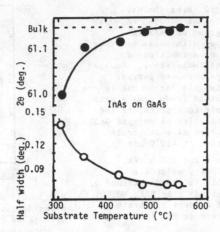

Fig.4. The variation of lattice constants (shown by X-ray diffraction angles) and half-widths of diffraction lines of InAs grown on GaAs as a function of substrate temperature.

Fig.5 shows the Hall mobilities and carrier concentrations of undoped PAE-InAs layers grown on semi-insulating GaAs, as a function of substrate temperature. The variation of the electronic properties has the same tendency with the crystallographic properties.

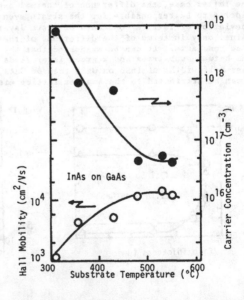

Fig.5. Hall mobilities and carrier concentrations of PAE-InAs grown on semi-insulating GaAs, as a function of substrate temperature.

Layers of similar electronic property were obtained in a wider range of supply ratio, compared with MBE [9,10,11], as shown in Fig.6. For example, Miggitt et al. complained that in MBE growth of InAs on GaAs, As/In supply ratio larger than 20/1 is required for mirror-like surface [9], but in PAE As/In ratio around 5/1 is sufficient as long as the surface morphology concerns. It is also observed that the electronic properties of InAs layers grown on InP are less sensitive to the growth conditions including this supply ratio, compared with InAs on GaAs.

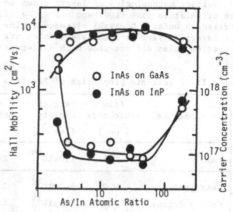

Fig.6. Hall mobilities and carrier concentrations of PAE-InAs grown at 450°C, as a function of supply ratio As/In.

The removal of native oxide from substrate surfaces at low temperatures with the help of plasma is one of the essential advantages of PAE which enables the low temperature epitaxial growth. Fig.7 shows RHEED patterns of InAs layers grown on (100)Si by PAE in hydrogen plasma, as a function of substrate temperature. It is possible to grow InAs layers even on Si below 630°C after the surface cleaning is achieved in hydrogen plasma at 530°C without the process of surface cleaning at higher temperatures before growth, while poly-crystalline layers grow on Si without this treatment.

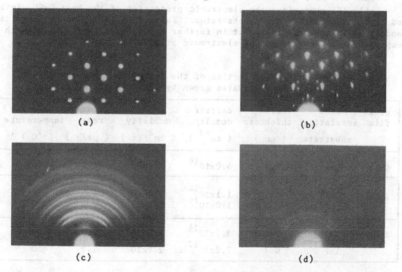

Fig.7. RHEED patterns of InAs layers grown on (100)Si (a) at 625°C by PAE, (b) at 350°C by PAE, (c) at 270°C by PAE and (d) at 610°C without hydrogen plasma cleaning.

Surface morphologies of layers grown on lattice mis-matched substrates with orientation just <100> and off <100> did not show remarkable differences under a Nomarski microscope. The electronic properties of very thin layers on various substrates, which are very sensitive to the interface properties, also did not show much difference as shown in Table II.

Table II Electronic properties of PAE-InAs layers on various substrates.

substrate	film thickness (μm)	carrier density (cm^{-3})	Hall mobility (cm^2/Vs)	growth rate (μm/h)	substrate temperature (°C)
InP	0.2	1.0×10^{19}	2.6×10^3	0.6	420
GaAs just	0.2	8.3×10^{18}	2.5×10^3	0.6	420
GaAs 3 degree off	0.2	1.5×10^{19}	1.8×10^3	0.6	420
GaAs 8 degree off	0.2	6.3×10^{18}	2.8×10^3	0.6	420

Undoped PAE-InSb layers showed n-type conduction and the similar variation of crystallographic and electronic properties as PAE-InAs were obtained as described elsewhere [6]. The optimum substrate temperature with applied rf power of 20W in this case is about 380°C for InSb and 500°C for InAs on GaAs.

Table III summarizes the electronic properties of the best GaSb, InSb and InAs layers on different substrates. Further optimization of deposition conditions is expected to result in further reduction of epitaxial growth temperature and improvement of electronic properties.

Table III Electronic properties of the best GaSb, InAs and InAs on different substrates grown by plasma-assisted epitaxy.

grown film	semi-insulating substrate	film thickness (μm)	carrier density (cm^{-3})	Hall mobility (cm^2/Vs)	growth rate (μm/h)	substrate temperature (°C)
GaSb	GaAs	1.0	6.0×10^{16}	750	1.0	410
InSb	GaAs	3.5	1.1×10^{16}	3.9×10^4	0.5	390
	sapphire	3.5	1.0×10^{16}	1.7×10^4	0.5	390
InAs	GaAs	2.0	3.6×10^{16}	1.5×10^4	2.0	550
	InP	0.7	7.2×10^{17}	2.1×10^4	1.2	450

PAE-p-GaSb/n-GaAs heterojunction photodiodes

Fig.8 shows an indepth profile of carrier concentration in a PAE-GaSb on semi-insulating GaAs grown at 390°C, revealed by differential van der Pauw measurement. The carrier concentration of a GaSb layer with average carrier concentration of 4×10^{16} cm^{-3} increases toward the interface between GaSb and GaAs. This tendency seems to result from large lattice mismatch of about 8% between GaSb and GaAs, rather than the doping effect due to the inter-diffusion between GaAs and GaSb. The similar phenomenon was observed on GaSb/GaAs structure grown by MBE by Yano et al. [7]. The carrier concentration of MBE-GaSb layers grown at 500-550°C increases by one order of magnitude over a wider region near the interface, probably due to the larger effect of the difference of thermal expansion coefficients or the enhanced inter-diffusion. The comparison of the result with ours clearly shows the advantage of low temperature epitaxy.

Table IV shows the comparison with the backward current densities of a GaSb homojunction and GaAs/GaSb heterojunction photodiodes, fabricated by LPE and PAE, respectively. The backward current density are drastically reduced by using a GaAs/GaSb heterojunction structure, which results mainly from cut-off of the current path through the surface inversion layer of n-GaSb substrate.

Fig. 9 shows photoresponse in the wavelength region between 0.85 and 1.7μm for the light incident from the GaAs substrate. Lower quantum efficiency will be improved by optimizing the structure with AR coating.

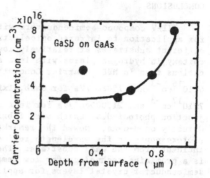

Fig.8. Indepth profile of carrier concentration in a PAE-GaSb on semi-insulating GaAs grown at 390°C, revealed by differential van der Pauw measurement.
The thickness of the GaSb layer is 1.05μm.

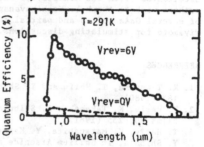

Fig.9. External quantum-efficiency spectra of photocurrent in a PAE-p-GaSb/n-GaAs photodiode at various bias voltages.

Table IV Comparison with backward current densities of n-GaSb/p-GaSb homojunction and n-GaAs/p-GaSb heterojunction photodiodes.

p-type carrier epi-layer density (cm^{-3})	n-type carrier substrate density (cm^{-3})	reverse current density (at Vrev=4V) (A/cm^2)
LPE-GaSb 7.0×10^{16}	n-GaSb 1.7×10^{18}	2.6×10^{-1}
PAE-GaSb 4.1×10^{16}	n-GaAs 2.8×10^{17}	8.7×10^{-4}
PAE-GaSb 4.1×10^{16}	n-GaAs 4.2×10^{15}	5.7×10^{-6}

CONCLUSIONS

III-V compound semiconductors with mirror surface, which are attractive for application to infrared optoelectronic devices and IC's, were grown on different substrates at relatively low temperatures by plasma-assisted epitaxy in hydrogen plasma with less supply ratio of anions relative to cations than in MBE. Carrier Concentrations and Hall mobilities of $6 \times 10^{16} \mathrm{cm}^{-3}$ and $750 \mathrm{cm}^2/\mathrm{Vs}$ for GaSb, $1 \times 10^{16} \mathrm{cm}^{-3}$ and $39,000 \mathrm{cm}^2/\mathrm{Vs}$ for InSb, and $7 \times 10^{17} \mathrm{cm}^{-3}$ and $21,000 \mathrm{cm}^2/\mathrm{Vs}$ for InAs were obtained. P-GaSb/n-GaAs heterojunction photodiodes, which were fabricated by growing epitaxial GaSb directly on n-GaAs, showed the remarkable reduction of dark current and the photoresponse in the wavelength region between 0.85 and 1.7μm for the light incident from the GaAs substrate. The present results demonstrate that PAE is a promising technology for low temperature epitaxial growth of compound semiconductor crystal layers for application to optoelectronic devices in the infrared region.

ACKNOWLEDGEMENT

This work was supported in part by the Scientific Research Grant-in-Aid #61114004 for Special Project Research on "Alloy Semiconductor Physics and Electronics" from the Ministry of Education, Science and Culture. The authors wish to thank Y. Kashiwayanagi of Furukawa Electric for the supply of several GaAs wafers and partial support, and Professors S. Ono and N. Miyamoto for stimulating discussions.

REFERENCES

1. K. Takenaka, T. Hariu and Y. Shibata, Jpn. J. Appl. Phys. Suppl.19-2, 183 (1980).
2. T. Hariu, K. Takenaka, S. Shibuya, Y. Komatsu and Y. Shibata, Thin Solid Films 80, 235 (1981).
3. T. Hariu, K. Matsushita, Y. Komatsu, S. Shibuya, S. Igarashi and Y. Shibata, in Gallium Arsenide and Related Compounds, edited by G. E. Stillman, Inst. Phys. Ser. 65, 141 (1982).
4. Y. Sato, K. Matsushita, T. Hariu and Y. Shibata, Appl. Phys. Lett. 44, 592 (1984).
5. K. Matsushita, T. Sato, Y. Sato, Y. Sugiyama, T. Hariu and Y. Shibata, IEEE Trans. Electron Devices ED-31, 1092 (1984).
6. T. Hariu, S. F. Fang, K. Shida, K. Matsushita and Q. Z. Gao, in Gallium Arsenide and Related Compounds, Inst. Phys. Ser. (1986) (to be published).
7. M. Yano, Y. Suzuki, T. Ishii, Y. Matsushima and M. Kimata, Jpn. J. Appl. Phys. 17, 2091 (1978).
8. L. Bernstein and R. J. Beals, J. Appl. Phys. 32, Letter to Editor, 122 (1961).
9. B. T. Miggitt, E. H. Parker, and R. M. King, Appl. Phys. Lett. 33, 528 (1978).
10. M. Yano, T. Takase and M. Kimata, phys. stat. sol.(a) 54, 707 (1979).
11. R. A. A. Kubiak, E. H. C. Parker, S. Newstead and J. J. Harris, Appl. Phys. A (Germany) 35, 61 (1984).

Author Index

Subject Index

MATERIALS RESEARCH SOCIETY CONFERENCE PROCEEDINGS

VLSI-I—Tungsten and Other Refractory Metals for VLSI Applications, R. S. Blewer, 1986; ISSN: 0886-7860; ISBN: 0-931837-32-4

VLSI-II—Tungsten and Other Refractory Metals for VLSI Applications, E.K. Broadbent, 1987; ISSN: 0886-7860; ISBN: 0-931837-66-9

TMC—Ternary and Multinary Compounds, S. Deb, A. Zunger, 1987; ISBN:0-931837-57-x

Printed in the United States
By Bookmasters